T0092406

Lecture Notes in Physics

The Lecture Notes in Physics

The series Lecture Notes in Physics (LNP), founded in 1969, reports new developments in physics research and teaching – quickly and informally, but with a high quality and the explicit aim to summarize and communicate current knowledge in an accessible way. Books published in this series are conceived as bridging material between advanced graduate textbooks and the forefront of research and to serve three purposes:

- to be a compact and modern up-to-date source of reference on a well-defined topic

- to serve as an accessible introduction to the field to postgraduate students and nonspecialist researchers from related areas

- to be a source of advanced teaching material for specialized seminars, courses and schools

Both monographs and multi-author volumes will be considered for publication. Edited volumes should, however, consist of a very limited number of contributions only. Proceedings will not be considered for LNP.

Volumes published in LNP are disseminated both in print and in electronic formats, the electronic archive being available at springerlink.com. The series content is indexed, abstracted and referenced by many abstracting and information services, bibliographic networks, subscription agencies, library networks, and consortia.

Proposals should be sent to a member of the Editorial Board, or directly to the managing editor at Springer:

Christian Caron
Springer Heidelberg
Physics Editorial Department I
Tiergartenstrasse 17
69121 Heidelberg / Germany
christian.caron@springer.com

A.V. Mikhailov (Ed.)

Integrability

 Springer

Alexander V. Mikhailov
University of Leeds
School of Mathematics
Applied Mathematics Department
Woodhouse Lane
Leeds
United Kingdom LS2 9JT
A.V.Mikhailov@leeds.ac.uk

Mikhailov, A.V. (Ed.), *Integrability*, Lect. Notes Phys. 767 (Springer, Berlin Heidelberg 2009), DOI 10.1007/978-3-540-88111-7

ISBN: 978-3-540-88110-0   e-ISBN: 978-3-540-88111-7

DOI 10.1007/978-3-540-88111-7

Lecture Notes in Physics ISSN: 0075-8450   e-ISSN: 1616-6361

Library of Congress Control Number: 2008936509

Cover design: Integra Software Services Pvt. Ltd.

Printed on acid-free paper

9 8 7 6 5 4 3 2 1

springer.com

Preface

The principal aim of the book is to give a comprehensive account of the variety of approaches to such an important and complex concept as Integrability. Developing mathematical models, physicists often raise the following questions: whether the model obtained is integrable or close in some sense to an integrable one and whether it can be studied in depth analytically. In this book we have tried to create a mathematical framework to address these issues, and we give descriptions of methods and review results.

In the Introduction we give a historical account of the birth and development of the theory of integrable equations, focusing on the main issue of the book – the concept of integrability itself. A universal definition of *Integrability* is proving to be elusive despite more than 40 years of its development. Often such notions as "exact solvability" or "regular behaviour" of solutions are associated with integrable systems. Unfortunately these notions do not lead to any rigorous mathematical definition. A constructive approach could be based upon the study of hidden and rich algebraic or analytic structures associated with integrable equations. The requirement of existence of elements of these structures could, in principle, be taken as a definition for integrability. It is astonishing that the final result is not sensitive to the choice of the structure taken; eventually we arrive at the same pattern of equations. The relationship between the different approaches is often far from obvious and needs to be understood better.

Integrable equations possess hidden symmetries and actually possess infinite hierarchies of local symmetries. This property is taken as a definition of integrability in the symmetry approach. A detailed introduction and review of the modern state of the symmetry approach is given in Chap. 1, written by A.V. Mikhailov and V. Sokolov. The symmetry approach provides powerful necessary conditions for the existence of local higher symmetries and/or conservation laws for systems of differential equations. For a given system of equations these conditions are easily verifiable and eventually can serve as a criterion of integrability. Chapter 1 also contains an account of classification results obtained and an extensive bibliography.

For evolutionary equations whose right-hand side is a homogeneous differential polynomial, the symbolic representation and powerful results of number theory allow us to achieve global classification results (Chap. 2, written by J. Sanders and J.P. Wang). One of the most spectacular results of this theory can be formulated as follows: any scalar integrable evolutionary equation whose right-hand side is a homogeneous differential polynomial (with a positive weight) belongs to one of the infinite hierarchies of equations of order 2, 3 or 5 and all these integrable hierarchies are explicitly listed. It is shown that for a scalar evolutionary equation the existence of one higher symmetry implies the existence of an infinite hierarchy of hidden symmetries and therefore the integrability of the equation. For systems of equations a similar statement is not valid: there are examples of systems which have only a finite number of higher local symmetries. Chapter 2 is an excellent introduction to the symbolic method and contains relevant number theory results in applications to the theory of integrable equations.

In Chap. 3, written by S.P. Novikov, the phenomenon of integrability is associated with hidden symmetries of linear spectral problems. Darboux and Laplas transformations for one- and two-dimensional Schrödinger operators are famous examples of the spectral symmetries. The proper discretisation of these operators, the corresponding discrete Darboux and Laplas transformations and their relation to integrable equations and finite gap solutions are discussed. Chapter 4, written by A. Shabat is devoted to a detailed study of continuous and discrete spectral symmetries in the one-dimensional case. A connection of these symmetries with the famous list of Painlevé equations and with dressing chains is discussed.

Chapters 5 and 6 explore perturbative and asymptotic aspects of integrable equations. The concept of approximate integrability, approximate symmetries and conservation laws are discussed in Chap. 5, written by Y. Hiraoka and Y. Kodama. It is an attempt to extend the classical theory of normal forms to the case of partial differential equations. If the main approximation is given by an integrable equation, the higher order corrections often violate integrability and give rise to new effects, such as inelasticity in soliton interaction, creation of new solitons as a result of soliton collisions, etc. Chapter 6, written by A. Degasperis, addresses multiscaling expansion and universal equations, i.e. nonlinear equations which determine the leading term in the asymptotic expansion. Francesco Calogero gave a simple explanation for why integrable equations, which are rather exceptional, are widely applicable. Universal equations have a good chance to be integrable, since the multiscaling expansion preserves the main attributes of integrability, such as symmetries, local conservation laws, etc. The analysis of higher order corrections in a multiscale expansion of a given system provides necessary conditions for integrability of the system.

In the analytical theory of differential equations we study the structure of singularities of the solutions. The absence of movable critical singularities can be taken as a criteria for isolation of integrable systems. This is at the heart of the Painlevé approach and its generalisations described in Chap. 7 written by A. Hone.

Chapter 8, written by J. Hietarinta, describes the modern development of the Hirota approach and bi-linear representation of integrable equations. This kind of representation proved to be very useful for construction and analysis of explicit multi-soliton solutions. It can also be used for a classification of integrable equations of special form.

Quantum integrability is a separate and well-developed subject. It deserves a separate volume. We include lectures of T. Miwa (Chap. 9) in order to give a flavour of quantum integrability and to highlight the symmetry aspects of quantum integrable systems in the example of XXZ model.

This book is a unique collection of articles which could serve as the core material for a number of graduate lecture courses. The chapters in the book are independent and self-contained. They can be read in any order. Chapter 1 is probably more pedagogical than others and can be recommended for those wishing to become acquainted with the subject. The book was specifically designed to be accessible to graduate students and post-docs.

Leeds, UK,   *Alexander V. Mikhailov*
September 2008

Acknowledgements

I would like to thank all the authors of the lecture courses for their excellent contributions, continuous support and patience. I would like to thank J.A. Sanders for his help in preparing the index for this book and A.N. Gorban, whose help and advice were very valuable at the final stage of this project. Also I am grateful to the Springer publishing team responsible for Lecture Notes in Physics for enthusiastic support and assistance.

Contents

Introduction

A.V. Mikhailov

In the introduction I would like to give a brief historical background and some aspects of the modern development of the theory of integrable systems with main focus on the problem of integrability of partial differential equations. I have been a witness to this development since 1972 when I joined Zakharov's seminar in Novosibirsk and later on working in L.D. Landau Institute for Theoretical Physics. I had been very lucky indeed to be in the right place at the right time. My account here does not intend to be complete – its only purpose is to give a general background of, motivations for and a smooth introduction to the topics covered in this book.

Inverse Scattering Transform and Integrable Equations

There is no doubt that the modern theory of integrable systems was inspired by the discovery of the inverse transform method. Over 40 years ago, Gardner, Green, Kruskal and Miura [33] proposed a method for solving the initial value problem for the Korteweg–de Vries (KdV) equation

$$u_t + u_{xxx} - 6uu_x = 0 \tag{1}$$

using the inverse scattering problem for the Schrödinger operator

$$L = -\frac{d^2}{dx^2} + u(x,t) - \lambda^2.$$

At that time it looked like a miracle or simply a trick. It was not at all clear whether such a trick could be applied to any other nonlinear differential equation (partial or ordinary). In 1968, Lax gave [53] an elegant and neat interpretation of results in [33] by representing the KdV equation in terms of two commuting operators

A.V. Mikhailov (✉)
School & Mathematics, University of Leeds, Leeds LS2 9JT, UK,
A.V.Mikhailov@leeds.ac.uk

Mikhailov, A.V: *Introduction*. Lect. Notes Phys. **767**, 1–18 (2009)
DOI 10.1007/978-3-540-88111-7_0

$$\left[\frac{d}{dx} - A, L\right] = 0, \quad A = 4\frac{d^3}{dx^3} - 3\left(u\frac{d}{dx} + \frac{d}{dx}u\right),$$

which we now call a *Lax pair*. This observation opened the way to many further generalisations and applications of the method.

The breakthrough was the discovery by Zakharov and Shabat in 1971 [88]: the nonlinear Schrödinger (NLS) equation

$$i\psi_t = \psi_{xx} + 2|\psi|^2\psi \tag{2}$$

can be solved by the inverse transform method. The Lax operator L corresponding to Eq. (2) is a Dirac-type operator. Soon after, Ablowitz, Kaup, Newell and Segur (the famous AKNS group from the Clarkson University, NY) made a number of extensions of the method to other equations including the sine-Gordon equation [2]

$$u_{tt} - u_{xx} + \sin u = 0 . \tag{3}$$

Then there was an entire avalanche of discoveries of integrable equations. People were "fishing" for integrable equations by commuting different types of operators, often without any clear idea about which equation will emerge as the result of the commutation.

Gradually the original Lax pair and spectral transform have been replaced by a more general approach based on a zero-curvature representation

$$U_t - V_x + [U, V] = 0$$

or a compatibility condition for two linear problems [89]

$$L\Psi = 0, M\Psi = 0, \quad \text{where} \quad L = \frac{d}{dx} - U, \ M = \frac{d}{dt} - V,$$

where $U = (x, t, \lambda), V = V(x, t, \lambda)$ are $N \times N$ matrices (or elements of a Lie algebra) depending on the spectral parameter λ in a certain way (i.e. polynomial, rational, elliptic or even meromorphic functions of λ). I remember, after the paper of D. Kaup [42] (which we discussed in Zakharov's seminar in Novosibirsk), where he proposed replacing the spectral problem by a more general problem with a polynomial dependence on the spectral parameter,[1] that I added a simple idea to control the Lorentz invariance of the resulting equations and discovered the Lax representation [59] for the two-dimensional massive Thirring model

$$i\gamma^\mu \partial_\mu \psi + m\psi + \gamma^\mu \psi(\bar{\psi}\gamma_\mu \psi) = 0$$

and the proof of its integrability [52]. More general rational dependence on the spectral parameter (of the Lax operator) enabled us to integrate the two-dimensional principal chiral field model [85]

---

[1] Later on I have realised that a polynomial dependence on the spectral parameter and even zero-curvature type representations had been introduced earlier by S. P. Novikov [72] in the context of the finite gap integration.

$$\partial^\mu(g^{-1}\partial_\mu g) = 0, \quad g(x,t) \in G,$$

for various Lie groups G [69] and its Grassmanian reductions

$$[P, P_{\xi\eta}] = 0.\ g = I - 2P,\ P^2 = P.$$

Elliptic dependence on the spectral parameter was inherented from the quantum Yang–Baxter theory and applied to integration of the classical Landau–Lifshitz equation describing anisotropic ferromagnets by Sklyanin [77]. A straightforward extension of λ dependence to rational functions on algebraic curves of higher genus meets certain obstacles due to the Riemann–Roch theorem (see discussion in [86]), but here there has been some recent promising progress [48]. A generalisation to a nonisospectral setup enables us to integrate the Ernst equation (a reduction of the Einstein equations of gravitation) and a number of other systems [11, 14].

The range of integrable systems was rapidly extending. The inverse transform method has been applied to differential difference equations, such as the Toda lattice

$$\frac{d^2 v_n}{dt^2} = \exp(v_{n+1}) - \exp(v_{n-1}),$$

the Volterra chain and the differential difference nonlinear Schrödinger equation [3, 28, 41, 57]. The Lax representation has been found and thus the complete integrability has been proven for the N-dimensional Euler top [58] (in 2006, Manakov and Sokolov were awarded the Kowalewskii prize for their achievements in the study of integrable tops). A number of important integro-differential equations and totally discrete systems proved to be integrable by the inverse transform method. The inverse transform method has been extended to multi-dimensional equations. Maybe the most famous examples are the Kadomtsev–Petviashvili equation [22, 89]

$$(u_t + u_{xxx} + 6uu_x)_x = \pm u_{yy}, \tag{4}$$

the Veselov–Novikov equation [79]

$$u_t + u_{xxx} + u_{yyy} + (uv)_x + (uw)_y = 0, \quad v_y = u_x,\ w_x = u_y$$

and the (anti) self-dual Yang–Mills equation [10]

$$F_{\mu\nu} = \pm F_{\mu\nu}^*, \quad F_{\mu\nu} = \partial_\mu A_\nu - \partial_\nu A_\mu + [A_\mu, A_\nu].$$

Simultaneously the methods of solutions were developing too. They are based on the solution of the inverse scattering problems for the corresponding Lax operators. Inverse scattering problems (or more generally, inverse spectral transforms) were vastly generalised. Zakharov and Shabat proposed the "dressing method" [89, 90]. They expressed solutions of integrable equations in terms of solutions of a matrix Riemann–Hilbert problem. In scattering theory soliton solutions correspond to reflectionless potentials (the Bargman potentials in the case of the Schrödinger operator L). In the dressing method exact multisoliton solutions can be found via a solution of a rational matrix Riemann–Hilbert problem. The latter can be tackled using linear algebra alone and therefore bypassing the need to use the full machinery of the inverse problem. The inverse transform based on the nonlocal Riemann–Hilbert problem [84] and the $\bar{\partial}$-problem [1] was developed for integration of multidimensional equations (i.e. the Kadomtsev–Petviashvili equation (4)).

Methods of algebraic geometry have been brought to the theory of integrable equations by S. P. Novikov [72]. He constructed a new class of solutions corresponding to algebraic spectral curves. Novikov has discovered quasi-periodic potentials, the continuous spectrums of which have a finite number of forbidden gaps. These potentials are useful and fruitful generalisation of the reflectionless (or Bargman) potentials in quantum scattering theory. That was a huge boost to the theory of integrable equations and likewise it has set a number of new pithy problems in classical algebraic geometry.

Self-similar solutions of integrable partial differential equations are solutions of ordinary differential equations satisfying the Painlevé property [5]. In particular, all six Painlevé equations can be obtained as symmetry reductions of integrable PDEs. The theory of iso-monodromic deformations, based on the Lax representation, enables us to find connection formulas for asymptotics of the Painlevé transcendents.

Gradually it has been becoming clear that the theory of integrable systems has many relations with almost all parts of mathematics. The inverse transform and the Gel'fand–Levitan–Marchenko equation are related to functional analysis. The matrix Riemann–Hilbert problem and the $\bar{\partial}$-problem are related to complex analysis. The theory of integrable systems uses methods of the theory of Lie algebras and representation theory, group theory, algebraic geometry, commutative and noncommutative algebra, number theory, etc. Dressing transformations, which add solitons to a solution, are known in differential geometry as Bäcklund transformations; on the level of the corresponding Lax operators it is Darboux transformations [75]. Darboux and Laplace transformations can be viewed as discrete spectral symmetries of Lax operators, while integrable hierarchies correspond to continuous spectral symmetries. Lectures of Novikov (Chap. 3) and Shabat (Chap. 4) give a seminal introduction to this area of research.

Hirota has discovered a quite simple and surprisingly very effective method for construction of exact multisoliton solutions of integrable equations [38, 39]. His method was based on the observation that KdV, NLS, sine-Gordon and other integrable equations can be rewritten in a special bi-linear form. Then the problem can be split, and partial "multisoliton" solutions correspond to special solutions (a finite sum of exponentials) of linear partial differential equations with constant coefficients. Moreover, many nonintegrable equations can be re-cast into the Hirota bi-linear form, but in the latter case the method is guaranteed to produce exact two-soliton solutions only. Later on the Hirota method received a deep mathematical interpretation in terms of the τ-function and representation theory [18]. Lectures of Hietarinta (Chap. 8) give a comprehensive introduction to the Hirota method and its applications. There is a very interesting recent development. Chakravarty and Kodama have given a classification of multisoliton solutions for the Kadomtsev–Petviashvili equation (4) relating the τ-function with the Schubert decomposition and totally nonnegative Grassmanian cells [17].

Initially the quantum field theory of exactly solvable models was developing rather independently from the theory of integrable equations. Currently many concepts from the theory of integrable systems have found their quantum analog and are well understood on the quantum level. The joint development of the quantum and classical theories of integrable systems proved to be very fruitful for both. We have already mentioned that some important classical integrable systems, such as the Landau–Lifshitz equation, have been inherited from the quantum theory. The reduction group describing the discrete symmetries of the Lax representation [60–62] helped to find new solutions of the Yang–Baxter equation [9], etc. The quantum theory of integrable systems is a big and well-established area of research. We do not plan to cover this area of research in this book–this topic deserves a separate issue of the Lecture Notes. Lectures of T. Miwa (Chap. 9), though, give a flavour of the quantum theory of integrable systems and show some links with the classical theory. We believe it is a useful complement to the main part of the book.

Testing for Integrability and Classification

How do we test whether a given system is integrable and if so, how can we integrate it? What are the integrability conditions? Can we describe all integrable systems of a certain type (classification problem)? Can we give a complete picture of all possible integrable systems of all orders (global classification)? To answer these challenging questions we ought to decide what integrability is. In order to classify equations we have to define the equivalence relation and ideally give a method to check whether two given equations are equivalent or not.

In the previous section we related integrability with the Lax representations (or their generalisations). Unfortunately the problem of whether a given equation has a Lax representation is still unsolved. A useful development in this direction was due to H. D. Wahlquist and F. B. Estabrook [80]. They proposed a theory of pseudo-potentials which in certain cases could lead to a construction of a Lax representation for a given equation. Their method is based on a number of assumptions on the structure of the Lax operator L and leads to a quite nontrivial, but purely Lie algebraic problem to find a representation of a Lie algebra with a given subset of commutation relations. The representation should nontrivially depend on a spectral parameter and satisfy some technical conditions. If the assumptions were correct and we have managed to represent the algebra, then we get the Lax representation. Actually, many useful equations have been shown to be integrable and the corresponding Lax representations have been found using this method. For example the Landau–Lifshitz equation for magnetics with uni-axial anisotropy has been integrated in this way [13] (see [66] for other examples). The Wahlquist–Estabrook approach, when it works, gives sufficient conditions for integrability (by the inverse transform method) for a given equation. A good introduction into this method and discussion can be found in [71]. Another idea would be to classify all possible Lax representations (of a particular type) and their reductions. That would provide us with a rather long (and complete in a certain sense) list of integrable equations, but would not answer the above questions for a given system.

Partial differential equations integrable by the inverse transform method (1+1) dimensional, like the KdV or NLS) enjoy one exceptional property – they all possess an infinite hierarchy of local conservation laws. By a local conservation law we mean that there exists a function of the dependent variables and their derivatives (called a *conserved density*) which, when differentiated in time, results in a total spatial derivative of another function of the dependent variables and their derivatives (sometime called a *flux*) (accurate definitions will be given in Chap. 1). For example, the first three local conservation laws for the Korteweg–de Vries equation (1) are

$$u_t + \left(u_{xx} - 3u^2\right)_x = 0, \qquad (u^2)_t + \left(2uu_{xx} - u_x^2 - 4u^3\right)_x = 0,$$
$$(u_x^2 + 2u^3)_t + \left(2u_{xxx}u_x - u_{xx}^2 + 6u^2u_{xx} - 12uu_x^2 - 9u^4\right)_x = 0,$$

where the first simply restates the KdV equation (1) in conserved form. The fourth and fifth conservation laws were found by Kruskal and Zabusky before the discovery of the inverse transform method. Using the Lax representation for the KdV it is easy to derive a recursion relation for an infinite hierarchy of conservation laws. Shabat told me that nontrivial local conservation laws for the nonlinear Schrödinger equation (2) were found before the discovery of the Lax representation for the NLS equation.

Thus, the existence of nontrivial local conservation laws is a good indicator of integrability and can even be taken as a definition of integrability. There were a few attempts to classify equations which possess extra (beyond momentum and energy) conservation laws. Maybe the first one was due to Kulish [51]. He listed all nonlinear Klein–Gordon type equations

$$u_{tt} - u_{xx} + a_1 u + a_2 u^2 + a_3 u^3 + \cdots, \qquad a_1 \neq 0,$$

which possess a conserved density of order four. He has shown that any nonlinear equation satisfying this property can be reduced to the sine-Gordon equation (3) by a linear transformation. That is a correct result, but one integrable Klein–Gordon type equation

$$u_{tt} - u_{xx} + e^u - e^{-2u} = 0 \tag{5}$$

has been missed. Equation (5) has been missed because it does not possess a conserved density of order four. Its first nontrivial conserved density is of order six. Nontrivial conservation laws for this equation have been found by Bullough and Dodd [19] in their attempt to study equations with higher conservation laws, but in the paper there is a wrong statement: that Eq. (5) possesses only a finite number of local conservation laws and thus is not integrable! The Lax representation for (5) has been found in [60] and it has been shown [62] that Eq. (5) has local conserved densities of orders $6n - 4$ and $6n$, $n \in \mathbb{N}$. The first indication that Eq. (5) is integrable was due to Zhiber and Shabat [91] in their development of the Symmetry Approach to classification of integrable equation (see the discussion of this approach later in this chapter and also in Chap. 1). Digging around in the literature it was found that Eq. (5) has been known in differential geometry for some time: it was derived and studied by Tzitzeica [78] almost a century ago. Thus we call it now the Tzitzeica equation. Possible gaps in the sequence of conservation laws together with certain technical problems make it difficult to develop a classification theory for equations possessing only a few nontrivial conserved densities. If we assume the existence of an infinite hierarchy of local conservation laws, then we can get a more advanced theory which is insensitive to the gaps in the sequence. This theory enables us to produce a classification of integrable systems (see for example [66, 68]). It is a part of the Symmetry Approach, which we will discuss later in this section, as well as in Chap. 1.

Every integrable equation possesses a rich Lie algebra of symmetries. By symmetries we mean infinitesimal transformations which map solutions of the equation into solutions. The symmetries are generated by the flows commuting with the equation (for a precise definition see Chap. 1). For example the KdV equation (1) has symmetries generated by

$$u_{t_G} = 1 + 6tu_x, \quad u_{t_S} = 2u + xu_x + 3tu_t, \quad u_{\tau_1} = u_x, \quad u_{\tau_3} = u_t,$$
$$u_{\tau_5} = u_{xxxxx} - 10uu_{xxx} - 20u_xu_{xx} + 30u^2u_x.$$

The first four transformations correspond to classical continuous Lie point symmetries of the KdV equation, namely the Galilean, scaling transformations and shifts in space and time:

$$t_G : u(x,t) \mapsto u(x+6\lambda t,t) + \lambda, \quad t_S : u(x,t) \mapsto e^{-2\lambda} u\left(e^\lambda x, e^{3\lambda} t\right),$$
$$\tau_1 : u(x,t) \mapsto u(x+\lambda,t), \quad \tau_3 : u(x,t) \mapsto u(x,t+\lambda), \quad \lambda \in \mathbb{R}.$$

The transformation generated by u_{τ_5} does not correspond to any classical (point or contact) symmetry. It is called a higher symmetry and the KdV equation possesses infinitely many higher symmetries. Having symmetries one can study symmetry-invariant solutions. Scaling-invariant solutions are self-similar and satisfy the Painlevé property, as we mentioned above (see also the end of this section and the discussion in Chap. 7). According to S. P. Novikov, solutions of the KdV equation invariant with respect to higher symmetries satisfy completely integrable finite-dimensional dynamical systems which can be solved in terms of hyperelliptic Θ-functions and correspond to finite gap spectral curves.

There are partial differential equations which can be related to linear equations by a differential substitution (a transformation which is not invertible in the classical sense). F. Calogero proposed to call such equations C-integrable, while equations integrable by means of the inverse transform method he proposed to call S-integrable. Maybe the most famous example of a C-integrable equation is the Burgers equation

$$u_t = u_{xx} + 2uu_x, \tag{6}$$

which can be related to the linear heat equation $v_t = v_{xx}$ by the Cole–Hopf substitution $u = v_x/v$. The Burgers equation has only one local conserved density, but possesses infinitely many symmetries. Symmetries of (6) can be easily found from the symmetries of the heat equation generated by $v_{\tau_n} = \partial_x^n v$ and the Cole–Hopf transformation. Thus existence of symmetries is a common property of both "C"-and "S"-integrable equations.

Symmetries of integrable equations can be found using the Lax representation [53]. Another elegant way is to construct symmetries with the help of a recursion operator (or Lenard's recursion operator). Acting on a generator of a symmetry the recursion operator produces a new generator of a symmetry and thus a hierarchy of symmetries. In the case of the KdV equation (1) the recursion operator is of the form

$$\Lambda = D_x^2 - 4u - 2u_x D_x^{-1}, \tag{7}$$

where D_x is the operator of total derivative in x and D_x^{-1} is the inverse total derivative (which is well defined on the image space of D_x). Starting from the *seed symmetry* $f_1 = u_x$, which is a total derivative and a generator of a spatial shift we can construct an infinite sequence of generators $f_{2k+1} = \Lambda^k(u_x)$:

$$f_3 = \Lambda(u_x) = u_{xxx} - 6uu_x,$$
$$f_5 = \Lambda^2(u_x) = u_{xxxxx} - 10uu_{xxx} - 20u_x u_{xx} + 30u^2 u_x, \dots.$$

In 1974 there were two publications [2] and [34]. In the last section of [34] there is a construction attributed to Lenard (as a private communication) which gives an elegant way to generate higher symmetries of the KdV equation. The recursion operator (7) is not written out explicitly, but it is obvious from the construction. Lenard's scheme has had a great impact on the theory of integrable equations. It is the corner stone of the bi-Hamiltonian (multi-Hamiltonian) theory, the framework of which has been laid down in the fundamental works of F. Magri [55], I. M. Gel'fand and I. Ya. Dorfman [20, 36] and then developed in hundreds of publications and a few monographs. There is a very interesting article [74] where a letter from Andrew Lenard is reproduced. According to Lenard himself, in 1967 he made his construction when he tried to answer a question from Kruskal in the common room discussion in Princeton and it took "only fifteen minutes or so". Lenard writes that he never published anything nor concerned himself with the subject since. In [2] the authors have constructed recursion operators for the nonlinear Schrödinger, modified KdV, sine-Gordon equations and operator (7) for the KdV equation. Their construction has not been related to any multi-Hamiltonian splitting, but originated from known Lax operators and squares of their eigenfunctions. Three years later, in 1977, Olver [73] arrived at the concept of recursion operator in a symmetry algebraic setup, not related to a bi-Hamiltonian splitting nor to a Lax representation. He coined the name *recursion operator*, re-proved the result of Lenard (7) and some results of [2] and found a recursion operator for the Burgers equation (6) (Eq. (6) is not Hamiltonian nor does it have a Lax representation). Recently it has become clear that recursion operators may have other splittings, suitable for non-Hamiltonian equations [67]. For C-integrable systems the recursion operator can be found using the linearising differential substitution. For S-integrable systems, when the Lax representation is known, there is a neat construction which helps to find it purely algebraically [37]. For multi-dimensional integrable equations the concept of the recursion operator and Lenard's scheme have to be considerably modified (see discussion in [21, 32, 56]).

Existence of a nontrivial (or higher order) symmetry can be taken as another definition of integrability. This idea has been put forward by A. S. Fokas [29, 30]. He applied it to the classification of KdV-type equations of the form

$$u_t = u_{xxx} + f(u, u_x).$$

Moreover, in his paper he stated a folklore conjecture that the existence of one higher symmetry implies infinitely many. In the case of homogeneous differential polynomial equations this conjecture has been proven by Sanders and Wang. It has been disproved for systems of equations: there are systems of two equations with only one or two higher symmetries (see discussion and references in Chap. 2).

From a technical point of view it is easier to study the existence of symmetries than local conservation laws, but both approaches have the same weakness – the sequence of higher symmetries, similar to the sequence of higher conservation laws, may have lacunae, which are not known in advance. This problem has been overcome in the Symmetry Approach proposed by Shabat with his group and co-authors. It has been shown (see Theorem 24, Chap. 1) that if an evolutionary equation

$$u_t = f(u, u_x, \ldots, \partial_x^n u), \quad n > 1, \tag{8}$$

possesses symmetries of arbitrarily high order, then there exists a *formal recursion operator*, i.e. a formal series

$$\Lambda = a_1 D_x + a_0 + a_{-1} D_x^{-1} + a_{-2} D_x^{-2} + \cdots$$

with local coefficients a_k (i.e. functions of u and its derivatives), satisfying the equation

$$\Lambda_t = [f_*, \Lambda], \tag{9}$$

where f_* is the differential operator corresponding to the Fréchet derivative of f. The same is true if the evolutionary equation possesses two local conservation laws of arbitrarily high order (see Theorem 32, Chap. 1). If the order of symmetries or conservation laws is high, but not arbitrarily high, then Eq. (9) has an approximate solution (i.e. a few first coefficients of formal series Λ are guaranteed to be local). The Lenard recursion operator (7) satisfies equation (9).

The concept of a formal recursion operator proved to be a very flexible and powerful tool for testing for integrability and for classification of integrable systems (see Chap. 1). Necessary conditions for the existence of a formal recursion operator can be formulated in a simple and useful form of a sequence of canonical densities for the evolutionary equation (8). The first necessary condition states that the function (the first canonical density)

$$\rho = \left(\frac{\partial f}{\partial u^{(n)}} \right)^{-\frac{1}{n}}, \qquad u^{(n)} = \partial_x^n u,$$

must be a conserved density (trivial or nontrivial) for Eq. (8). Higher canonical densities can be found recursively. If they are nontrivial, then they are local conservation laws for the equation. If almost all canonical densities are trivial (i.e. total x-derivatives), it indicates that the equation is likely to be C-integrable. Existence of a formal recursion operator does not guarantee integrability but provides us with a tight test for integrability.

Thinking about a classification of integrable equations we have to decide which equations we consider to be equivalent, what is the equivalence relation and how to verify in practice whether two given equations belong to the same equivalence class or not. Obviously, equations related by classical invertible transformations (point or contact) should be treated as equivalent in our algebraic approach. Some analytical properties of solutions (such as the Painlevé property) may not survive with invertible transformations, but local conservation laws, symmetries, recursion and Lax operators can be easily transformed. In the Symmetry Approach based on the existence of a formal recursion operator the integrability conditions are invariant with respect to invertible changes of variables. There are some "weakly" noninvertible differential substitutions, such as potentiation (i.e. introduction of a potential) and the inverse potentiation and their generalisations which respect symmetries and conservation laws.

For example the potential version of the Korteweg–de Vries equation

$$v_t + v_{xxx} - 3v_x^2 = 0 \qquad (10)$$

can be reduced to the KdV equation (1) by a substitution $u = v_x$. The generators of the infinite hierarchy of KdV symmetries are total derivatives, they can be pulled back and give us generators of the potential KdV. Conserved densities of the potential KdV can be found from the conserved densities of KdV using the substitution directly. The recursion operator $\hat{\Lambda}$ for potential KdV can be found from the Lenard recursion operator (7) in an obvious way

$$\hat{\Lambda} = D_x^{-1} \cdot \Lambda|_{u \to v_x} \cdot D_x.$$

R. Miura has discovered a remarkable nonlinear differential substitution [70]

$$u = w_x + w^2 \qquad (11)$$

which maps solutions of the modified KdV equation

$$w_t + w_{xxx} - 6w^2 w_x = 0$$

into solutions of the KdV equation (1). The Miura transformation can also be used for re-calculation of symmetries and conserved densities.

Thus there are several levels of equivalence relation and corresponding classification. We consider equations to be equivalent if they are related by

(i)   classical invertible transformations;
(ii) the above in composition with potentiation and inverse potentiation;
(iii) all of the above in composition with Miura-type transformations.

For example,

1. All C-integrable equations of the form

$$u_t = F(u, u_x, u_{xx}, x, t)$$

by transformations (i) can be reduced to a list of four nonlinear equations or a linear equation (see Sect. 5.2, Chap. 1); by transformations (ii) they all can be linearised.

2. All S-integrable equations of the form

$$u_t = u_{xxx} + F(u, u_x, u_{xx})$$

by transformations (ii) can be reduced to a list of five equations (see Theorem 73, Chap. 1) and by transformations (iii) to a list of two equations, namely either to the Korteweg–de Vries equation (1) or to the Krichever–Novikov [47] equation.

The sequence of canonical conserved densities carry valuable information on the class that particular equation belongs to and helps to find transformations between equivalent equations. Similarly, it is also true for systems of integrable equations (see [66, 68] and references therein).

It has become possible to study the global structure of integrable hierarchies using the symbolic method and some results from Number Theory [81]. In the symbolic representation (the Gel'fand–Dikii symbolic calculus [35]) the existence of an infinite hierarchy of higher symmetries can be related to common factors for an infinite sequence of commutative multi-variate polynomials (see Chap. 2). Periodicity of the factors in the sequence of polynomials (originally studied in 1836 by Cauchy and Liouville [16])

$$G_n = (x + y)^n - x^n - y^n$$

explains the lacunae in hierarchies of symmetries of integrable evolutionary equations. Using the symbolic method it has been shown that (under certain technical conditions, which I believe can be relaxed) there are only a finite number of integrable hierarchies of evolutionary equations (see Sect. 4.2, Chap. 2 and [76, 81]) and it is sufficient to study integrable equations of orders $2, 3$ and 5: all other equations with nontrivial symmetries are members of their hierarchies. Thus, one nontrivial symmetry implies infinitely many. The latter in general is not true for systems of equations: there are examples of systems of two equations which possess only one or two higher symmetries (the proof is number-theoretical and based on the p-adic analysis, see Chap. 2). The symbolic method can be extended to nonevolutionary, nonlocal, multi-component and multi-dimensional systems [12, 63–65]. I think that the symbolic method has great potential for development.

In the analytic theory of ordinary differential equations there is an important class of equations which enjoy the property that the location of any algebraic, logarithmic or essential singularity of their solutions is independent of the initial conditions. Sofia Kowalevsi was the first to test this (what we now call Painlevé) property to isolate the integrable case of a system describing a heavy top with a fixed point

[45, 46]. Classification of second-order nonlinear equations with these properties was studied in exhaustive detail by Painlevé (see [40] and references).

In 1977, Ablowitz and Segur [5] noted that classical symmetry reductions of integrable equations, such as the KdV (1), mKdV (10) and sine-Gordon equations, result in ordinary differential equations which can be transformed to ones on the Painlevé list. Together with Ramani they formulated a conjecture [4] that *all ordinary differential equations derived from completely integrable partial differential equations have the Painlevé property.* In 1983, Weiss, Tabor and Carnevale [82] proposed an extension of the Painlevé approach to partial differential equations without any use of symmetry reductions or the Painlevé classification results. The obvious advantage of the Painlevé test is its simplicity. The test is very useful and can be applied to a rather wide class of equations. A disadvantage is that the result may depend on the choice of the dependent variables. If an equation does satisfy the test, the Painlevé approach in general does not tell us how to integrate the equation. Chapter 7 serves as a good introduction to the Painlevé theory of ordinary and partial differential equations and its relation to the problem of integrability.

Asymptotic Expansions and Normal Forms

Why are certain nonlinear PDEs both widely applicable and integrable? That is the title of the paper [15], where Francesco Calogero pointed out the universality of the Korteweg–de Vries, nonlinear Schrödinger and other integrable equations. These equations are widely applicable because they represent a dominant balance in a multi-scaling expansion. Asymptotic expansions preserve the property of integrability, so the dominant balance is likely to be integrable (see the discussion in Chap. 6). Trying to resolve the Fermi–Pasta–Ulam paradox [27], Zabusky and Kruskal used a long wave asymptotic expansion to reduce the nonlinear chain of oscillators

$$\frac{d^2 y_n}{dt^2} = -\frac{\partial}{\partial y_n} \sum_{k=0}^{N-1} U(y_{k+1} - y_k), \quad U(x) = \frac{x^2}{2} + \alpha \frac{x^4}{4}, \quad y_0 = y_N = 0 \qquad (12)$$

to the Korteweg–de Vries equation (1), which was known to be "universal" and widely applicable in hydrodynamics, plasma physics, etc. [49, 50, 83]. That was the motivation to study the Korteweg–de Vries equation which led to remarkable discoveries of solitons, nontrivial conservation laws, the inverse transform method and opened up the whole new and rich area of research – the theory of integrable equations.

The FPU nonlinear chain (12) is not an integrable system, but for relatively small amplitudes of oscillations, in the continuous limit ($N \to \infty$) together with a long wave asymptotic expansion it leads to the Korteweg–de Vries equation which represents the dominant balance. Thus, the non-integrable effects, such as a stochastisation of the chain, are hidden in the corrections to the integrable dominant balance.

In the asymptotic theory we can define a group of asymptotically invertible near-identity transformations and use them to transform the corrections to a simple and managable form. That is very similar to the main idea of the classical normal form theory for finite-dimensional dynamical systems. Analysing the higher corrections to an integrable dominant balance we can find obstacles to (asymptotic) integrability which could serve as the integrability test for the system and help to estimate the time required for nonintegrable effects to become visible.

The multi-scale asymptotic theory for partial differential equations depends on the class of initial data (or expected solutions). There are three consistent choices:

(i) the small amplitude limit: the dominant balance is represented by a linear system;

(ii) nonlinear terms and linear dispersion are of the same order: the dominant balance is a nonlinear "soliton" equation;

(iii) the hydrodynamic limit: the dominant balance is a hydrodynamic-type equation,

which lead to three different normal form theories.

In the first case (i) it is a generalisation of the Poincare–Dulac theory of normal forms or of the Birkhoff theory for the Hamiltonian systems. It originates from a seminal work of Zakharov and Schulman [87] and has received further rigorous development in [7, 8]. In this case the terms in the asymptotic expansion are ordered according the power of the dependent variable (and its derivatives) and perturbation theory is an adequate tool. In the application to the FPU chain (12) in the long wave approximation this case would correspond to solutions with $|y_{k+1} - y_k| \simeq N^{-3}$ or smaller.

In the second case (ii), the corresponding normal form theory aims to account for effects of nonintegrability on the soliton properties and soliton interaction [43]. The dominant balance is a nonlinear homogeneous equation which is assumed to be integrable. If the first correction is a symmetry of the dominant balance, or can be transformed into a symmetry, then the correction does not violate integrability at that level of the asymptotic approximation. Obstacles to transforming a correction to a symmetry of the principal balance are obstacles to integrability, so they could serve for testing of integrability [44]. Some interesting further development and applications of the Kodama theory of normal forms can be found in [26, 31].

To illustrate it in the example of the FPU chain (12) we should assume that $|y_{k+1} - y_k| \simeq N^{-2}$. In this case the nonlinear term and a ("numerical") dispersion are of the same order. Taking the continuous limit and performing a long wave asymptotic expansion we would get

$$u_\tau = \alpha u u_x + \frac{1}{24} u_{xxx}$$
$$+ N^{-2} \left(\frac{u_{xxxxx}}{1920} + \frac{\alpha}{24} u u_{xxx} + \frac{5\alpha}{48} u_x u_{xx} - \frac{\alpha^2}{2} u^2 u_x \right) + O(N^{-4}),$$

where $\tau = N^{-3}t$. Thus for $t \ll N^5$ the system is well approximated by the Korteweg–de Vries equation and demonstrates integrable regular behaviour. The first correction is not a symmetry of the principal balance, but it can be transformed to a symmetry by a near-identity transformation (see Theorem 10 and its proof in Chap. 5). Thus, corrections of order N^{-2} will not result in nonintegrable effects and the system will demonstrate integrable behaviour for $t \ll N^7$. The next correction (of order N^{-4}) cannot be transformed to a symmetry and is an obstacle to integrability (it is easy to check that the invariant of the transformation $\mu_1^{(2)} \neq 0$, in the notation of Chap. 5). It leads to an inelasticity in soliton collisions and eventual stochastisation of the chain. The time scale corresponding to the first nonintegrable correction is $t \simeq N^7$, where N is the number of particles in the chain (12) ($N = 64$ in [27]).

The third case (iii) corresponds to Dubrovin's normal form theory [23–25, 54]. It is a very recent development and not covered in this book. For "big" amplitudes the dominant balance of the multi-scale asymptotic expansion is often of the hydrodynamic type and is integrable. In the case of one dependent variable the hydrodynamic-type equation can be written in the Riemann form

$$v_t = v v_x. \tag{13}$$

The corrections are polynomials in derivatives of the dependent variable with smooth coefficients and the order of a term is defined as the total number of derivatives

$$u_t = u u_x + \epsilon(b_1(u)u_{xx} + b_2(u)u_x^2) + \epsilon^2(b_3(u)u_{xxx} + b_4(u)u_x u_{xx} + b_5(u)u_x^3) + \cdots. \tag{14}$$

For example, in the case of the FPU chain (12), assuming $|y_{k+1} - y_k| \simeq N^{-1}$ in the continuous limit and for long simple waves we get

$$u_t = \alpha u u_x + O(N^{-1}).$$

One of the remarkable results is that any equation (14) can be transformed in the "normal form" (13) by a near-identity transformation (invertible in the asymptotic sense) [54]. Such transformations generalise the famous Miura transformations (11) and are thus called the *quasi-Miura transformations*. For example the Korteweg–de Vries equation

$$u_t = u u_x + \frac{\epsilon^2}{12} u_{xxx}$$

can be transformed to the form (13) ($+O(\epsilon^6)$) by the transformation [6]:

$$v = u + \frac{\epsilon^2}{24}\left(\frac{u_2^2}{u_1^2} - \frac{u_3}{u_1}\right) + \frac{\epsilon^4}{24 \cdot 240} D^2\left(5\frac{u_4}{u_1^2} - 9\frac{u_2 u_3}{u_1^3} + 4\frac{u_2^3}{u_1^4}\right) + O(\epsilon^6).$$

Another remarkable observation is that near a gradient catastrophe point of solutions of the Riemann equation (13) the corresponding solutions of (Hamiltonian) Eq. (14) have a universal behaviour (the Dubrovin Conjecture [23]).

Acknowledgments I would like to thank my friends and colleagues with whom I have had numerous discussions of various aspects of the theory of integrable equations: E.A. Kuznetsov, V.E. Zakharov, A.B. Shabat, V.V. Sokolov, S.V. Manakov, S.P. Novikov, A.P. Veselov, Y. Kodama, M.D. Kruskal, A.C. Newell, F. Calogero, A. Degasperis and A.P. Fordy. It is my pleasure to thank G.Sh. Fridman, who long ago (in 1972) advised me, then a second year undergraduate student, to join Zakharov's group in Novosibirsk and study the recently discovered inverse scattering method.

References

1. M.J. Ablowitz, D. Bar Yaacov, and A.S. Fokas, On the inverse scattering transform for the Kadomtsev–Petviashvili equation, Studies Appl. Math. 69, 135–143, 1983.
2. M.J. Ablowitz, D.J. Kaup, A.C. Newell, and H. Segur, The inverse scattering transform – Fourier analysis for nonlinear problems, Stud. Appl. Math. 53, 249–315, 1974.
3. M.J. Ablowitz and J.F. Ladik, Nonlinear differential–difference equations. J. Math. Phys. 16, 598–603, 1975.
4. M.J. Ablowitz, A. Ramani, and H. Segur, Nonlinear evolutionary equations and ordinary differential equations of Painlevé type., Lett. Nuovo Cimento 23, 333–338, 1978.
5. M.J. Ablowitz and H. Segur, Exact linearisation of a Painlevé transcendent, Phys. Rev. Lett. 38, 1103–1106, 1977.
6. V.A. Baikov, R.K. Gazizov and N.Kh. Ibragimov, Approximate symmetries and formal linearisation. PMTF 2, 40–49, 1989 (In Russian).
7. D. Bambusi, Birkhoff normal form for some nonlinear PDEs, Comm. Math. Phys. 234(2), 253–285, 2003.
8. D. Bambusi, Galerkin averaging method and Poincaré normal form for some quasilinear PDEs, Ann. Scula Norm. Sup. Pisa, (2006, to appear).
9. A.A. Belavin, Hidden symmetry of an integrable system. Pisma ZETP 32(2), 182–186, 1980.
10. A.A. Belavin and V.E. Zakharov, Yang-Mills equations and inverse scattering problem, Phys. Lett. B 73(1) 53–57, 1978.
11. V.A. Belinskii and V.E. Zakharov, Einstein equations by the inverse scattering problem technique and construction of exact soliton solutions, Zh. Eksp. Teor. Fiz. 75(6), 1955–1971, 1978 [Sov. Phys. JETP, 48 (6) 985–994 (1978)].
12. F. Beukers, J.A. Sanders, and J.P. Wang. On integrability of systems of evolution equations. J. Differ. Equations, 172(2), 396–408, 2001.
13. A.E. Borovik, N-soliton solutions of the nonlinear Landau-Lifshitz equation, JETF Lett. 28, 629, 1978.
14. S.P. Burtsev, V.E. Zakharov, and A.V. Mikhailov, The Inverse Scattering Method with Variable Spectral Parameter, Theor. Math. Phy., 70, 232–241, 1987.
15. F. Calogero, Why are certain nonlinear PDEs both widely applicable and integrable?, in "What is Integrability?" V.E. Zakharov (ed.), Springer series in Nonlinear Dynamics, 1–62, 1991.
16. A. Cauchy and J. Liouville. Rapport sur un mémoire de M. Lamé relatif au dernier théoréme de Fermat. C. R. Acad. Sci. Paris, 9, 359–363, 1839.
17. S. Chakravarty and Y. Kodama, Classification of the line-soliton solutions of KPII, arXiv:0710.1456v1 [nlin.SI] 8 Oct 2007.
18. E. Date, M. Jimbo, M. Kashiwara, and T. Miwa, Transformation groups for soliton equations, Proc. RIMS Symposium on Nonlinear Integrable Systems – Classical and Qantum Theory, M. Jimbo and T. Miwa (eds.), Word Scientific Press, 1983.
19. R.K. Dodd and R.K. Bullough, Polynomial conserved densities for the sine-Gordon equation, Proc. R. Soc. Lond. A 352, 481–503, 1977.
20. I.Ya. Dorfman, Dirac Structures and Integrability of Nonlinear Evolution Equations, John Wiley&Sons, Chichester, 1993.

21. I.Ya. Dorfman and A.S. Fokas. Hamiltonian theory over noncommutative rings and integrability in multidimensions. J. Math. Phys. 33(7), 2504–2514, 1992.

22. V.S. Druma, Analytic solutions of the two-dimensional Korteweg–de Vries equation, Sov Phys. JETP Lett, 19, 381–388, 1974.

23. B. Dubrovin, On Hamiltonian perturbations of hyperbolic systems of conservation laws, II: universality of critical behaviour, Commun. Math. Phys. 267, 117–139 (2006).

24. B. Dubrovin and Y. Zhang, Normal forms of hierarchies of integrable PDEs, Frobenius manifolds and Gromov – Witten invariants. Math.DG/0108160, 2001.

25. B. Dubrovin, T.Grava, and C.Klein, On universality of critical behaviour in the focusing nonlinear Schrödinger equation, elliptic umbilic catastrophe and the tritronquée solution to the Painlevé I equation, arXiv:0704.0501v3 [math.AP], 2007.

26. H.R. Dullin, G.A. Gotteald and D.D. Holm, On asymptotically equivalent shallow water equations, Physica D. 10, 1–14, 2004.

27. E. Fermi, J. Pasta, and S. Ulam, Studies of nonlinear problems, I, Los Alamos Rep. LA1940, 1955; reprod. in Nonlinear Wave Motion, A.C. Newell (ed.), Lectures in Applied Mathematics, 15, Ameriacn Math.Soc., Providence, RI, 143–196, 1974.

28. H. Flashka, The Toda Lattice II. Inverse scattering solution. Progr. Theor Phys 51, 703–716, 1974.

29. A.S. Fokas, A symmetry approach to exactly solvable evolution equations, J. Math. Phys. 21(6), 1318–1325, 1980.

30. A.S. Fokas, Symmetries and integrability, Stud. Appl. Math. 77, 253–299, 1987.

31. A.S. Fokas and Q.M. Liu, Asylptotic integrability of water waves. Phys. Rev. Lett. 77(12), 2347–2351, 1996.

32. A.S. Fokas and P.M. Santini. Recursion operators and bi-Hamiltonian structures in mul- tidimensions 2, Commun. Math. Phys. 116(3), 449–474, 1988.

33. C.S. Gardner, J.M. Green, M.D. Kruskal, and R.M. Miura, Method for solving the Korteweg–de Vries equation, Phys. Rev. Lett. 19, 1095–1097, 1967.

34. C.S. Gardner, J.M. Green, M.D. Kruskal, and R.M. Miura, Korteweg–de Vries equation and generaliztions. IV. Methods for exact solution, Comm. Pure Appl. Math. 27, 97–133, 1974.

35. I. M. Gel'fand and L. A. Dickii. Asymptotic properties of the resolvent of Sturm-Liouville equations, and the algebra of Korteweg-de Vries equations, Uspehi Mat. Nauk, 30(5(185)), 67–100, 1975. English translation: Russian Math. Surveys, 30(5), 77–113, 1975.

36. I.M. Gel'fand and I.Ya. Dorfman, Hamiltonian operators and algebraic structures related to them, Funct. Anal. Appl. 13, 248–262, 1979.

37. M. Gurses, A. Karasu and V.V. Sokolov, On construction of recursion operator from Lax representation, JMPh, 40(12), 6473–6490, 1999.

38. R. Hirota, Exact solutions of the Korteweg–de Vries equation for multiple collisions of solitons, Phys. Rev Lett. 27, 1192–1194, 1972.

39. R. Hirota, Direct methods in soliton theory, in Topics in Current Physics, 17, R. Bullough and P. Caudrey (eds.), Springer-Verlag, New York, 157–175, 1980.

40. E.L. Ince, Ordinary Differential Equations, 1927, reprinted by Dover, New York, 1956.

41. M. Kac and P. van Moerbeke, On an explicitly soluble system of nonlinear differential equations related to certain Toda lattices, Advances Math. 16, 160–169, 1975.

42. D.J. Kaup, A Higher-OrderWaterWave Equation and Its Method of Solution, Prog. Theor. Phys. 54, 396–408 (1975).

43. Y.Kodama, Normal Form and Solitons, 319–340, in Topics in Soliton Theory and Exactly Solvable Nonlinear Equations, M. J. Ablowitz et al. (eds.), World Scientific, 1987.

44. Y. Kodama and A.V. Mikhailov, Obstacles to Asymptotic Integrability, 173–204, in Algebraic aspects of Integrability, I.M. Gel'fand and A.S Fokas (eds.), Birkhäuser, Boston, 1996.

45. S. Kowalewski, Sur le problème de la rotation d'un corps solide autour d'un point fixe. Acta. Math. 12, 177–232, 1889.

46. S. Kowalewski, Sur une propriètè du système d'équations différentielles qui défini la rotation d'un corps solide autour d'un point fixe. Acta. Math. 14, 81–93, 1889.

47. I.M. Krichever and S.P. Novikov, Holomorphic bundles and non-linear equations. Finite-zone solutions of rank 2. Dokl. Akad. Nauk SSSR 247, 33–36, 1979 (Trans.: Soviet Math. Dokl. 20, 650–654, 1979).

48. I.M. Krichever and O.K. Sheinman, Lax operator algebras, arXiv:math/0701648v3 [math.RT] 15 May 2007.

49. M.D. Kruskal and N.J. Zabusky, Progress on the Fermi-Pasta-Ulam nonlinear string problem, Princeton Plasma Physics Laboratory Annual Report, MATTQ-21, Princeton, NJ, 301–308, 1963.

50. M.D. Kruskal, Asymptotology in numerical computation: Progress and plans on the Fermi-Pasta-Ulam problem, Proc. IBM Scientific Computing Simposium on Large Scale Problems in Physics, IBM Data Processing Division, White Plans, NY, 43–62, 1965.

51. P.P. Kulish, Factorization of classical and quantum S-matrices and conservation laws,. Teor. Mat. Fiz. 26, No. 2, 198–205, 1976.

52. E.A. Kuznetsov and A.V. Mikhailov, On the Compete Integrability of the Two Dimensional Classical Thirring Model, Theoretical and Mathematical Physics, 30, 303, 1977.

53. P. Lax, Integrals of nonlinear equations of evolution and solitary waves, Comm. Pure. Appl. Math. 21(5), 467–490, 1968.

54. S.Q. Liu and Y. Zhang, On Quasitriviality and Integrability of a Class of Scalar Evolutionary PDEs, J. Geom. Phys. 57, 101–119, 2006; nlin.SI/0510019, 2005.

55. F. Magri, A simple model of the integrable Hamiltonian equation, J. Math. Phys. 19, 1156–1162, 1978.

56. F. Magri, C. Morosi, and G. Tondo. Nijenhuis g-manifolds and lenard bicomplexes – a new approach to kp systems, Commun. Math. Phys. 115(3), 457–475, 1988.

57. S.V. Manakov, Complete Integrability and Stochastization in Discrete Dynamical Systems, Zh. Eksp. Teor. Fiz. 67, 543–555, 1974.

58. S.V. Manakov, Remarks on the integrals of the Euler equations of the n–dimensional heavy top, Func. Anal. Appl. 10, 93–94, 1976.

59. A.V. Mikhailov, Integrability of the Two Dimensional Thirring Model, Lett. J. Exp. Theor. Phys., 23, 356–358, 1976.

60. A.V. Mikhailov, On the Integrability of two-dimensional Generalization of the Toda Lattice, Lett. J. Exp. Theor. Phys., 30, 443–448, 1979.

61. A.V. Mikhailov, Reduction in Integrable Systems. The Reduction Group. Pisma ZETP 32(2), 187–192, 1980.

62. A.V. Mikhailov, Reduction Problem and the Inverse Scattering method, Physica 3D, N1&2, p73–117, 1981.

63. A. V. Mikhailov and V. S. Novikov, Perturbative symmetry approach, J. Phys. A 35(22): 4775–4790, 2002.

64. A.V. Mikhailov and V.S. Novikov, Classification of integrable Benjamin-Onotype equations. Moscow Math. J. 3(4), 1293–1305, 2003.

65. A.V. Mikhailov, V.S. Novikov, and J.P. Wang. On classification of integrable non-evolutionary equations. Stud. Appl. Math. 118, 419–457, 2007.

66. A.V. Mikhailov, A.B. Shabat, and R.I. Yamilov. Extension of the module of invertible transformations. Classification of integrable systems. Comm. Math. Phys. 115(1), 1–19, 1988.

67. A.V. Mikhailov and V.V. Sokolov, Integrable ODEs on Associative Algebras, Comm. Math. Phys. 211(1), 231–251, 2000.

68. A.V. Mikhailov, V.V. Sokolov, A.B. Shabat, The symmetry approach to classification of integrable equations, in What is Integrability? V.E. Zakharov (ed.), Springer series in Nonlinear Dynamics, 115–184, 1991.

69. A.V. Mikhailov and V.E. Zakharov, On the Integrability of Classical Spinor Models in Two Dimensional Space-time, Comm. Math. Phys., 74, 21–40, 1980

70. R.M. Miura, Korteweg–de Vries equation and generaliztions. I. A remarkable explicit nonlinear terasnformation, J. Math. Phys. 9, 1202–1204, 1968.

71. A.C. Newell, Solitons in Mathematics an Physics, SIAM, Philadelphia, 1985.

72. S.P. Novikov, Periodic problem for the Korteweg–de Vries equation, Func. Anal. Appl.4, 54–66, 1974.

73. P.J. Olver, Evolution equations possessing infinitely many symmetries, J. Math. Phys. 18(6), 1212–1215, 1977.

74. J. Praught and R.G. Smirnov, Andrew Lenard: A Mistery Unraveled, SIGMA, 1, Paper 005, 7 pages, 2005.

75. C. Rogers and W.K. Schief, Bäcklund and Darboux Transformations Geometry and modern applications in soliton theory, Cambridge texts in Applied Mathematics, CUP, 2002.

76. J.A. Sanders and J.P. Wang. On the integrability of homogeneous scalar evolution equations. J. Differ. Equations 147(2), 410–434, 1998.

77. E.K. Sklyanin, On complete integrability of the Landau-Lifschitz equation, preprint LOMI, E-3-79, Leningrad, 1979.

78. G. Tzitzeica, Sur une nouvelle classe de surfaces, C.R.Acad. Sci. Paris 144, 1257–1259, 1907.

79. A.P. Veselov and S.P. Novikov, Finite–gap two–dimensional potential Schrödinger operators. Explicit formulas and evolution equations. Dokl. Akad. Nauk SSSR, 279, 20–24, 1984.

80. H.D. Wahlquist and F.B. Estabrook, Prolongation structures in nonlinear evolution equations J. Math. Phys. 16,. 1–7, 1975.

81. J.P. Wang. Symmetries and Conservation Laws of Evolution Equations. PhD thesis, Vrije Universiteit/Thomas Stieltjes Institute, Amsterdam, 1998.

82. J.Weiss, M. Tabor, and G. Carnevale, The Painlevé property for partial differential equations J. Math. Phys. 24, 522–526, 1983.

83. N.J. Zabusky and M.D. Kruskal, Interaction of solitonsin a collisionless plasma and the recurrence of initial states, Phys. Rev. Lett. 15, 240–243, 1965.

84. V.E. Zakharov and S.V. Manankov, Multidimensional nonlinear integrable systems and methods for constructing their solutions, Zap. Nauchn. Sem. LOMI, 133, 77–91, 1984.

85. V.E. Zakharov and A.V. Mikhailov, Relativistically Invariant Models of Field Theory, Integrable by Inverse Transform Method, J. Exper. Theor. Phys., 74, 1953–1973, 1978.

86. V.E. Zakharov and A.V. Mikhailov, The Inverse Scattering Method with the Spectral Parameter on Algebraic Curves, Funct. Anal. Appl. 17, 1–6, 1983.

87. V.E. Zakharov and E.I. Schulman, Integrability of Nonlinear Systems and Perturbation Theory. 185–250, in What is Integrability, V.E. Zakharov (ed.), Springer–Verlag, Berlin Heidelberg, 1991.

88. V.E. Zakharov and A.B. Shabat, Exact theory of two-dimensional self-focusing and one-dimensional self-modulation of waves in nonlinear media, Zh. Eksp. Teor. Fiz. 61(1), 118–134, 1971, [Sov. Phys. JETP 34 (1) 62–69 (1972)].

89. V.E. Zakharov and A.B. Shabat, A scheme for integrating the nonlinear equations of mathematical physics by the method of the inverse scattering problem, Func. Annal. Appl. 8(3), 226–235, 1974.

90. V.E. Zakharov and A.B. Shabat, A scheme for integrating the nonlinear equations of mathematical physics by the method of the inverse scattering problem II, Func. Annal. Appl. 13(3), 13–22, 1979.

91. A.V. Zhiber and A.B. Shabat, Klein-Gordon equations with a nontrivial group, Sov. Phys. Dokl. 247(5), 1103–1107, 1979.

Chapter 1
Symmetries of Differential Equations and the Problem of Integrability

A.V. Mikhailov and V.V. Sokolov

1.1 Introduction

The goal of our lectures is to give an introduction to the symmetry approach to the problems of integrability. The basic concepts are discussed in the first two chapters where we give definitions and formulate statements in a simple way with complete proofs. In the other chapters we attempt to make a brief account of the results obtained in more than 20 years of the development and give references to original articles as well as to comprehensive review papers. We illustrate the achievements in the description and classification of integrable equations and discuss a variety of the problems associated with the Symmetry Approach and modern trends.

Many people have contributed to the development of the Symmetry Approach. The main credits here have to be given to Alexei Shabat whose pioneer works often determined principal directions of the research. The major results in the solution of specific classification problems had been obtained by Serguei Svinolupov with his remarkable abilities to exercise and structuralize very complex algebraic computations. We would also like to mention significant contributions of Ravil Yamilov and Anatoli Zhiber and others. We are very grateful to the above-mentioned colleagues for numerous discussions and mutual collaborations which casted our understanding and vision of the Symmetry Approach to the testing and classification of integrable equations.

In our lecture course we do not include the recent works of V. Adler, V. Marikhin, A. Shabat and R. Yamilov related to integrable chains, Bäcklund transformations and Lagrangian aspects of the Symmetry Approach (see [1] and references), nor the works of I. Habibullin devoted to a symmetry approach to initial-boundary problems for integrable equations (see for instance [23]).

A.V. Mikhailov (✉)
School of Mathematics, University of Leeds, Leeds LS2 9JT, UK,
A.V.Mikhailov@leeds.ac.uk

Mikhailov, A.V., Sokolov, V.V.: *Symmetries of Differential Equations and the Problem of Integrability.* Lect. Notes Phys. **767**, 19–88 (2009)
DOI 10.1007/978-3-540-88111-7_1

1.2 Symmetries and First Integrals of Finite-Dimensional Dynamical Systems

We will consider differential equations in the spirit of elementary differential algebra. In this approach the derivations associated with differential equations become the central objects of the theory and such notions as symmetries, first integrals, transformations, etc. can be naturally defined in their terms. The derivations are usually represented by vector fields. While for ordinary differential equations these vector fields are finite dimensional, in the case of partial differential equations they become infinite dimensional and certain accuracy is required for formulation of correct definitions. We found that the usage of vector fields proved to be more suitable for actual computations than the dual approach based on the theory of differential forms.

1.2.1 Dynamical Systems and Vector Fields

In this section we remind the standard definition of finite-dimensional vector fields and list some of their properties. One of the purposes of this section is to make the book self-contained. From the other side we prepare the ground for less standard consideration of infinite-dimensional vector fields in the theory of partial differential equation.

Suppose we have a dynamical system

$$\frac{d u_i}{dt} = F_i(u_1, \dots, u_n), \quad i = 1, \dots, n. \tag{1.1}$$

Using the chain rule any function $G(u_1, \dots, u_n)$ can be differentiated in time in virtue of the system (1.1)

$$\frac{dG}{dt} = \sum_{k=1}^{n} F_k(u_1, \dots, u_n) \frac{\partial G}{\partial u_k}. \tag{1.2}$$

Now we can forget that $u_1, \dots, u_n$ are functions of time t and regard them as the set of *independent* variables. In order to emphasize the independence of symbols $u_1, \dots, u_n$ we call them the *dynamical* variables. We denote the set of all functions[1] of dynamical variables as $\mathscr{F}$.

The expression (1.2) defines a derivation, i.e. a linear map $D_t : \mathscr{F} \to \mathscr{F}$, which satisfies the Leibnitz rule $D_t(ab) = D_t(a)b + aD_t(b)$ (usually the derivation D_t is called the *operator of total t-derivative*). This linear map can be represented by a vector field

$$D_F = \sum_{k=1}^{n} F_k \frac{\partial}{\partial u_k}. \tag{1.3}$$

---

[1] In what follows we need $\mathscr{F}$ to have the structure of a differential field. The field of locally meromorphic functions is suitable in many cases.

Definition 1. Linear homogeneous differential operator of the form

$$X = \sum_{k=1}^{n} X_k \frac{\partial}{\partial u_k}, \quad X_i \in \mathscr{F}, \tag{1.4}$$

is called a vector field on $\mathscr{F}$.

The most important operation on vector fields is their Lie brackets. If X, Y are vector fields

$$X = \sum_{k=1}^{n} X_k \frac{\partial}{\partial u_k},$$

$$Y = \sum_{k=1}^{n} Y_k \frac{\partial}{\partial u_k},$$

then their Lie bracket $[X, Y]$ is defined as the commutator of the differential operators:

$$[X, Y] = X \circ Y - Y \circ X. \tag{1.5}$$

It is easy to verify that $[X, Y]$ is a first-order differential operator of the form (1.3). Indeed,

$$X \circ Y(f) = X \left(\sum_{k=1}^{n} Y_k \frac{\partial f}{\partial u_k} \right) = \sum_{k=1}^{n} X(Y_k) \frac{\partial f}{\partial u_k} + \sum_{k,m=1}^{n} X_k Y_m \frac{\partial^2 f}{\partial u_k \partial u_m},$$

$$Y \circ X(f) = Y \left(\sum_{k=1}^{n} X_k \frac{\partial f}{u_k} \right) = \sum_{k=1}^{n} Y(X_k) \frac{\partial f}{u_k} + \sum_{k,m=1}^{n} Y_k X_m \frac{\partial^2 f}{\partial u_k \partial u_m},$$

The second derivatives cancel out and the result of commutation is a first-order homogeneous differential operator, i.e. a vector field

$$[X, Y] = Z, \quad Z = \sum_{k=1}^{n} Z_k \frac{\partial}{\partial u_k},$$

where

$$Z_k = X(Y_k) - Y(X_k). \tag{1.6}$$

It follows from definition (1.5) that the Lie bracket has the following properties:

- Bilinearity
$$[X, \alpha Y + \beta Z] = \alpha [X, Y] + \beta [X, Z], \tag{1.7}$$

- Skew-Symmetry
$$[X, Y] = -[Y, X], \tag{1.8}$$

- Jacobi Identity
$$[X, [Y, Z]] + [Y, [Z, X]] + [Z, [X, Y]] = 0, \tag{1.9}$$

where X, Y, Z are vector fields and $\alpha, \beta \in \mathcal{K}$ are constants. The above identities (1.7), (1.8), (1.9) mean that the set of vector fields form a Lie algebra over the field of constants $\mathcal{K}$.

Apart from the Lie bracket, there is yet another important operation, namely a multiplication of vector fields on functions from $\mathcal{F}$ from the left. It is easy to see that for any vector fields X, Y and $a \in \mathcal{F}$ the following identity

$$[X, aY] = X(a)Y + a[X, Y]$$

holds.

Let us consider a transformation to a new set $(\hat{u}_1, \ldots, \hat{u}_n)$ of dynamical variables given by

$$\hat{u}_1 = \phi_1(u_1, \ldots, u_n), \ldots, \hat{u}_n = \phi_n(u_1, \ldots, u_n). \qquad (1.10)$$

We assume that transformation (1.10) is locally invertible, i.e. the corresponding Jacobi matrix

$$J_{i,j} = \frac{\partial \hat{u}_i}{\partial u_j} \qquad (1.11)$$

is not singular $\det(J) \neq 0$. The inverse transformation can be written in the form

$$u_1 = \hat{\phi}_1(\hat{u}_1, \ldots, \hat{u}_n), \ldots, u_n = \hat{\phi}_n(\hat{u}_1, \ldots, \hat{u}_n). \qquad (1.12)$$

Transformation (1.12) defines a map $\sigma : \mathcal{F} \to \hat{\mathcal{F}}$, where $\hat{\mathcal{F}}$ is a set of all functions of dynamical variables $\hat{u}_1, \ldots, \hat{u}_n$. The map is given by

$$\sigma : a(u_1, \ldots, a_n) \to a(\hat{\phi}_1(\hat{u}_1, \ldots, \hat{u}_n), \ldots, \hat{\phi}_n(\hat{u}_1, \ldots, \hat{u}_n))$$

for any $a(u_1, \ldots, u_n) \in \mathcal{F}$. Similarly, transformation (1.10) defines the inverse map $\sigma^{-1} : \hat{\mathcal{F}} \to \mathcal{F}$ by

$$\sigma^{-1} : b(\hat{u}_1, \ldots, \hat{u}_n) \to b(\phi_1(u_1, \ldots, u_n), \ldots, \phi_n(u_1, \ldots, u_n)),$$

for any $b(\hat{u}_1, \ldots, \hat{u}_n) \in \hat{\mathcal{F}}$. The meaning of σ is obvious: in arguments of all functions we simply express $u_1, \ldots, u_n$ in terms of new variables.

If we have an operator $A : \mathcal{F} \to \mathcal{F}$, then the corresponding operator $\sigma(A) : \hat{\mathcal{F}} \to \hat{\mathcal{F}}$ is defined by

$$\sigma(A) = \sigma \circ A \circ \sigma^{-1}, \qquad (1.13)$$

and $\circ$ means the composition.

It follows from (1.13) that for any operators A, B

$$\sigma(A \circ B) = \sigma(A) \circ \sigma(B),$$

and in particular, for the commutator of vector fields on $\mathcal{F}$

$$\sigma([X, Y]) = [\sigma(X), \sigma(Y)].$$

Lemma 2. *Let X be a vector field of the form* (1.4) *on $\mathscr{F}$. Then*

$$\sigma(X) = \sum_{k=1}^{n} \hat{X}_k \frac{\partial}{\partial \hat{u}_k}$$

is a vector field on $\hat{\mathscr{F}}$ with

$$\hat{X}_k = \sigma(X(\phi_k)). \tag{1.14}$$

Proof. It follows from the chain rule for a differentiation of the composition of functions that

$$\sigma\left(\frac{\partial}{\partial u_k}\right) = \sum_{j=1}^{n} \sigma\left(\frac{\partial \phi_j}{\partial u_k}\right) \frac{\partial}{\partial \hat{u}_j}.$$

Therefore $\sigma(X)$ is a vector field of the form

$$\sigma(X) = \sum_{k=1}^{n} \sigma(X_k)\sigma\left(\frac{\partial}{\partial \hat{u}_k}\right) = \sum_{k,j=1}^{n} \sigma(X_k)\sigma\left(\frac{\partial \phi_j}{\partial u_k}\right) \frac{\partial}{\partial \hat{u}_j} = \sum_{j=1}^{n} \sigma(X(\phi_j)) \frac{\partial}{\partial \hat{u}_j}.$$

∎

Let us denote that the coefficients $\hat{\mathbf{X}} = (\hat{X}_1, \ldots, \hat{X}_n)$ of the transformed vector field (1.14) are nothing but $J\mathbf{X}$ with a subsequent re-expression of $u_1, \ldots, u_n$ via (1.12) and J is the Jacobi matrix (1.11) of the transformation. Obviously, the coefficients of the vector field $\hat{X}_F$ give the right-hand side for the system (1.1) in new variables.

The following statement is one of the basic theorems in the theory of ODEs (see, for example, Proposition 1.29 in Olver [49]).

Theorem 3. *Suppose at some point $u_1^0, \ldots, u_n^0$ the vector field $X_F \neq 0$, then there exists a nonsingular transformation of the form* (1.12) *such that in an open vicinity of this point the transformed vector field takes the form $\hat{X} = \partial/\partial u_1$.*

Unfortunately this theorem does not provide a constructive way to find this transformation in a closed form. The situation changes dramatically if we have exactly n linearly independent over $\mathscr{F}$ vector fields $X^1, \ldots, X^n$ such that $[X^i, X^j] = 0$ for all i and j. In this case there exists a transformation of the form (1.12) such that in new variables $\hat{X}^i = \partial/\partial \hat{u}_i$. Moreover this transformation can be found in *quadratures*. In other words all n corresponding dynamical systems can be simultaneously integrated and the answer can be written in quadratures. We shall discuss the issue of integrability of dynamical systems in quadratures and prove the above statement.

One of the most important notion in the local theory of differential equations is the notion of linearized equations. Suppose $\mathbf{u} = (u_1, \ldots, u_n)$ satisfies equation (1.1). Let us replace $\mathbf{u}$ in (1.1) by $\mathbf{u} + \epsilon \mathbf{v}$ and assuming $\epsilon \to 0$ find the equation for $\mathbf{v}$ in the first order of ϵ. It has the form

$$\mathbf{v}_t = \mathbf{F}_* \mathbf{v},$$

where $\mathbf{F}_*$ is a matrix with entries

$$F_{*i,j} = \frac{\partial F_i}{\partial u_j}. \tag{1.15}$$

The matrix $\mathbf{F}_*$ is called the Fréchet derivative of the vector-function $\mathbf{F} = (F_1, \ldots, F_n)$. Formal definition of the Fréchet derivative in more general situations will be given later.

With any vector field (1.4) on $\mathscr{F}$ we can assign a similar matrix[2] X_* of the form

$$X_{*i,j} = \frac{\partial X_i}{\partial u_j}. \tag{1.16}$$

If $Z = [X, Y]$, then it follows from (1.6) that

$$\mathbf{Z}_* = X(\mathbf{Y}_*) - Y(\mathbf{X}_*) + [\mathbf{X}_*, \mathbf{Y}_*],$$

where $[\mathbf{X}_*, \mathbf{Y}_*] = \mathbf{X}_* \mathbf{Y}_* - \mathbf{Y}_* \mathbf{X}_*$ is a commutator of the matrices.

Under the change of variables the transformation rule for X_* (1.16) is given by

$$\hat{X}_* = J X_* J^{-1} + X(J) J^{-1},$$

where in the right-hand side we have to re-express $\mathbf{u}$ in terms of $\hat{\mathbf{u}}$ according to (1.12).

1.2.2 First Integrals

First integrals of a dynamical system can be defined as elements of the kernel space for the corresponding vector field.

Definition 4. A function $I = I(u_1, \ldots, u_n) \in \mathscr{F}$ is a first integral of the dynamical system (1.1) if $X_F(I) = 0$.

Any function of first integrals is a first integral. It is important to count only functionally independent first integrals.

Definition 5. First integrals $\phi_k(u_1, \ldots, u_n)$, $k = 1, \ldots, m$, are called functionally independent if the Jacobi matrix

$$\frac{D(\phi_1, \ldots, \phi_m)}{D(u_1, \cdots, u_n)} = \begin{vmatrix} \dfrac{\partial \phi_1}{\partial u_1} & \cdots & \dfrac{\partial \phi_1}{\partial u_n} \\ \vdots & & \vdots \\ \dfrac{\partial \phi_m}{\partial u_1} & \cdots & \dfrac{\partial \phi_m}{\partial u_n} \end{vmatrix}$$

has the maximal rank.

---

[2] As a matter of fact the matrix $\mathbf{X}_*$ defines the action of the vector field on the cotangent space. Namely, it follows from the standard definition $X(a\,db) = X(a)\,db + a\,d(X(b))$ that $X(du_k) = \sum_{m=1}^{n} X_{*km} du_m$.

Example 6. The Euler–Poinsot equations for a rigid body with a fixed point can be written in the form

$$
\begin{cases}
\dot{p}_1 = (c-b)p_2 p_3 + zq_2 - yq_3 \\
\dot{p}_2 = (a-c)p_1 p_3 + xq_3 - zq_1 \\
\dot{p}_3 = (b-a)p_1 p_2 + yq_1 - xq_2 \\
\dot{q}_1 = cq_2 p_3 - bq_3 p_2 \\
\dot{q}_2 = aq_3 p_1 - cq_1 p_3 \\
\dot{q}_3 = bq_1 p_2 - aq_2 p_1,
\end{cases}
\tag{1.17}
$$

where a,b,c,x,y,z are constant parameters of the problem. For any values of these parameters, equation (1.17) has the following first integrals

$$
I_1 = ap_1^2 + bp_2^2 + cp_3^2 + 2xq_1 + 2yq_2 + 2zq_3,
\tag{1.18}
$$
$$
I_2 = q_1 p_1 + q_2 p_2 + q_3 p_3,
\tag{1.19}
$$
$$
I_3 = q_1^2 + q_2^2 + q_3^2.
\tag{1.20}
$$

It is easy to see that the corresponding Jacobi matrix

$$
J = \begin{pmatrix}
2ap_1 & 2bp_2 & 2cp_3 & 2x & 2y & 2z \\
q_1 & q_2 & q_3 & p_1 & p_2 & p_3 \\
0 & 0 & 0 & 2q_1 & 2q_2 & 2q_3
\end{pmatrix}
$$

has rank 3 in a generic point $(p_1, p_2, p_3, q_1, q_2, q_3)$, and therefore these first integrals are functionally independent.

Making the change of variables (1.10) we have to replace $u_1, \ldots, u_n$ by $\hat{\phi}_1, \ldots, \hat{\phi}_n$ in first integrals:

$$
\hat{I}(\hat{u}_1, \ldots, \hat{u}_n) = \sigma\left(I(u_1, \ldots, u_n)\right) = I\left(\hat{\phi}_1, \ldots, \hat{\phi}_n\right).
\tag{1.21}
$$

Function $\hat{I}$ belongs to the kernel space of the vector field $\hat{X}$. Indeed, it follows from (1.13) and (1.21) that

$$
\hat{X}(\hat{I}) = \sigma(X(\sigma^{-1}(\sigma(I)))) = \sigma(X(I)) = 0.
$$

The existence of functionally independent first integrals follows immediately from Theorem 3.

Proposition 7. *Let the coefficients of the vector field X_F be continuous and continuously differentiable functions which do not vanish simultaneously at a point $u_1^0, \ldots, u_n^0$. Then in some neighbourhood of this point there exist $n-1$ functionally independent first integrals for the corresponding dynamical system (1.1).*

Proof. It follows from Theorem 3 that there exist such new variables that $\hat{X}_F = \partial/\partial\hat{u}_1$. In this variables $\hat{u}_2, \hat{u}_3, \ldots, \hat{u}_n$ are $n-1$ functionally independent first integrals for the vector field $\hat{X}_F$. In the old variables (1.10), functions $\phi_2(u_1, \ldots, u_n), \ldots,$ $\phi_2(u_1, \ldots, u_n)$ are $n-1$ independent first integrals for (1.1).

∎

1.2.3 Symmetries

Another fundamental concept of the local theory of nonlinear ODEs is the infinitesimal symmetry.

Definition 8. A vector field

$$X_G = \sum_{k=1}^{n} G_k(u_1, u_2, \ldots, u_n) \frac{\partial}{\partial u_k}$$

is called (infinitesimal) symmetry of dynamical system (1.1) iff

$$[X_F, X_G] = 0. \tag{1.22}$$

Condition (1.22) is equivalent to the fact that the dynamical systems (1.1) and

$$\frac{d u_i}{d \tau} = G_i(u_1, \ldots, u_n), \quad i = 1, \ldots, n, \tag{1.23}$$

are compatible. It means that for any initial data $\mathbf{u}_0$ there exists a common solution $\mathbf{u}(t, \tau)$ of systems (1.1) and (1.23) such that $\mathbf{u}(0,0) = \mathbf{u}_0$.

Sometimes the dynamical system (1.23) itself is called a symmetry (instead of the corresponding vector field X_G). Notice also that identity (1.22) can be rewritten in the following two equivalent forms:

$$\frac{dG}{dt} = \mathbf{F}_*(G) \tag{1.24}$$

or

$$\mathbf{F}_*(G) - \mathbf{G}_*(F) = 0.$$

Relation (1.24) means that the vector-function $\mathbf{G}$ satisfies the linearization of dynamical system (1.1).

Example 9. Let us consider the system

$$\frac{du_1}{dt} = 1, \quad \frac{du_2}{dt} = 0, \ \ldots, \ \frac{du_n}{dt} = 0. \tag{1.25}$$

The corresponding vector field is $X_F = \frac{\partial}{\partial u_1}$ and condition (1.22) is equivalent to

$$X_G = \sum_{k=1}^{n} G_k(u_2, \ldots, u_n) \frac{\partial}{\partial u_k}.$$

According to Theorem 3 a generic symmetry for arbitrary dynamical system (1.1) also depends on n functions of $(n-1)$ dependent variables.

1.2.4 Lie's Theorem

Both first integrals and symmetries are very useful if we want to integrate dynamical system (1.1) by quadratures. Suppose we know $n-1$ functionally independent first integrals $I_1,\dots,I_{n-1}$ of (1.1). Making a change of variables

$$\hat{u}_1 = \phi_1(u_1,\dots,u_n), \hat{u}_2 = I_1(u_1,\dots,u_n),\dots,\hat{u}_n = I_{n-1}(u_1,\dots,u_n),$$

for any ϕ_1, we get a system of the form

$$\frac{d\hat{u}_1}{dt} = \hat{f}_1(\hat{u}_1,\dots,\hat{u}_n), \quad \frac{d\hat{u}_2}{dt} = 0, \ \dots, \ \frac{d\hat{u}_n}{dt} = 0,$$

which can be easily integrated in quadratures.

The procedure of integrating (1.1) if $n-1$ symmetries

$$X^1 = \sum_{k=1}^{n} G_k^1 \frac{\partial}{\partial u_k}, \quad X^2 = \sum_{k=1}^{n} G_k^2 \frac{\partial}{\partial u_k}, \dots, \quad X^{n-1} = \sum_{k=1}^{n} G_k^{n-1} \frac{\partial}{\partial u_k} \tag{1.26}$$

are given is not so standard. For an efficient use of symmetries (1.26) we have to impose some restrictions on the structure of the Lie algebra generated by the vector fields X^i. The simplest version of a statement of such a sort reads as follows.

Theorem 10. *Suppose dynamical system (1.1) has $(n-1)$ symmetries (1.26) such that*

- *the matrix*

$$\begin{pmatrix} F_1 & F_2 & \dots & F_n \\ G_1^1 & G_2^1 & \dots & G_n^1 \\ \dots & \dots & \dots & \dots \\ G_1^{n-1} & G_2^{n-1} & \dots & G_n^{n-1} \end{pmatrix} \tag{1.27}$$

is nondegenerate
- *and*

$$[X_i, X_j] = 0, \qquad 1 \le i, j \le n-1.$$

Then (1.1) can be integrated in quadratures.

Proof. It turns out that we can explicitly find a transformation (1.10) such that

$$\sigma(X_F) = \frac{\partial}{\partial \hat{u}_1}, \qquad \sigma(X_i) = \frac{\partial}{\partial \hat{u}_i}, \quad i = 1,\dots,n-1.$$

Indeed, it follows from (1.14) that the unknown functions $\phi_i(u_1,\dots,u_n)$ must satisfy the following system of equations

$$X_F(\phi_1) = 1, \qquad X_F(\phi_i) = 0, \quad i > 1,$$

$$X_i(\phi_j) = \delta_j^i.$$

In particular, the function ϕ_1 satisfies the following conditions

$$X_F(\phi_1) = 1, \quad X_1(\phi_1) = 0, \quad \dots, X_{n-1}(\phi_1) = 0.$$

Let us consider these relations as a system of algebraic linear equations with respect to unknowns $z_i = \dfrac{\partial \phi_1}{\partial u_i}$. Since the determinant of matrix (1.27) is not zero, the system has a unique solution. It follows from the Frobenious theorem that $\dfrac{\partial z_j}{\partial u_i} = \dfrac{\partial z_i}{\partial u_j}$. Now, to reconstruct the function ϕ_1 one has to perform a sequence of integrations with respect to variables $u_1, u_2, \dots, u_n$. In a similar way we can find the functions $\phi_2, \dots, \phi_n$. Thus we can bring our system to the canonical form (1.25), solve (1.25) and perform the inverse transformation (1.12) in order to find the general solution of the initial dynamical system (1.1).   ∎

A more general statement which involves both first integrals and symmetries can be formulated as follows:

Theorem 11. *Suppose dynamical system* (1.1) *has k symmetries of the form* (1.26) *and* $(n - k - 1)$ *functionally independent first integrals* $I_1, \dots, I_{n-k-1}$ *such that*

- *the matrix*

$$\begin{pmatrix} F_1 & F_2 & \dots & F_n \\ G_1^1 & G_2^1 & \dots & G_n^1 \\ \dots & \dots & \dots & \dots \\ G_1^k & G_2^k & \dots & G_n^k \end{pmatrix}$$

  has the maximal rank;

-

$$[X_i, X_j] = 0, \qquad 1 \le i, j \le n - 1;$$

- *and*

$$X_i(I_j) = 0, \qquad 1 \le i \le k, \quad 1 \le j \le n - k - 1.$$

Then (1.1) *can be integrated in quadratures.*

Proof. Under assumptions of Theorem 11 all symmetries can be restricted to the surface of the common level of all first integrals $I_i = c_i$ and we can apply Theorem 10.   ∎

Notice that the above statements are *local* and do not give us any information about the global behaviour of solutions, Liouville tori, etc. But on the local level Theorem 11 is a very useful generalization of the famous Liouville theorem from the Hamiltonian mechanics. In this theorem $n = 2m$ and the $m - 1$ integrals $I_1, \dots, I_{m-1}$, which are in involution with respect to the corresponding nondegenerate Poisson bracket $\{\cdot, \cdot\}$, automatically provide $m - 1$ symmetries of the form

$$\frac{d\,u_i}{d\tau_j} = \{u_i, I_j\}, \quad i = 1, \ldots, n, \quad j = 1, \ldots, m-1. \tag{1.28}$$

Taking together with the Hamiltonian H, they form a set of integrals and symmetries satisfying the conditions of Theorem 11 and provide the integrability of dynamical system (1.1) in quadratures.

1.2.5 Classification Problems in Rigid Body Dynamics

Let us consider the Euler–Poinsot equations (1.17) from the viewpoint of Theorem 11. These equations are Hamiltonian. One of the possible Hamiltonian structures is defined by the following Poisson brackets:

$$\{p_i, p_j\} = \varepsilon_{ijk}\, p_k, \qquad \{p_i, q_j\} = \varepsilon_{ijk}\, q_k, \qquad \{q_i, q_j\} = 0, \tag{1.29}$$

where ε_{ijk} is the signum of the permutation (i, j, k) if i, j, k are distinct and $\varepsilon_{ijk} = 0$ otherwise. The Poisson brackets between any two functions $f(p_1, p_2, p_3, q_1, q_2, q_3)$ and $f(p_1, p_2, p_3, q_1, q_2, q_3)$ is defined by

$$\{f, g\} = \sum \{p_i, p_j\} \frac{\partial f}{\partial p_i} \frac{\partial g}{\partial p_j} + \sum \{p_i, q_j\} \frac{\partial f}{\partial p_i} \frac{\partial g}{\partial q_j} + \sum \{q_i, q_j\} \frac{\partial f}{\partial q_i} \frac{\partial g}{\partial q_j}.$$

The function $H = I_1$ defined by formulas (1.18) is the Hamiltonian for (1.17). Integrals I_2 and I_3 are the Casimirs for brackets (1.29) and therefore they produce trivial symmetries (1.28). Thus, for generic equations (1.17) we need only one additional first integral I_4 such that $\{H, I_4\} = 0$. This integral provides one more symmetry and Theorem 11 could be applied.

Note that Eqs. (1.17) are homogeneous if we assign the weight 1 to variables p_i, weight 2 to q_i and weight 1 to t-derivation. Let us try to find an additional homogeneous polynomial first integrals for (1.17).

The simplest classification problem is to find all possible sets of constants a, b, c, x, y, z such that function

$$I_4 = \lambda_1 p_1 + \lambda_2 p_2 + \lambda_3 p_3$$

is the first integral. A trivial calculation leads to the following system of bilinear equations:

$$\lambda_1(b-c) = \lambda_2(a-c) = \lambda_3(a-b) = 0,$$

$$x\lambda_2 - y\lambda_1 = x\lambda_3 - z\lambda_1 = y\lambda_3 - z\lambda_2 = 0.$$

For example we can choose $a = b \neq c$ and $\lambda_3 \neq 0$. Then $x = y = 0$ and we get the Lagrange case with additional integral $I_4 = p_3$.

It is not difficult to verify that an additional integral of weight 2 exists only for the Euler case for which a, b, c are arbitrary and $x = y = z = 0$. The additional integral has the form $I_4 = p_1^2 + p_2^2 + p_3^2$.

Further computations can be performed with the help of special software for study of overdetermined systems of algebraic equations. It turns out that there are no cases with additional integral of weight 3 and there exists only one case with additional integral of weight 4. This is the famous Kowalewski case defined by relations $a = b = 1$, $c = 2$ and $z = 0$. The additional integral has the form

$$I_4 = \left(p_1^2 - p_2^2 - 2xq_1 + 2yq_2\right)^2 + 4\left(p_1p_2 - yq_1 - zq_2\right)^2.$$

It has been shown in [26] that there are no more cases with polynomial additional integrals. Moreover, if (1.17) possesses a single-valued meromorphic integral, then it belongs to one of the above three cases [82].

The classification of integrable cases for the Kirchhoff equations describing the motion of a rigid body in an ideal fluid is much more complicated. These equations are Hamiltonian with respect to the same Poisson brackets (1.29), but here the Hamiltonian function is a generic second-degree homogeneous polynomial of variables p_i, q_i:

$$H = <P, A(P)> + <P, B(Q)> + <Q, C(Q)>,$$

where $P = (p_1, p_2, p_3)$, $Q = (q_1, q_2, q_3)$. Without loss of generality we can assume that the matrices $A = \{a_{ij}\}$ and $C = \{c_{ij}\}$ are symmetric. Together with an arbitrary matrix $B = \{b_{ij}\}$ there are 21 parameters in the Hamiltonian. To reduce the number of parameters we have to use linear transformations which preserve the Poisson brackets. Such transformations are called canonical. The canonical transformations for brackets (1.29) form a six-parameter Lie group consisting of

- orthogonal transformations $\hat{P} = S(P)$, $\hat{Q} = S(Q)$, where $SS^T = E$;
- transformations of the form $\hat{q}_i = q_i$,

$$\hat{p}_1 = p_1 - \mu_1q_2 + \mu_2q_3, \quad \hat{p}_2 = p_2 + \mu_1q_1 + \mu_3q_3, \quad \hat{p}_3 = p_3 - \mu_2q_1 - \mu_3q_2, \tag{1.30}$$

where μ_i are arbitrary parameters.

Using the orthogonal transformations one can bring the matrix A to the diagonal form:

$$A = \begin{pmatrix} a_1 & 0 & 0 \\ 0 & a_2 & 0 \\ 0 & 0 & a_3 \end{pmatrix}.$$

With the help of (1.30) we can transform the matrix B to the symmetric (or to the upper triangular) form.

There are classical integrable cases found by Kirchhoff, Clebsch and Steklov–Lyapunov [63]. For all these cases the matrices B and C are diagonal and the Hamiltonian is of the form

$$H = a_1p_1^2 + a_2p_2^2 + a_3p_3^2 + 2b_{11}p_1q_1 + 2b_{22}p_2q_2 + 2b_{33}p_3q_3 \\ + c_{11}q_1^2 + c_{22}q_2^2 + c_{33}q_3^2.$$

The Kirchhoff case is described by the relations

$$a_1 = a_2, \qquad b_{11} = b_{22}, \qquad c_{11} = c_{22}.$$

This the only case with a linear additional integral $I_4 = p_3$.

For the Clebsch and Steklov–Lyapunov cases the coefficients a_i are arbitrary and the remaining parameters satisfy the following conditions:

$$b_{11} = b_{22} = b_{33},$$
$$\frac{c_{11} - c_{22}}{a_3} + \frac{c_{33} - c_{11}}{a_2} + \frac{c_{22} - c_{33}}{a_1} = 0$$

and

$$\frac{b_{11} - b_{22}}{a_3} + \frac{b_{33} - b_{11}}{a_2} + \frac{b_{22} - b_{33}}{a_1} = 0,$$
$$c_{11} - \frac{(b_{22} - b_{33})^2}{a_1} = c_{22} - \frac{(b_{33} - b_{11})^2}{a_2} = c_{33} - \frac{(b_{11} - b_{22})^2}{a_3},$$

respectively. For each of these cases there exists an additional quadratic integral. It can be proven that for the case $a_i > 0$ any Kirchhoff equations with an additional quadratic integral are equivalent (up to linear canonical transformations) to one of these two cases.

Very recently a new integrable case [57] has been found with the Hamiltonian

$$H = p_1^2 + p_2^2 + 2 p_3^2 + 2 (\mu_1 q_1 + \mu_2 q_2) p_3 - (\mu_1^2 + \mu_2^2) q_3^2. \qquad (1.31)$$

The additional integral I_4 is of fourth degree and can be written in a factorized form $I_4 = k_1 k_2$, where the factors are given by $k_1 = p_3$ and

$$k_2 = \left(p_1^2 + p_2^2 + p_3^2\right) p_3 + 2 \left(\mu_1 p_1 + \mu_2 p_2\right) \left(p_1 q_1 + p_2 q_2\right)$$
$$+ 2 \left(\mu_1 q_1 + \mu_2 q_2\right) p_3^2 + \left(\mu_1 q_1 + \mu_2 q_2\right)^2 p_3$$
$$- \left(\mu_1^2 + \mu_2^2\right) \left(2 p_1 q_1 + 2 p_2 q_2 + p_3 q_3\right) q_3.$$

It turns out that both k_1 and k_2 are invariant relations:

$$\dot{k}_1 = 2 (\mu_2 q_1 - \mu_1 q_2) k_1, \qquad \dot{k}_2 = -2 (\mu_2 q_1 - \mu_1 q_2) k_2.$$

According to a classification theorem from [57], the Kirchhoff equations with Hamiltonian (1.31) is the only integrable case with $a_1 = a_2 \neq a_3$ and an additional first integral of fourth degree.

It turns out that there exists a remarkable nonhomogeneous integrable combination of the Kowalewski Hamiltonian and the Hamiltonian (1.31):

$$\tilde{H} = p_1^2 + p_2^2 + 2 p_3^2 + 2\varepsilon (\mu_1 q_1 + \mu_2 q_2) p_3 - \varepsilon^2 (\mu_1^2 + \mu_2^2) q_3^2 + 2\lambda (\mu_2 q_1 - \mu_1 q_2).$$

One can check that in this case there also exists an additional integral of fourth degree. If we put $\varepsilon = 0$ then the Hamiltonian reduces just to the Kowalewski case. The case $\lambda = 0$ coincides with (1.31).

1.3 Basic Concepts of the Symmetry Approach

1.3.1 Dynamical Variables for Partial Differential Equations

The case of single ODE of nth order can be regarded as a particular case of general dynamic system (1.1) considered in the previous section. We identify equation

$$u_n = f(x, u, u_x, \ldots, u_{n-1}) \tag{1.32}$$

with the vector field

$$D = \frac{\partial}{\partial x} + \sum_{i=0}^{n-2} u_{i+1} \frac{\partial}{\partial u_i} + f \frac{\partial}{\partial u_{n-1}}.$$

This vector field acts on the set of all functions depending on dynamical variables $x, u, u_1 = u_x, \ldots, u_{n-1}$. Usually D is called a derivation in virtue of Eq. (1.32) or total derivative operator with respect to x.

Suppose we have an evolution partial differential equation

$$u_t = F(u, u_1, \ldots, u_n, x, t), \quad n \geq 2, \tag{1.33}$$

where $u_0 = u(x,t), u_1 = u_x(x,t), u_2 = u_{xx}(x,t), \ldots, u_n = \partial_x^n u(x,t)$ and F is an analytic function (often a polynomial) of its arguments. Following Sophus Lie, we shall assume the variables $u = u_0, u_1, \ldots, u_n, \ldots$ to be independent and will call them the *dynamical variables*. In these variables the vector field

$$D = \frac{\partial}{\partial x} + u_1 \frac{\partial}{\partial u_0} + u_2 \frac{\partial}{\partial u_1} + u_3 \frac{\partial}{\partial u_2} + \cdots \tag{1.34}$$

represents the total derivative operator with respect to x.

We will apply the operator D to functions of finite number of dynamical variables and therefore only a finite number of terms in the sum (1.34) is required.

The most adequate to this viewpoint is the language differential algebra where it is assumed that all functions such as F belong to a proper differential field $\mathscr{F}$ [30] generated by u and the derivation D (1.34). We shall assume that $\mathbb{C} \in \mathscr{F}$. Evolution partial differential equation (1.33) defines another derivation of the field $\mathscr{F}$

$$D_t = \frac{\partial}{\partial t} + F_0 \frac{\partial}{\partial u_0} + F_1 \frac{\partial}{\partial u_1} + F_2 \frac{\partial}{\partial u_2} + \cdots, \qquad F_k \in \mathscr{F},$$

where

$$F_0 = F(u,\dots,u_n,x,t), \qquad F_1 = D(F_0), \qquad \dots, \qquad F_n = D^n(F_0), \qquad \dots .$$

The vector field D_t represents the total derivative with respect to time t due to evolutionary equation (1.33). Derivations D_t and D commute:

$$[D_t,D] = \sum_{s=0}(D_t(u_{s+1}) - D(F_s))\frac{\partial}{\partial u_s} = \sum_{s=0}(F_{s+1} - D(F_s))\frac{\partial}{\partial u_s} = 0.$$

Equation (1.33) can be represented by two compatible infinite-dimensional dynamical systems

$$D(u_s) = u_{s+1}, \qquad D_t(u_s) = F_s, \qquad s = 0,1,2,\dots . \tag{1.35}$$

Example 12. For the Korteweg–de Vries equation

$$u_t = u_{xxx} + 6uu_x, \tag{1.36}$$

function $F_0 = u_3 + 6uu_1$ and first few equations of the system (1.35) are of the form

$$\begin{aligned}
D(u) &= u_1, & D_t(u) &= u_3 + 6uu_1, \\
D(u_1) &= u_2, & D_t(u_1) &= u_4 + 6uu_2 + 6u_1^2, \\
D(u_2) &= u_3, & D_t(u_2) &= u_5 + 6uu_3 + 18u_1u_2,
\end{aligned}$$

$$\dots$$

1.3.2 Fréchet Derivative, Euler's Operator and Formal Pseudo-differential Series

Definition 13. For any element $a \in \mathscr{F}$ the Fréchet derivative is defined as a linear differential operator of the form

$$a_* = \sum_k \frac{\partial a}{\partial u_k}D^k.$$

The order of function a is defined as the order of the differential operator a_* (i.e. the maximal power of D). We denote a_*^+ the formally conjugated operator

$$a_*^+ = \sum_k(-1)^k D^k \circ \frac{\partial a}{\partial u_k}.$$

Definition 14. The Euler operator or the variational derivative of $a \in \mathscr{F}$ is defined as

$$\frac{\delta a}{\delta u} = \sum_k (-1)^k D^k \left(\frac{\partial a}{\partial u_k}\right) = a_*^+(1).$$

If function a is a total derivative $a = D(b), b \in \mathscr{F}$ (we say that $a \in \mathrm{Im}(D)$, and $\mathrm{Im}(D)$ is defined as the image $D : \mathscr{F} \to \mathrm{Im}(D)$ of the derivation D) then the variational derivative vanishes. Moreover the vanishing of the variational derivative is almost a criteria that the function belongs to $\mathrm{Im}(D)$ [21]:

Theorem 15. *For $a \in \mathscr{F}$ the variational derivative vanishes*

$$\frac{\delta a}{\delta u} = 0$$

if and only if $a \in \mathrm{Im}(D) + \mathbb{C}$.

Here we list a few useful identities:

$$(ab)_* = ab_* + ba_*, \tag{1.37}$$

$$(D(a))_* = D \circ a_* = D(a_*) + a_* \circ D, \tag{1.38}$$

$$(D_t(a))_* = D_t(a_*) + a_* \circ F_*, \tag{1.39}$$

$$(a_*(b))_* = D_b(a_*) + a_* \circ b_*, \tag{1.40}$$

$$\left(\frac{\delta a}{\delta u}\right)_* = \left(\frac{\delta a}{\delta u}\right)_*^+, \tag{1.41}$$

$$\frac{\delta}{\delta u}(D_t(a)) = D_t\left(\frac{\delta a}{\delta u}\right) + F_*^+\left(\frac{\delta a}{\delta u}\right), \tag{1.42}$$

which are valid for any $a, b, F \in \mathscr{F}$.

For further consideration we will need formal pseudo-differential series, which for simplicity we shall call formal series (of order $m = \mathrm{ord}A$)

$$A = a_m D^m + a_{m-1} D^{m-1} + \cdots + a_0 + a_{-1} D^{-1} + a_{-2} D^{-2} + \cdots, \quad a_k \in \mathscr{F}. \tag{1.43}$$

The product of two formal series is defined by

$$aD^k \circ bD^m = a\left(bD^{m+k} + C_k^1 D(b)D^{k+m-1} + C_k^2 D^2(b)D^{k+m-2} + \cdots\right), \tag{1.44}$$

where $k, m \in \mathbb{Z}$ and C_n^j is the binomial coefficient

$$C_n^j = \frac{n(n-1)(n-2)\cdots(n-j+1)}{j!}.$$

This product is associative.

A conjugated formal series A^+ is defined as

$$A^+ = (-1)^m D^m \circ a_m + \cdots + a_0 - D^{-1} \circ a_{-1} + D^{-2} \circ a_{-2} + \cdots.$$

Example 16. Let

$$A = uD^2 + u_1 D, \qquad B = -u_1 D^3, \qquad C = uD^{-1},$$

then

$$A^+ = D^2 \circ u - D \circ u_1 = A, \qquad B^+ = D^3 \circ u_1 = u_1 D^3 + 3u_2 D^2 + 3u_3 D + u_4,$$
$$C^+ = -D^{-1} \circ u = -uD^{-1} + u_1 D^{-2} - u_2 D^{-3} + \cdots.$$

Formal series form a skew-field. For any element (1.43) we can find uniquely the inverse element

$$B = b_{-m} D^{-m} + b_{-m-1} D^{-m-1} + \cdots, \qquad b_k \in \mathscr{F},$$

such that $A \circ B = B \circ A = 1$. Indeed, multiplying A and B and equating the result to 1 we find that $a_m b_{-m} = 1$, i.e. $b_{-m} = 1/a_m$, then at D^{-1} we have

$$m a_m D(b_{-m}) + a_m b_{-m-1} + a_{m-1} b_{-m} = 0$$

and therefore

$$b_{-m-1} = -\frac{a_{m-1}}{a_m^2} - mD\left(\frac{1}{a_m}\right), \qquad \text{etc.}$$

First k coefficients of the series B can be uniquely determined in terms of the first k coefficients of A.

Moreover we can find the mth root of the series A (1.43), i.e. a series

$$C = c_1 D + c_0 + c_{-1} D^{-1} + c_{-2} D^{-2} + \cdots$$

such that $C^m = A$ and if we know first k coefficients of the series A we can find the first k coefficients of the series C.

Example 17. Let $A = D^2 + u$. Assuming

$$C = c_1 D + c_0 + c_{-1} D^{-1} + c_{-2} D^{-2} + \cdots$$

we compute (using (1.44))

$$C^2 = C \circ C = \left(c_1 D + c_0 + c_{-1} D^{-1} + \cdots\right) \circ \left(c_1 D + c_0 + c_{-1} D^{-1} + \cdots\right) =$$
$$c_1^2 D^2 + (c_1 D(c_1) + c_1 c_0 + c_0 c_1)D + c_1 D(c_0) + c_0^2 + c_1 c_{-1} + c_{-1} c_1 + \cdots,$$

and compare the result with A. At D^2 we find $c_1^2 = 1$ or $c_1 = \pm 1$. Let us choose the positive root $c_1 = 1$. Now at D we have $2c_0 = 0$, i.e. $c_0 = 0$. At D^0 we have $2c_{-1} = u$, at D^{-1} we find $c_{-2} = -u_1/4$, etc.

$$C = D + \frac{u}{2} D^{-1} - \frac{u_1}{4} D^{-2} + \cdots.$$

We can easily find as many coefficients of C as required.

Definition 18. The residue of a formal series $A = \sum_{k \leq n} a_k D^k$, $a_k \in \mathscr{F}$, is the coefficient at D^{-1}

$$res(A) = a_{-1}.$$

The logarithmic residue of A is defined as

$$res \log A = \frac{a_{n-1}}{a_n}.$$

For any two formal series A, B of order n and m, respectively, the logarithmic residue satisfies the following identity:

$$res \log(A \circ B) = res \log(A) + res \log(B) + nD(\log(b_m)).$$

For any derivation D_t of the field $\mathscr{F}$ and any formal series A we have

$$D_t(res \log(A)) = res(D_t(A) \circ A^{-1}). \tag{1.45}$$

We will use the following important Adler's Theorem [10].

Theorem 19. *For any two formal series A, B the residue of the commutator belongs to $Im(D)$:*

$$res[A, B] = D(\sigma(A, B)),$$

where

$$\sigma(A, B) = \sum_{p \leq ord(B),\, q \leq ord(A)}^{p+q+1 > 0} C_q^{p+q+1} \sum_{s=0}^{p+q} (-1)^s D^s(a_q) D^{p+q-s}(b_q).$$

1.3.3 Infinitesimal Symmetries of Evolution PDEs

Here we will give a few equivalent definitions of infinitesimal symmetries of Eq. (1.33). The definition of symmetries for PDEs is very similar to the definition for ODEs (compare with Sect. 1.3).

Traditionally symmetries of equations are defined as transformations which map solutions of the equation into solutions. Suppose u is an arbitrary solution of Eq. (1.33). Let us consider an infinitesimal transformation

$$\hat{u} = u + \tau G(u, \ldots, u_m, x, t) \tag{1.46}$$

which depends on small parameter τ. We say that the transformation (1.46) defines an infinitesimal symmetry of Eq. (1.33) if $\hat{u}$ satisfies equation

$$\hat{u}_t = F(\hat{u}, \ldots, \hat{u}_n, x, t) + O(\tau^2).$$

If we substitute $\hat{u}$ (1.46) in (1.33) and request the cancellation of terms of order τ, we receive

$$D_t(G(u,\ldots,u_m,x,t)) = F_*(G(u,\ldots,u_m,x,t)), \tag{1.47}$$

where F_* denotes the Fréchet derivative of F.

The generator of a symmetry of Eq. (1.33) can be defined as a function $G(u,\ldots,u_m,x,t)$ which satisfies the corresponding linearized equation (1.47).

A symmetry of Eq. (1.33) can also be defined as a derivation

$$X = g_0\frac{\partial}{\partial u_0} + g_1\frac{\partial}{\partial u_1} + g_2\frac{\partial}{\partial u_2} + \cdots, \qquad g_k \in \mathscr{F}, \tag{1.48}$$

of $\mathscr{F}$, which commutes with the derivations D and D_t (compare with Sect. 2.1). It follows from $[D,X] = 0$ that $g_k = D^k(g_0)$.

We call derivation (1.48) which commutes with D *evolutionary*. Any evolutionary derivation has the following form:

$$D_G = G\frac{\partial}{\partial u_0} + D(G)\frac{\partial}{\partial u_1} + D^2(G)\frac{\partial}{\partial u_2} + \cdots, \qquad G_k \in \mathscr{F}, \tag{1.49}$$

for some function $G \in \mathscr{F}$. We call this function G *generator* of evolutionary derivation (1.49). All evolutionary derivations form a Lie algebra with respect to the standard commutator

$$D_K = D_G \circ D_H - D_H \circ D_G \tag{1.50}$$

of vector fields. The generator K of the commutator is given by

$$K = H_*(G) - G_*(H). \tag{1.51}$$

Formula (1.51) defines a Lie bracket on our differential field $\mathscr{F}$.

Suppose we have two symmetries D_G and D_H, then the commutator (1.50) commutes with D_t due to the Jacobi identity for vector fields and therefore also corresponds to a symmetry. A linear combination of symmetries with constant coefficients is also a symmetry. In other words, infinitesimal symmetries of an equation form a Lie algebra over $\mathbb{C}$, which is a subalgebra of the Lie algebra of all evolutionary derivations.

It is easy to verify that the condition $[D_t, D_G] = 0$ is equivalent to (1.47). Since $D_t = \frac{\partial}{\partial t} + D_F$ we see that D_F is a symmetry for Eq. (1.33) iff the function F does not depend on t explicitly.

There is one-to-one correspondence between the evolutionary derivations and evolution partial differential equations. The derivation D_G (1.49) corresponds to equation

$$u_\tau = G(u,\ldots,u_m,x,t). \tag{1.52}$$

Since D_t and D_G commute, Eqs. (1.33) and (1.52) are compatible. Often evolution equations (1.52) which are compatible with (1.33) are called *symmetries*. The order of a symmetry is the order of its generator G.

The new time τ in (1.52) plays the role of a group parameter. In order to find a one-parameter family $u(x,t,\tau)$ of solutions including a given solution $u(x,t)$ of Eq. (1.33), we have to solve Eq. (1.52) with initial data $u(x,t)$.

Example 20. For the KdV equation (1.36) we have obvious symmetries with generators

$$G_1 = u_1, \qquad G_3 = u_t = u_3 + 6uu_1$$

of order 1 and 3 which correspond to the shifts in space and time. The Galilean and scaling transformations are generated by

$$G_g = 1 + 6tu_1, \qquad G_s = 2u + xu_1 + 3t(u_3 + 6uu_1).$$

There are infinitely many high-order symmetries for the KdV equation; the first nontrivial one has order 5 and is of the form

$$G_5 = u_5 + 10uu_3 + 20u_1u_2 + 30u^2u_1.$$

It is easy to verify that all these functions are indeed generators of symmetries according to the definitions given above.

Symmetries of PDEs help a lot to find partial solutions. Suppose Eq. (1.33) has a symmetry with a generator $G(u, \ldots, u_m, x, t)$. There is a subclass of solutions of (1.33) which is invariant with respect to this symmetry, i.e. corresponding symmetry transformation (1.46) does not change the solution, or in other words $G(u, \ldots, u_m, x, t) = 0$ for such solutions. The condition $G = 0$ preserves with time; indeed, function G satisfies linear equation (1.47) and $G = 0$ is the obvious solution and if the condition $G = 0$ was true at initial time it remains to be true at any time. The condition $G = 0$ allows us to reduce the PDE or the corresponding infinite-dimensional dynamical system (1.35) to an ODE (a finite-dimensional dynamical system). In order to do the corresponding symmetry reduction we should resolve equation $G = 0$ and express the highest derivative u_m in terms of the lower derivatives $u_m = g(u, u_1, \ldots, u_{m-1}, x, t)$. Now we have only m-independent dynamical variables $u, \ldots, u_{m-1}$ and all other variables can be expressed in their terms. For example

$$u_{m+1} = D(u_m) = D(g) = g_x + \sum_{k=0}^{m-2} u_{k+1}\frac{\partial g}{\partial u_k} + g\frac{\partial g}{\partial u_{m-1}}.$$

After such reduction the system (1.35) gives two compatible m-dimensional dynamical systems.

1.3.4 Formal Recursion Operator

For simplicity here and in the sequel we assume that the right-hand side of equation

$$u_t = F(u, \ldots, u_n, x), \qquad n \geq 2, \tag{1.53}$$

and also its symmetries and conservation laws do not depend on time explicitly. Our theoretical constructions are based on the fundamental concept of *formal recursion operator*,[3] which is in the core of the Symmetry Approach.

Definition 21. A formal series

$$\Lambda = l_1 D + l_0 + l_{-1} D^{-1} + \cdots, \qquad l_k \in \mathscr{F}, \tag{1.54}$$

is called a formal recursion operator for Eq. (1.53) if it satisfies equation

$$D_t(\Lambda) - [F_*, \Lambda] = 0. \tag{1.55}$$

Here we have to make one important stipulation that in our approach we will never consider Λ as an operator, i.e. we will never act by Λ to any function. Relation (1.55) is regarded as an infinite system of equations for the coefficients l_i. In the cases when the formal series (1.54) or its power is finite or can be summed up and represented in the form of a pseudo-differential operator, it gives us a *recursion operator* which maps symmetries of Eq. (1.53) into symmetries. Indeed, if the action of Λ is properly defined on a symmetry generator G (i.e. $\Lambda(G) \in \mathscr{F}$), then $H = \Lambda(G)$ satisfies Eq. (1.47):

$$\begin{aligned}
D_t(H) = D_t(\Lambda G) &= D_t(\Lambda)(G) + \Lambda D_t(G) \\
&= [F_*, \Lambda](G) + \Lambda F_*(G) = F_* \Lambda(G) = F_*(H)
\end{aligned}$$

and according to our definition H is a generator of a symmetry.

Proposition 22. *If Λ is a recursion operator for Eq. (1.53), then any power $\tilde{\Lambda} = \Lambda^k$ also satisfies Eq. (1.55). In particular,*

$$\hat{\Lambda} = c_1 \Lambda + c_0 + c_{-1} \Lambda^{-1} + c_{-2} \Lambda^{-2} + \cdots$$

is a formal recursion operator for (1.53) for any $c_k \in \mathbb{C}$.

The coefficients of the formal recursion operator can be found from Eq. (1.55).

Example 23. Let us consider equations of the KdV type

$$u_t = u_3 + f(u_1, u) \tag{1.56}$$

and find a few coefficients $l_1, l_0, \ldots$ of the formal recursion operator Λ. We substitute

$$F_* = D^3 + \frac{\partial f}{\partial u_1} D + \frac{\partial f}{\partial u}, \qquad L = l_1 D + l_0 + l_{-1} D^{-1} + \cdots$$

in (1.55) and collect coefficients at $D^3, D^2, \ldots$.

---

[3] In our previous publications [42, 45, 55] we called it a formal symmetry. We think that the term *formal recursion operator* is more adequate for many reasons.

We obtain

$$D^3: \quad 3D(l_1) = 0; \qquad D^2: \quad 3D^2(l_1) + 3D(l_0) = 0;$$
$$D: \quad D^3(l_1) + 3D^2(l_0) + 3D(l_{-1}) + \frac{\partial f}{\partial u_1} D(l_1) = (l_1)_t + l_1 D\left(\frac{\partial f}{\partial u_1}\right).$$

From the first equation it follows that l_1 is a constant and we set $l_1 = 1$. Now, from the second equation, it follows that l_0 is a constant and we choose $l_0 = 0$ (any constant is a trivial solution of Eq. (1.55)). It follows from the third equation that

$$D(l_{-1}) = D\left(\frac{1}{3}\frac{\partial f}{\partial u_1}\right),$$

and therefore

$$l_{-1} = \frac{1}{3}\frac{\partial f}{\partial u_1} + c_{-1}, \qquad c_{-1} \in \mathbb{C}.$$

The constant of integration c_{-1} can be set equal to zero without loss of generality (Proposition 22). Therefore

$$\Lambda = D + \frac{1}{3}\frac{\partial f}{\partial u_1}D^{-1} + \cdots. \tag{1.57}$$

The concept of formal recursion operator is very universal in the theory of integrable equations. If Eq. (1.53) possesses an infinite hierarchy of symmetries [28] or conservation laws [69] of arbitrary high order or can be linearized by a differential substitution [70] the formal series Λ satisfying Eq. (1.55) exists and the sequence of its coefficients $l_1, l_0, \ldots \in \mathscr{F}$ can be found explicitly. Below we formulate and illustrate some main results of the theory, the details of proofs one can be found in original papers or in reviews [42, 45, 55].

Theorem 24. *If Eq. (1.53) possesses an infinite hierarchy of higher symmetries of infinitely increasing order then it has a formal recursion operator.*

The main idea of the proof of Theorem 24 and the relation between the structure of the formal recursion operator and symmetries can be illustrated by the following consideration. Suppose Eq. (1.53) has a symmetry with a generator G. Function G satisfies Eq. (1.47). Let us compute the Fréchet derivative from this equation. Using identities (1.37), (1.39), (1.40) we get equation

$$D_t(G_*) + G_*F_* = D_G(F_*) + F_*G_*,$$

which can be rearranged in the form

$$D_t(G_*) - [F_*, G_*] = D_G(F_*). \tag{1.58}$$

Now let us assume that the order of Eq. (1.53) is fixed, say $n = 3$ (i.e. $F = F(u, u_1, u_2, u_3, x)$ in (1.53)), and the symmetry G has a very high order (say, for example, $m = 125$, i.e. $G = G(u, u_1, \ldots, u_{125}, x)$). Equation (1.58) for operators (the

Fréchet derivative is a differential operator) is understood as equations for the coefficients of the operators at each power D^k. In the right-hand side of Eq. (1.58) we have operator $D_G(F_*)$ of order 3 (or less, if the leading coefficient of F_* is a constant). The product F_*G_* in the left-hand side of the equation has the order $n + m = 3 + 125 = 128$. It means that in first $128 - 3 = 125$ equations the right-hand side does not contribute and first 124 terms of operator G_* satisfy the same equation (1.55) as the formal recursion operator Λ. We can use $G_*^{1/125}$ as an approximate for Λ or, more precisely,

$$\Lambda = (G_*)^{1/m} + \tilde{l}_{-123}D^{-123} + \tilde{l}_{-124}D^{-124} + \cdots. \tag{1.59}$$

If Eq. (1.53) has an infinite hierarchy of symmetries G_s and the order of symmetries is going to infinity as $s \to \infty$ then one can show that there exists a formal series Λ, such that Eq. (1.55) is satisfied at any order D^k, $k = n, n-1, \ldots, 0, -1, \ldots$. That is the basic idea for the proof of Theorem 24.

1.3.5 Conservation Laws

In contrast to symmetries, the notion of first integrals cannot be generalized to the case of PDEs. It is replaced by the concept of local conservation laws, which can also be related to constants of motion.

Definition 25. A function $\rho \in \mathscr{F}$ is called a density of a local conservation law of Eq. (1.33) if there exists a function $\sigma \in \mathscr{F}$ such that

$$D_t(\rho) = D(\sigma). \tag{1.60}$$

Equation (1.60) is evidently satisfied if $\rho = D(h)$ for any $h \in \mathscr{F}$. In this case $\sigma = D_t(h)$. We call such "conservation laws" trivial.

Definition 26. Two conserved densities ρ_1, ρ_2 are called equivalent $\rho_1 \sim \rho_2$ if the difference $\rho_1 - \rho_2$ is a trivial density (i.e. $\rho_1 - \rho_2 \in \mathrm{Im}(D)$).

Definition 27. The order $\mathrm{ord}(\rho)$ of a conserved density ρ is defined as the order of the differential operator

$$R = \left(\frac{\delta \rho}{\delta u} \right)_*.$$

For trivial densities $\delta \rho / \delta u = 0$ (see Theorem 15) and therefore equivalent densities have the same order. For example, densities $\rho_1 = u_1^2 + u_3$ and $\rho_2 = -uu_2$ are equivalent and according to our Definition 27, we have $\mathrm{ord}(\rho_1) = \mathrm{ord}(\rho_2) = 2$. In literature the order of a conserved density ρ is often defined as the minimal order of densities equivalent to ρ. Using this definition we get $\mathrm{ord}(\rho_1) = 1$. It is easy to prove that for the scalar equations (1.33) the latter order always differs by a factor

2 from ours. We prefer our definition because it is more suitable for the systems of evolution equations.

A linear combination of conserved densities with constant coefficients is also a conserved density. Therefore the set of conserved densities forms a linear space, actually a factor space over $\text{Im}(D)$.

Example 28. Functions $\rho_1 = u$, $\rho_2 = u^2$, $\rho_3 = -u_1^2 + 2u^3$, $\hat{\rho}_3 = uu_2 + 2u^3$ are conserved densities of the Korteweg–de Vries equation (1.36). Indeed,

$$D_t(u) = D\left(u_2 + 3u^2\right), \qquad D_t(u^2) = D\left(2uu_2 - u_1^2 + 4u^3\right),$$
$$D_t(\rho_3) = D\left(9u^4 + 6u^2u_2 + u_2^2 - 12uu_1^2 - 2u_1u_3\right).$$

Densities ρ_3 and $\hat{\rho}_3$ are equivalent, $\hat{\rho}_3 - \rho_3 = D(u_1)$. Densities ρ_1, ρ_2 are of zero order, $\text{ord}\rho_3 = \text{ord}(\hat{\rho}_3) = 2$. Function u^3 is not a density of a conservation law for the Korteweg–de Vries equation. Indeed, $D_t(u^3) = 3u^2u_3 + 18u^3u_1$. In order to check that the right-hand side is not a total derivative we apply the Euler operator (Theorem 15)

$$\frac{\delta}{\delta u}\left(3u^2u_3 + 18u^3u_1\right) = -18u_1u_2 \neq 0.$$

If u is a periodic (in x) function with period L, then $I_k = \int_0^L \rho_k\,dx$ do not depend on time and are constants of motion.

If we substitute a density of a conservation law in the identity (1.42) we find that $\frac{\delta\rho}{\delta u}$ is a co-symmetry (i.e. it satisfies the equation conjugated to (1.47)):

$$D_t\left(\frac{\delta\rho}{\delta u}\right) + F_*^+\left(\frac{\delta\rho}{\delta u}\right) = 0. \tag{1.61}$$

Theorem 29. *Evolution equation of even order*

$$u_t = F(u, u_1, \ldots, u_{2n}, x) \tag{1.62}$$

cannot have a conserved density ρ of order higher than $2n$.

Proof. We prove the theorem by a contradiction. Let us assume that Eq. (1.62) has a conserved density ρ and $\text{ord}(\rho) = k > 2n$. The variational derivative of ρ satisfies Eq. (1.61). Let us compute the Fréchet derivative from Eq. (1.61) using identities (1.37), (1.38), (1.39). The result can be represented as

$$D_t(R) + R \circ F_* + F_*^+ \circ R = Q, \tag{1.63}$$

where

$$R = \left(\frac{\delta\rho}{\delta u}\right)_*, \qquad Q = -\sum_{k-1}^{2n}(-1)^k D^k \circ \left(\frac{\delta\rho}{\delta u}(F_k)_*\right)$$

and F_k are the coefficients of the Fréchet derivative

$$F_* = F_{2n}D^{2n} + F_{2n-1}D^{2n-1} + \cdots + F_0, \quad F_{2n} \neq 0.$$

It is easy to see that the differential operator Q has order $4n$ or less. Order of $R \circ F_*$ and $F_*^+ \circ R$ is equal to $k + 2n$. Substituting $R = r_k D^k + r_{k-1} D^{k-1} + \cdots$ $(r_k \neq 0)$ in (1.63) and collecting terms at D^{k+2n} we get a contradiction

$$2 r_k F_{2n} = 0.$$

∎

The proof of Theorem 29, and in particularly relation (1.63), motivates the following definition:

Definition 30. A formal series

$$S = s_m D^m + s_{m-1} D^{m-1} + \cdots + s_0 + s_{-1} D^{-1} + \cdots, \qquad s_m \neq 0, \qquad (1.64)$$

is called a formal symplectic operator of order m for Eq. (1.33) if it satisfies equation

$$D_t(S) + S \circ F_* + F_*^+ \circ S = 0. \qquad (1.65)$$

Notice that if Eq. (1.33) has a Hamiltonian structure, then the symplectic operator (inverse to the Hamiltonian operator) satisfies Eq. (1.64) [13]. It is easy to verify that the ratio $S_1^{-1} S_2$ of any two solutions of (1.65) satisfies Eq. (1.55) for the formal recursion operator.

Identity (1.63) is a key point in the proof [69] of the following statements:

Theorem 31. *If Eq. (1.33) possesses an infinite hierarchy of conserved densities of infinitely increasing order, then it has a formal symplectic operator.*

Theorem 32. *If Eq. (1.53) possesses an infinite hierarchy of higher conserved densities of infinitely increasing order, then it has a formal recursion operator.*

1.3.6 Canonical Densities and Necessary Integrability Conditions

In this section we formulate necessary conditions for the existence of high-order symmetries or conservation laws. These conditions are formulated in terms of a sequence of *canonical conservation laws*.

For Eq. (1.53) with formal recursion operator Λ we define a sequence of *canonical densities*.

Definition 33. The functions

$$\rho_i = \text{res}(\Lambda^i), \qquad i = -1, 1, 2, \ldots, \text{ and} \qquad \rho_0 = \text{res} \log(\Lambda)$$

are called canonical densities for Eq. (1.53).

Theorem 34. *If Eq. (1.53) has a formal recursion operator, then canonical densities*

$$\rho_i = \text{res}(\Lambda^i), \quad i = -1, 1, 2, \ldots, \text{and} \quad \rho_0 = \text{res}\log(\Lambda)$$

are defined local conservation laws

$$D_t(\rho_i) = D(\sigma_i), \quad \sigma_i \in \mathscr{F}, \quad i = -1, 0, 1, 2, \ldots, \tag{1.66}$$

for Eq. (1.53).

Proof. If a formal series Λ satisfies Eq. (1.55), so does a formal series Λ^k, $k = -1, 1, 2, 3, \ldots$. Using Adler's Theorem 19 we get

$$D_t(\rho_k) = D_t(\text{res}(\Lambda^k)) = \text{res}([F_*, \Lambda^k]) = D(\sigma_k) \in \text{Im}(D), \quad k = -1, 1, 2, 3, \ldots.$$

It follows from identity (1.45) and Theorem 19 that

$$D_t(\rho_0) = \text{res}(D_t(\Lambda)\Lambda^{-1}) = \text{res}([F_*, \Lambda]\Lambda^{-1})$$
$$= \text{res}([F_*\Lambda^{-1}, \Lambda]) = D(\sigma_0) \in \text{Im}(D).$$

∎

Theorem 35. *Under the assumptions of Theorem 32 all even canonical densities ρ_{2j} are trivial.*

Example 36. The Korteweg–de Vries equation $u_t = u_3 + 6uu_1$ has a recursion operator

$$\hat{\Lambda} = D^2 + 4u + 2u_1 D^{-1},$$

which satisfies Eq. (1.55). The formal recursion operator for the Korteweg–de Vries equation can be represented as $\Lambda = \hat{\Lambda}^{1/2}$. The infinite hierarchy of commutative symmetries of KdV can be obtained as

$$G_{2k+1} = \hat{\Lambda}^k(u_1). \tag{1.67}$$

The first five canonical densities for the KdV equation (Example 36) are

$$\rho_{-1} = 1, \quad \rho_0 = 0, \quad \rho_1 = 2u, \quad \rho_2 = 2u_1, \quad \rho_3 = 2u_2 + u^2.$$

Example 37. The Burgers equation $u_t = u_2 + 2uu_1$ has the (formal) recursion operator

$$\Lambda = D + u + u_1 D^{-1}.$$

Functions $G_n = \Lambda^n(u_1)$ are generators of symmetries of the Burgers equation. The canonical densities for the Burgers equation are

$$\rho_{-1} = 1, \quad \rho_0 = u, \quad \rho_1 = u_1, \quad \rho_2 = u_2 + 2uu_1, \ldots.$$

Since ρ_0 is not trivial, the Burgers equation cannot possess an infinite series of conservation law. This fact also follows from Theorem 29.

Thus, if we have an equation of the form (1.53) and know the (formal) recursion operator, we can construct a sequence of canonical densities, which gives us a sequence of conservation laws (some of them, or even all, may be trivial).

Question 38. How to check that for a given equation (1.53) a formal recursion operator exists? What are the obstacles for the existence of a formal recursion operator for a given equation?

The coefficients of a formal recursion operator Λ can be found directly from the linear equation (1.55). First $n-1$ coefficients $l_1, l_0, \ldots, l_{3-n}$ of Λ coincide with the first $n-1$ coefficients of the formal series $(F_*)^{1/n}$. Indeed, since the right-hand side F of Eq. (1.53) does not depend on time explicitly, it generates a symmetry (a time shift) and we can use the ansatz

$$\Lambda = (F_*)^{1/n} + \tilde{l}_{2-n} D^{2-n} + \tilde{l}_{1-n} D^{1-n} + \cdots$$

(compare with (1.59)).

Having first $n-1$ coefficients of Λ we can find $n-1$ canonical densities $\rho_{-1}, \rho_0, \ldots, \rho_{n-2}$ explicitly (in terms of the coefficients $F_i = \frac{\partial F}{\partial u_i}$ of the Fréchet derivative $F_* = F_n D^n + F_{n-1} D^{n-1} + \cdots + F_0$). Equating coefficients at D in Eq. (1.55) it can be shown that the first unknown coefficient l_{2-n} of Λ can be found (as element of $\mathscr{F}$) if and only if the first canonical density

$$\rho_{-1} = F_n^{-\frac{1}{n}} \tag{1.68}$$

is a density of a local conservation law for Eq. (1.53), i.e. there exists such function $\sigma_{-1} \in \mathscr{F}$ that $D_t(\rho_{-1}) = D(\sigma_{-1})$. Coefficient l_{2-n} can be expressed explicitly in terms of the coefficients $F_n, \ldots, F_0$ and σ_{-1}. If $D_t(\rho_{-1}) \notin \mathrm{Im}(D)$, which we can easily verify applying the Euler operator and checking that $\delta D_t(\rho_{-1})/\delta u \neq 0$ (Theorem 15), then the formal recursion operator does not exist and consequently Eq. (1.53) cannot have infinite hierarchy of higher symmetries or conservation laws. Similarly the next coefficient l_{1-n} can be found (as element of $\mathscr{F}$) if and only if the canonical density ρ_0 is conserved; then l_{1-n} can be explicitly expressed in terms of $F_n, \ldots, F_0, \sigma_{-1}, \sigma_0$, etc. In such a way we could obtain as many coefficients of the formal recursion operator Λ as we wish, unless we meet an obstacle: it may happen that we find a canonical density ρ_k such that it does not define a conservation law, i.e. $D_t(\rho_k) \notin \mathrm{Im}(D)$, and therefore element $\sigma_k \in \mathscr{F}$, such that $D_t(\rho_k) = D(\sigma_k)$ does not exist!

Example 39. Let us consider evolution equations of second order

$$u_t = F(x, u, u_1, u_2). \tag{1.69}$$

Computations described above show that the densities of three first canonical conservation laws (1.66) can be written in the form

$$\rho_{-1} = \left(\frac{\partial F}{\partial u_2}\right)^{-1/2},$$

$$\rho_0 = \left(\frac{\partial F}{\partial u_2}\right)^{-1/2} \sigma_{-1} - \left(\frac{\partial F}{\partial u_2}\right)^{-1} \left(\frac{\partial F}{\partial u_1}\right),$$

$$\rho_1 = \rho_{-1}\frac{\partial F}{\partial u} - \frac{\rho_0^2}{4\rho_{-1}} + \frac{\rho_0\sigma_{-1}}{2} - \frac{\rho_{-1}\sigma_0}{2}.$$

1.3.7 Simple Classification Problems

It follows from Theorem 24 or 32 that canonical conservation laws provide necessary conditions for the existence of higher symmetries or conservation laws. We shall call the fact that the function ρ_i is a density of a local conservation law *ith integrability condition* for Eq. (1.53). Using these conditions we can prove nonintegrability of given equations.

Example 40. It is known that partial differential equations

$$u_t = u^n u_n, \qquad n = 2, 3, \tag{1.70}$$

are integrable (i.e. possesses infinitely many symmetries). The question is: whether any equation of such type is integrable for $n > 3$? For Eq. (1.70) we have (1.68)

$$\rho_{-1} = \frac{1}{u}.$$

Let us verify the condition that ρ_{-1} is a conserved density. The right-hand side of expression

$$D_t(\rho_{-1}) = -u^{n-2}u_n$$

should be a total derivative of a function from $\mathscr{F}$, i.e. belong to $\text{Im}(D)$. Thus, if we apply the Euler operator (i.e. take the variational derivative) $\delta/\delta u$, we should receive zero. The result is

$$\frac{\delta}{\delta u}\left(u^{n-2}u_n\right) = (-1)^n D^n \left(u^{n-2}\right) + (n-2)u^{n-3}u_n.$$

It is zero for $n = 2, 3$ and different from zero for any $n > 3$. Conclusion: for $n > 3$ Eq. (1.70) does not pass the test for integrability and therefore it cannot possess higher symmetries or a hierarchy of conservation laws.

Let us consider equations of the KdV type (1.56) and find restrictions on the function $f(u_1, u)$ which follows from first two nontrivial integrability conditions.

It follows from (1.57) that the first nontrivial canonical density is

$$\rho_1 = l_{-1} = \frac{1}{3}\frac{\partial f}{\partial u_1}.$$

Thus for any integrable equation of the KdV type (1.56) we should have

$$D_t\left(\frac{\partial f}{\partial u_1}\right) = D(\sigma_1), \qquad \sigma_1 \in \mathcal{F}. \tag{1.71}$$

Example 41. For the mKdV equation $u_t = u_3 + 3u^2 u_1$ we have

$$\rho_1 = u^2$$

and it is indeed a conserved density

$$D_t(u^2) = 2uD_t(u) = 2u(u_3 + 3u^2 u_1) = D\left(2u_2 - u_1^2 + \frac{3}{2}u^4\right).$$

Applying the Euler operator $\delta/\delta u$ to (1.71) we find an explicit form

$$0 = \frac{\delta}{\delta u}D_t\left(\frac{\partial f}{\partial u_1}\right) = 3u_4\left(u_2\frac{\partial^4 f}{\partial u_1^4} + u_1\frac{\partial^4 f}{\partial u_1^3 \partial u}\right) + \cdots$$

of the first integrability condition. The identity

$$u_2\frac{\partial^4 f}{\partial u_1^4} + u_1\frac{\partial^4 f}{\partial u_1^3 \partial u} = 0$$

gives rise to

$$f(u_1, u) = \lambda u_1^3 + A(u)u_1^2 + B(u)u_1 + C(u),$$

where λ is a constant.

For such f the first condition turns out to be equivalent to

$$\lambda A' = 0, \qquad\qquad B''' + 8\lambda B' = 0,$$

$$(B'C)' = 0, \qquad\qquad AB' + 6\lambda C' = 0.$$

The second integrability condition for the KdV-type Eq. (1.56) has the form

$$D_t\left(\frac{\partial f}{\partial u}\right) = D(\sigma_2).$$

Using this fact we can derive a few more differential relations between $A(u)$, $B(u)$, $C(u)$. Solving them all together we obtain the following list of equations:

$$u_t = u_{xxx} + \left(c_1 u^2 + c_2 u + c_3\right)u_x,$$

$$u_t = u_{xxx} + c_1 u_x^3 + c_2 u_x^2 + c_3 u_x + c_4,$$

$$u_t = u_{xxx} - \frac{1}{2}u_x^3 + \left(c_1 e^{2u} + c_2 e^{-2u} + c_3\right)u_x,$$

where c_1, c_2, c_3, c_4 are arbitrary constants. In the latter equation we normalize λ to $-1/2$ by a scaling. Only these equations (of type (1.56)) have passed through the first two necessary integrability conditions $(D_t(\rho_1), D_t(\rho_2) \in \mathrm{Im}(D))$. Actually all these equations are integrable, i.e. possess infinitely many commuting symmetries, higher conservation laws, have Lax's representations, etc. In this particular case the first two integrability conditions proved to be sufficient for the classification.

1.3.8 Almost Invertible Transformations and Differential Substitutions of Miura Type

The first two equations from the above list are related in the following way. Let us differentiate equation

$$u_t = u_3 + c_1 u_1^3 + c_2 u_1^2 + c_3 u_1 + c_4$$

with respect to x and denote

$$\hat{u} = u_x.$$

Then

$$\hat{u}_t = \hat{u}_3 + 3c_1 \hat{u}^2 \hat{u}_1 + 2c_2 \hat{u} \hat{u}_1 + c_3 \hat{u}_1.$$

Obviously, the same transformation can be applied to any equation

$$u_t = F(u_1, u_2, \ldots, u_n) \tag{1.72}$$

with right-hand side, which does not depend on u.

In order to give an invariant description of such transformations we note that the initial equation (1.72) is invariant with respect to a one-parameter group of shifts $u \to u + \tau$. The invariants of the group are $t, x, u_1, u_2, \ldots$ and we simply take the simplest three invariants as new variables $\hat{t}, \hat{x}$ and $\hat{u}$.

The Cole–Hopf transformation

$$\hat{u} = \frac{u_x}{u}$$

between the heat equation $u_t = u_2$ and the Burgers equation $\hat{u}_t = \hat{u}_2 + 2\hat{u}\hat{u}_1$ admits the same algebraic interpretation. In this case the one-parameter symmetry group of the initial equation is the scaling group $u \to \tau u$.

We shall call such type of transformations related to a one-parameter group of symmetries the transformation of *differentiation*.

The inverse transformation can be applied if the right-hand side of the equation is a total x-derivative. Consider, for example, the equation

$$u_t = u_3 + 3c_1 u^2 u_1 + 2c_2 u_1 + c_3 u_1 = D(u_2 + c_1 u^3 + c_2 u^2 + c_3 u).$$

Integrating the equation with respect to x and introducing a new variable $\hat{u}$ such that $D(\hat{u}) = u$, we get

$$\hat{u}_t = \hat{u}_3 + c_1\hat{u}_1^3 + c_2\hat{u}_1^2 + c_3\hat{u}_1 + c_4.$$

Notice that the initial equation has the form of a conservation law $\rho_t = D(\sigma)$, where $\rho = u$, and we introduce the potential $\hat{u}$ such that $D(\hat{u}) = \rho$ and $(\hat{u})_t = \sigma$.

It is clear that if the equation has a conserved density of zero order $\rho = s(u)$ then the new variable $\hat{u} = D^{-1}(\rho)$ satisfies an equation of the form (1.72). We call a transformation related to a conserved density the *potentiation*. If two equations can be related via a finite chain of differentiations and potentiations, we say that these equations are equivalent up to *almost invertible transformations*.

The almost invertible transformations are extremely important in the theory of integrable equations. For example, using the fact that for any integrable equation the function (1.68) is a conserved density, we can prove the following general statement [55]:

Theorem 42. *Any integrable equation of the form*

$$u_t = f(u)u_3 + F(u_2,u_1,u), \qquad f'(u) \neq 0,$$

can be reduced by a potentiation and point transformations to the form

$$u_t = u_3 + G(u_2,u_1).$$

Proof. First, we make the point transformation $\tilde{u} = f(u)^{-1/3}$. It brings the equation to the form

$$\tilde{u}_t = \frac{\tilde{u}_3}{\tilde{u}^3} + \tilde{F}(\tilde{u}_2,\tilde{u}_1,\tilde{u}).$$

It follows from (1.68) that $\rho_{-1} = \tilde{u}$. According to Theorem 34, the function $\tilde{u}$ is a conserved density for any integrable equation of this form and therefore the equation can be written as

$$\tilde{u}_t = D\left(\frac{\tilde{u}_2}{\tilde{u}^3} + \Psi(\tilde{u}_1,\tilde{u})\right). \tag{1.73}$$

The next step is the potentiation $\hat{u} = D^{-1}(\tilde{u})$. As the result we get

$$\hat{u}_t = \frac{\hat{u}_3}{\hat{u}_1^3} + \Psi(\hat{u}_2,\hat{u}_1).$$

The last step is the point transformation

$$\hat{t} = t, \quad \hat{x} = u, \quad \hat{u} = x. \tag{1.74}$$

For any point transformation we have the following expressions for new derivations:

$$\hat{D}_{\hat{t}} = f_1 D_t + f_2 D, \qquad \hat{D} = f_3 D_t + f_4 D$$

for some functions $f_i \in \mathscr{F}$. In the case of transformations $\hat{t} = t$, $\hat{x} = p(x,u)$, $\hat{u} = q(x,u)$ applying these relations to $t' = t$, we get $f_1 = 1, f_3 = 0$ and hence

$$\hat{D}_{\hat{t}} = D_t + f_2 D, \qquad \hat{D} = f_4 D.$$

For transformation (1.74) we have $0 = D_t(u) + f_2 u_1, 1 = f_4 u_1$. This implies

$$f_2 = -\frac{u_t}{u_1}, \qquad f_4 = \frac{1}{u_1}$$

and

$$\hat{u}_1 = \hat{D}(\hat{u}) = f_4 D(x) = \frac{1}{u_1}, \qquad \hat{u}_2 = \hat{D}(\hat{u}_1) = -\frac{u_2}{u_1^3},$$

$$\hat{u}_3 = \hat{D}(\hat{u}_2) = -\frac{u_3}{u_1^4} + \frac{3u_2^2}{u_1^5}, \qquad \hat{u}_t = \hat{D}_t(\hat{u}) = -\frac{u_t}{u_1}.$$

Using these formulas we find that any equation

$$\hat{u}_t = \frac{\hat{u}_3}{\hat{u}_1^3} + \Psi(\hat{u}_2, \hat{u}_1)$$

transforms to an equation of the form

$$u_t = u_3 + G(u_2, u_1).$$

∎

Example 43. For the Harry–Dim equation $u_t = u^3 u_3$ we have to take $\tilde{u} = \frac{1}{u}$. Equation (1.73) is given by

$$\tilde{u}_t = D\left(\frac{\tilde{u}_2}{\tilde{u}^3} - \frac{3\tilde{u}_1^2}{2\tilde{u}^4}\right).$$

After the potentiation we get

$$\hat{u}_t = \frac{\hat{u}_3}{\hat{u}_1^3} - \frac{3\hat{u}_2^2}{2\hat{u}_1^4}.$$

Transformation (1.74) brings the latter equation to

$$u_t = u_3 - \frac{3u_2^2}{2u_1}.$$

Note that this so-called Schwartz–KdV equation admits a group of classical symmetries

$$u \to \frac{\alpha u + \beta}{\gamma u + \delta}.$$

The simplest three differential invariants of this group are $t, x, \frac{u_3}{u_1} - \frac{3u_2^2}{2u_1^2}$. The corresponding differential substitution

$$\hat{t} = t, \quad \hat{x} = x, \quad \hat{u} = \frac{u_3}{u_1} - \frac{3u_2^2}{2u_1^2}$$

reduces the Schwartz–KdV equation to the KdV equation $\hat{u} = \hat{u}_3 + 3\hat{u}\hat{u}_1$.

A different example of a differential substitution gives us the famous Miura transformation [47]

$$\hat{u} = u_1 - u^2,$$

which links the mKdV equation $u_t = u_3 - 6u^2u_1$ and the KdV equation $\hat{u}_t = \hat{u}_3 + 6\hat{u}\hat{u}_1$. Differential substitutions of Miura type are well known in the theory of integrable evolution PDEs (see [47]).

A relation

$$\hat{u} = P(x, u, u_1, \ldots, u_k) \tag{1.75}$$

is called a *differential substitution* of order k from the equation

$$u_t = f(x, u, u_1, \ldots, u_n) \tag{1.76}$$

to the equation

$$\hat{u}_t = g(x, \hat{u}, \hat{u}_1, \ldots, \hat{u}_n) \tag{1.77}$$

if for any solution $u(x, t)$ of Eq. (1.76) the function (1.75) satisfies (1.77).

Theorem 44. *If Eqs. (1.76) and (1.77) are related via a differential substitution (1.75) and Eq. (1.77) possesses the formal recursion operator, them Eq. (1.76) also possesses the formal recursion operator.*

To prove this theorem it suffices to verify that if $\hat{R}$ is a formal recursion operator for Eq. (1.77), then

$$\Lambda = P_*^{-1} \sigma(\hat{\Lambda}) P_*,$$

where σ denotes the substitution $\hat{u}_i \rightarrow D^i(P)$, is a formal recursion operator for (1.76).

Proposition 45. *Let ρ_i and $\hat{\rho}_i$ be the canonical densities of Eqs. (1.76) and (1.77) defined by the formal recursion operators Λ and $\hat{\Lambda}$. Then*

$$\rho_i \sim \sigma(\hat{\rho}_i), \qquad i = -1, 0, 1, 2, \ldots.$$

Corollary 46. *If Eq. (1.76) is related to a linear equation*

$$\hat{u}_t = \sum_{k=0}^{n} c_k \hat{u}_k, \ c_k \in \mathbb{C},$$

by a differential substitution (1.75), then Eq. (1.76) has the formal recursion operator and all its canonical densities are trivial.

A relation between special nonlocal symmetries of Eq. (1.76) and differential substitutions of first order has been discovered in [56]. Using this observation, in [80, 81] a classification of first-order substitutions has been done.

1.4 Modifications and Generalizations

1.4.1 Systems of Partial Differential Equations

All objects defined above can be easily extended to the case of system of equations. Let us consider a system of N evolutionary partial differential equations of the form

$$\mathbf{u}_t = \mathbf{F}(\mathbf{u}, \mathbf{u}_1, \ldots, \mathbf{u}_n, x, t), \qquad (1.78)$$

$$\mathbf{u} = \left(u^1, \ldots, u^N\right), \qquad \mathbf{u}_k = \partial_x^k(\mathbf{u}), \qquad \mathbf{F} = \left(F^1, \ldots, F^N\right).$$

Each entry F^s of the vector-function $\mathbf{F}$ belongs to $\mathscr{F}$, where $\mathscr{F}$ is a differential field generated by components of vector $\mathbf{u}$ and derivation

$$D = \frac{\partial}{\partial x} + \sum_{k=0}^{\infty} \sum_{s=1}^{N} u_{k+1}^s \frac{\partial}{\partial u_k^s}.$$

Equation (1.78) defines another derivation

$$D_t = \frac{\partial}{\partial t} + \sum_{k=0}^{\infty} \sum_{s=1}^{N} F_k^s \frac{\partial}{\partial u_k^s},$$

where $F_s^k = D^s(F^k)$.

For any vector-function $\mathbf{a}$ the Fréchet derivative is a differential operator with matrix coefficients

$$\mathbf{a}_* = \sum_k \frac{\partial \mathbf{a}}{\partial \mathbf{u}_k} D^k, \qquad \left(\frac{\partial \mathbf{a}}{\partial \mathbf{u}_k}\right)_{ij} = \frac{\partial a^i}{\partial u_k^j}.$$

The definition of symmetries and conservation laws for system of equations is exactly the same as in the scalar case. We note only that in this case the generator of symmetry is a vector function whereas the conserved density is still a scalar object.

Formal series

$$A = \mathbf{a}_n D^n + \mathbf{a}_{n-1} D^{n-1} + \cdots$$

and recursion operators also have matrix-valued coefficients. In the definition of the residue we have to add the trace operation:

$$\mathrm{res}(A) = \mathrm{trace}\left(\mathbf{a}_{-1}\right).$$

With such a definition of the residue Adler's Theorem 19 is valid.

If the leading coefficient $\mathbf{a}_n$ of the series A is nonsingular $(\det(\mathbf{a}_n) \neq 0)$, then we can find the inverse series B such that $AB = BA = I$.

Equation (1.55), which defines the formal recursion operator, can be rewritten in the form

$$[D_t - \mathbf{F}_*, \Lambda] = 0. \tag{1.79}$$

If the eigenvalues $\mu_1, \ldots, \mu_N$ of the leading matrix coefficient $\mathbf{F}_n$ of operator $\mathbf{F}_*$ are pair-wise distinct $\mu_i \neq \mu_j$, we can transform both operators $D_t - \mathbf{F}_*$ and Λ from (1.79) to a diagonal form [40, 45].

Theorem 47. *Suppose the leading matrix coefficient $\mathbf{F}_n$ of the Fréchet derivative*

$$\mathbf{F}_* = \mathbf{F}_n D^n + \mathbf{F}_{n-1} D^{n-1} + \cdots + \mathbf{F}_0, \qquad \mathbf{F}_k = \left(\frac{\partial \mathbf{F}}{\partial \mathbf{u}_k} \right),$$

has pair-wise distinct eigenvalues $\mu_1, \ldots, \mu_N$ ($\mu_i \neq \mu_j$), then there exists an unique formal series

$$T = T_0(I + t_1 D^{-1} + t_2 D^{-2} + \cdots)$$

such that all matrices t_k are off-diagonal,

$$\mathbf{F}_n = T_0^{-1} diag(\mu_1, \ldots, \mu_N) T_0,$$

and all coefficients Φ_k of the formal series

$$\Phi = T \circ \mathbf{F}_* \circ T^{-1} + D_t(T) \circ T^{-1} = \tag{1.80}$$
$$diag(\mu_1, \ldots, \mu_N) D^n + \Phi_{n-1} D^{n-1} + \Phi_{n-2} D^{n-2} + \cdots$$

are diagonal. Moreover, if Λ is a formal recursion operator satisfying Eq. (1.79) then

$$\hat{\Lambda} = T \circ \Lambda \circ T^{-1}$$

is a formal series with diagonal coefficients which satisfies equation

$$D_t(\hat{\Lambda}) = [\Phi, \hat{\Lambda}].$$

Example 48. Let us consider the following system of equations

$$\begin{array}{ll} u_t = u_{xx} + f(u, v, u_x, v_x) \\ -v_t = v_{xx} + g(u, v, u_x, v_x) \end{array}, \qquad \mathbf{F} = \left(\begin{array}{c} u_2 + f(u, v, u_1, v_1) \\ -v_2 - g(u, v, u_1, v_1) \end{array} \right).$$

The leading matrix $diag(1, -1)$ of the corresponding Fréchet derivative

$$\mathbf{F}_* = \left(\begin{array}{cc} 1 & 0 \\ 0 & -1 \end{array} \right) D^2 + \left(\begin{array}{cc} \partial f/\partial u_1 & \partial f/\partial v_1 \\ -\partial g/\partial u_1 & -\partial g/\partial v_1 \end{array} \right) D + \left(\begin{array}{cc} \partial f/\partial u & \partial f/\partial v \\ -\partial g/\partial u & -\partial g/\partial v \end{array} \right)$$

has distinct eigenvalues and is already in the diagonal form. Therefore $T_0 = I$. We rewrite Eq. (1.80) in the form

$$\Phi \circ T = T \circ \mathbf{F}_* + D_t(T), \qquad \Phi = diag(\hat{F}, -\hat{G}),$$

where
$$T = I + t_{-1}D^{-1} + t_{-2}D^{-2} + t_{-3}D^{-3} + \cdots,$$
$$\hat{F} = D^2 + F_1 D + F_0 + F_{-1}D^{-1} + \cdots,$$
$$\hat{G} = D^2 + G_1 D + G_0 + G_{-1}D^{-1} + \cdots,$$

and collect the coefficients at $D^k, k = 1, 0, -1, \ldots$. We obtain

$$t_{-1} = \frac{1}{2}\begin{pmatrix} 0 & \partial f/\partial v_1 \\ \partial g/\partial u_1 & 0 \end{pmatrix},$$

$$F_1 = \frac{\partial f}{\partial u_1}, \quad G_1 = \frac{\partial g}{\partial v_1}, \quad F_2 = \frac{\partial f}{\partial u} - \frac{1}{2}\frac{\partial f}{\partial v_1}\frac{\partial g}{\partial u_1}, \quad G_2 = \frac{\partial g}{\partial v} - \frac{1}{2}\frac{\partial f}{\partial v_1}\frac{\partial g}{\partial u_1}.$$

The formal diagonalization (Theorem 47) reduces the problem to find a formal recursion operator with matrix coefficients to a set of N scalar problems, simplifying computations, and enables to prove some nontrivial statements for integrable systems of equations (such as Theorem 50 below).

In the study of integrability for systems of equations we have to be more careful and exclude some degenerated cases of "partial" integrability. Consider for example that the following system consists of two de-coupled equations:

$$u_t = u_{xxx} + 6uu_x,$$
$$v_t = v_{xxx} + v^3 v_x.$$

One of these equations is integrable (KdV) and therefore possesses an infinite hierarchy of symmetries (1.67), but the second equation does not have any high-order symmetry. Such a "system" has an infinite hierarchy of symmetries of the form $\mathbf{G} = (G_{2k+1}, 0)$, where G_i are symmetries of the KdV equation. This system will also have a degenerated recursion operator. This example may look trivial, but after a change of variables one could receive a system and the corresponding hierarchy of symmetries which is difficult to recognize. Therefore we have to give an invariant definition of nondegenerate symmetry.

Definition 49. A symmetry $G(\mathbf{u}, \mathbf{u}_1, \ldots, \mathbf{u}_m)$ of order m of Eq. (1.78) is called nondegenerate if the leading term of its Fréchet derivative is nonsingular

$$\det\left(\frac{\partial \mathbf{G}}{\partial \mathbf{u}_m}\right) \neq 0.$$

For simplicity, let us restrict ourselves by considering nondegenerate systems

$$\mathbf{u}_t = A\mathbf{u}_n + \mathbf{F}(\mathbf{u}, \mathbf{u}_1, \ldots, \mathbf{u}_{n-1}), \quad n \geq 2, \tag{1.81}$$

with constant diagonal matrix $A = \text{diag}(a_1, \ldots, a_N)$, where $a_i \neq a_j$, $a_i \neq 0$. High-order symmetries of Eq. (1.81), if they exist, have a similar form

$$\mathbf{u}_\tau = B\mathbf{u}_m + \mathbf{G}(\mathbf{u}, \mathbf{u}_1, \ldots, \mathbf{u}_{m-1}), \quad m \geq 2, \quad B = \text{diag}(b_1, \ldots, b_N). \tag{1.82}$$

This symmetry is nondegenerate iff $b_i \neq 0$, $i = 1, \ldots, N$.

Theorem 50. *Suppose the system of equations* (1.81) *possesses an infinite hierarchy of nondegenerate symmetries* (1.82) *of arbitrary high order, then Eq.* (1.79) *for formal recursion operator* Λ *has a solution*

$$\Lambda = l_1 D + l_0 + l_{-1} D^{-1} + \cdots$$

for any constant diagonal matrix l_1.

This statement shows that any system (1.81) possessing a hierarchy of nondegenerate symmetries has N different formal recursion operators corresponding to $l_1 = \text{diag}(1,0,0...,0), \ldots, l_1 = \text{diag}(0,0,...0,1)$.

1.4.2 Integrable Polynomial Systems and Nonassociative Algebraic Structures

A complete classification of integrable cases for systems of evolution equations (1.78) becomes a very difficult problem even in the case $N = 2$. For $N > 2$ the general classification problem looks hopeless.

The only possibility here is to consider some specific classes of systems, which are interesting for applications and/or for pure mathematics. For example, some classes of polynomial N-component systems generalizing well-known integrable scalar models can be studied.

In the case of polynomial equations, integrability conditions yield an overdetermined system of algebraic equations for coefficients of the right-hand side. As a rule, it is very difficult to understand how many solutions such a system may have. Moreover, one should expect that the classification problem for N-component polynomial systems contains, as a sub-problem, a classical "unsolvable" classification problem of algebra, such as the description of all finite-dimensional Lie algebras.

However, the usage of the algebraic language usually allows us to reformulate the answer in componentless terms. After that we have a chance to use nontrivial algebraic classification results such as the classification of simple Lie algebras, simple Jordan algebras, etc.

In order to illustrate all the above points, let us consider Svinolupov's result [67] concerning multi-component generalizations

$$u_t^i = u_{xxx}^i + C_{jk}^i u^j u_x^k \tag{1.83}$$

of the Korteweg–de Vries equation. Here and below we assume that the summation is carried out over repeated indices. Since any linear transformation of **u** preserves the class (1.83), the description of integrable cases is to be invariant under these transformations.

To solve the problem of complication of computations, let us interpret C_{jk}^i as the structural constants of an (noncommutative and nonassociative) algebra J. Recall that if $e_1, \ldots, e_N$ be a basis of J, then the multiplication rule is uniquely defined by the formula $e_i \circ e_j = C^i_{jk} e_k$. The constants C^i_{jk} are called the structural constants of the algebra J.

The formula

$$X \circ Y = \lambda <X,C>Y + \mu <Y,C>X + \nu <X,Y>C, \qquad (1.84)$$

where $<\,,\,>$ is the standard scalar product in a vector space J and C is a given vector, gives us for different constants λ, μ, ν a number of interesting examples of nonassociative algebras. The so-called vector-integrable differential equations are closely related to those.

Using the notation $u = u^i e_i$, we can rewrite (1.83) in the form

$$u_t = u_{xxx} + u \circ u_x, \qquad (1.85)$$

where $u(x,t)$ is a J-valued function. It is easy to see that equations related by linear transformations correspond to isomorphic algebras.

Now the main question is: for which algebras Eq. (1.85) is integrable.

Theorem 51. *Suppose* $C^i_{jk} = C^i_{kj}$ *or, in other words, J is commutative. Then (1.85) possesses an infinite sequence of higher symmetries of the form*

$$u^i_\tau = u^i_n + P^i(\mathbf{u}, \mathbf{u}_x, \ldots, \mathbf{u}_{n-1})$$

iff J is a Jordan algebra.

Definition 52. A commutative algebra J is said to be Jordan if the following identity is fulfilled:

$$AS(X \circ X, Y, X) = 0,$$

where $AS(X,Y,Z)$ means the associator:

$$AS(X,Y,Z) = (X \circ Y) \circ Z - X \circ (Y \circ Z).$$

It is well known that the set of all matrices is a Jordan algebra with respect to the anticommutator operation

$$X \circ Y = \frac{1}{2}(XY + YX). \qquad (1.86)$$

Another example of Jordan multiplication is given by

$$X \circ Y = <X,C>Y + <Y,C>X - <X,Y>C. \qquad (1.87)$$

This operation turns a N-dimensional vector space J to a (simple) Jordan algebra.

Although there is no description of all the Jordan algebras this theorem allows one

  i) to check the integrability of a given system (1.83);
  ii) to classify all integrable cases for small dimensions;
  iii) to construct the most interesting examples of an arbitrary high dimension.

Let us explain what the term "most interesting" means. A system of equations (1.83) is called irreducible if it cannot be reduced to the block-triangular form by an appropriate linear transformation (in the case of the block-triangular system, the functions $u^1, \ldots, u^M$ $(M < N)$ satisfy an autonomous system of the form (1.83), and remaining equations are linear in $u^{M+1}, \ldots, u^N$). It turns out that irreducible systems are associated with the simple algebras. Thus, one can use a well-known algebraic result [29], namely the exhaustive description of all the simple Jordan algebras, to construct all irreducible systems. They are nothing but so-called vector and matrix Korteweg–de Vries equations [3, 58].

The matrix KdV equation, corresponding to the simple Jordan algebra (1.86), has the following form:

$$U_t = U_{xxx} + 3\left(UU_x + U_xU\right), \tag{1.88}$$

where $U(x,t)$ is an unknown $N \times N$-matrix. The simplest higher symmetry of this equation can also be written in the matrix form:

$$U_\tau = U_{xxxxx} + 5\left(UU_{xxx} + U_{xxx}U\right) +$$

$$10\left(U_xU_{xx} + U_{xx}U_x\right) + 10\left(U^2U_x + UU_xU + U_xU^2\right).$$

It is obvious that the reduction $U^T = U$, where superscript "T" stands for matrix transpose, is compatible with the structure of matrix equation (1.88). This reduction corresponds to another series of simple Jordan algebras.

One more interesting example is related to operation (1.87) which gives rise to the following vector KdV equation [58]:

$$\mathbf{u}_t = \mathbf{u}_{xxx} + <C, \mathbf{u}> \mathbf{u}_x + <C, \mathbf{u}_x> \mathbf{u} - <\mathbf{u}, \mathbf{u}_x> C. \tag{1.89}$$

Usually (see [3]), one refers to the system

$$\mathbf{u}_t = \mathbf{u}_{xxx} + <C, \mathbf{u}> \mathbf{u}_x + <C, \mathbf{u}_x> \mathbf{u} \tag{1.90}$$

as the vector KdV equation. However, the system (1.90) is reducible. Indeed, using an orthogonal transformation of the vector $\mathbf{u}$ we can bring the vector C to $(1, 0, \ldots, 0)$. After that the first equation in the system becomes separate. In contrast, system (1.89) is irreducible.

The description of simple Jordan algebras shows that only one exceptional simple Jordan algebra of dimension 27 leads to an irreducible integrable system (1.83) that essentially differs from (1.88) and (1.89).

It turns out that besides Jordan algebras such well-known nonassociative algebraic structures as the left-symmetric algebras, Jordan triple systems and Jordan pairs are closely connected to polynomial multi-component integrable systems. This connection allows one to clarify the nature of known vector and matrix generalizations (see, for instance [3, 18, 19]) of classical scalar integrable equations and to construct some new examples of this kind [58, 65, 66].

The left-symmetric algebras are related to systems of the form

$$u^i_t = u^i_{xx} + 2a^i_{jk}u^j u^k_x + b^i_{jkm}u^j u^k u^m, \qquad i = 1,\dots,N. \qquad (1.91)$$

Definition 53. An algebra J is called left-symmetric if

$$AS(X,Y,Z) - AS(Y,X,Z) = 0.$$

Any associative algebra is left-symmetric one. The formula

$$X \circ Y = <X,C> Y + <X,Y> C \qquad (1.92)$$

give us an example of left-symmetric algebra of the type (1.84).

Theorem 54. *System*(1.91) *has higher symmetries iff it can be written as*

$$u_t = u_{xx} + 2u \circ u_x + u \circ (u \circ u) - (u \circ u) \circ u, \qquad (1.93)$$

where $\circ$ denotes the multiplication in a left-symmetric algebra A.

Let us consider two simplest examples of the systems (1.93).

Example 55. The set of all the quadratic matrices forms an associative (and, therefore, left-symmetric) algebra. The corresponding equation (1.93) is the matrix Burgers equation

$$U_t = U_{xx} + 2U U_x.$$

Example 56. The left-symmetric algebra (1.92) generates the following vector Burgers equation

$$\mathbf{u}_t = \mathbf{u}_{xx} + 2 <\mathbf{u},\mathbf{u}_x> C + 2 <\mathbf{u},C> \mathbf{u}_x + \|\mathbf{u}\|^2 <\mathbf{u},C> C - \|C\|^2 \|\mathbf{u}\|^2 \mathbf{u}.$$

Multi-component equations of the nonlinear Schrödinger type and of the mKdV type are related to so-called Jordan triple systems.

Definition 57. A triple system $\{X,Y,Z\}$ is said to be Jordan if

$$\{X,Y,Z\} = \{Z,Y,X\}$$

and

$$\{X,\{Y,Z,V\},W\} - \{W,V,\{X,Y,Z\}\} + \{Z,Y,\{X,V,W\}\} - \{X,V,\{Z,Y,W\}\} = 0.$$

The set of $n \times n$-matrices equipped with the operation

$$\{X,Y,Z\} = \frac{1}{2}(XYZ + ZYX) \qquad (1.94)$$

is a Jordan triple system. The vector space of all $n \times m$-matrices is a Jordan triple system with respect to operation

$${X,Y,Z} = \frac{1}{2}(XY^tZ + ZY^tX),$$

where "t" stands for transposition. The following operations

$${X,Y,Z} = <X,Y>Z+<Y,Z>X-<X,Z>Y \tag{1.95}$$

and

$${X,Y,Z} = <X,Y>Z+<Y,Z>X \tag{1.96}$$

define two "vector" (cf. (1.84)) simple Jordan triple systems.

Theorem 58. *For any Jordan triple system the equation*

$$u_t = u_{xx} + 2\{u,v,u\}, \qquad v_t = -v_{xx} - 2\{v,u,v\}$$

possesses higher symmetries.

Theorem 59. *For any Jordan triple system the equation*

$$u_t = u_{xxx} + \{u,u,u_x\}$$

possesses higher symmetries.

Theorem 60. *For any Jordan triple system the equation*

$$u_t = u_{xx} + 2\{v,u,v\}_x, \qquad v_t = -v_{xx} - 2\{u,v,u\}_x$$

possesses higher symmetries.

The formulas (1.94), (1.95), (1.96) yield the following examples of corresponding integrable matrix and vector equations:
the matrix NLS equation

$$U_t = U_{xx} + 2UVU, \qquad V_t = -V_{xx} - 2VUV;$$

the vector NLS equation (S. Manakov)

$$\mathbf{u}_t = \mathbf{u}_{xx} + <\mathbf{u},\mathbf{v}>\mathbf{u}, \qquad \mathbf{v}_t = \mathbf{v}_{xx} - <\mathbf{u},\mathbf{v}>\mathbf{v};$$

a different vector NLS equation

$$\mathbf{u}_t = \mathbf{u}_{xx} + 2<\mathbf{u},\mathbf{v}>\mathbf{u} - <\mathbf{u},\mathbf{u}>\mathbf{v},$$
$$\mathbf{v}_t = -\mathbf{v}_{xx} - 2<\mathbf{u},\mathbf{v}>\mathbf{v} + <\mathbf{v},\mathbf{v}>\mathbf{u};$$

the matrix mKdV equation

$$U_t = U_{xxx} + U^2U_x + U_xU^2;$$

the vector mKdV equation

$$\mathbf{u}_t = \mathbf{u}_{xxx} + <\mathbf{u}, \mathbf{u}> \mathbf{u}_x; \qquad (1.97)$$

a different vector mKdV equation

$$\mathbf{u}_t = \mathbf{u}_{xxx} + <\mathbf{u}, \mathbf{u}> \mathbf{u}_x + <\mathbf{u}, \mathbf{u}_x> \mathbf{u}. \qquad (1.98)$$

Different results establishing relationships between multi-component integrable systems and nonassociative algebras are described in Sect. 5.

1.4.3 Integrable Nonabelian and Vector Equations

In the previous section we have seen that the most interesting examples of integrable systems, which come from general algebraic considerations, have very particular structure. They are matrix- or vector-integrable equations. In the next section we consider so-called nonabelian equations, which are natural generalization of the matrix equations.

1.4.3.1 Nonabelian Equations

In order to formalize the concept of matrix equations, let us consider evolution equations on free associative algebra $\mathscr{F}$. In the case of one-field non-abelian equations the generators of $\mathscr{F}$ are

$$U, \quad U_1 = U_x, \quad \ldots, \quad U_k, \quad \ldots, \qquad (1.99)$$

and the equation is of the form

$$U_t = F(U, U_1, \ldots, U_n), \qquad (1.100)$$

where F is a (noncommutative) polynomial. All definitions can be easily generalized to the case of several nonabelian variables.

Since $\mathscr{F}$ is assumed to be a free algebra no algebraic relations for the generators (1.99) are allowed. It is not true if we consider, for example, Eq. (1.100) for 2×2 matrix U. But if we want Eq. (1.100) to be integrable for the matrix U of arbitrary size, the assumption about absence of algebraic relations becomes adequate to the problem.

Actually, this formula does not mean that we consider an element of nonassociative algebra depending on time t. As usual, (1.100) defines a derivation D_t of $\mathscr{F}$ which commutes with

$$D = \sum_0^\infty U_{i+1} \frac{\partial}{\partial U_i}.$$

It is easy to check that this derivation is defined by the vector field

$$D_t = \frac{\partial}{\partial t} + \sum_0^\infty D^i(F) \frac{\partial}{\partial U_i}.$$

The generalization of the symmetry approach to differential equations on associative algebras requires proper definitions for such concepts as symmetry, first integral, Fréchet derivative and formal recursion operator.

As in the scalar case, the symmetry is an evolution equation

$$U_\tau = G(U, U_1, \ldots, U_m),$$

such that the vector field

$$D_G = \sum_0^\infty D^i(G) \frac{\partial}{\partial u_i}$$

commutes with D_t. The polynomial G is called *symmetry generator*.

The condition $[D_t, D_G] = 0$ is equivalent to $D_t(G) = D_G(F)$. The latter relation can be rewritten as

$$G_*(F) - F_*(G) = 0, \tag{1.101}$$

where the Fréchet derivative H_* for any $H \in \mathscr{F}$ is defined in the following standard way.

For any $a \in \mathscr{F}$ we denote by L_a and R_a the operators of left and, correspondingly, of right multiplication by a:

$$L_a(X) = aX, \qquad R_a(X) = X a, \qquad X \in \mathscr{F}.$$

The associativity of $\mathscr{F}$ is equivalent to the identity $[L_a, R_b] = 0$ for any a and b. Moreover,

$$L_{ab} = L_a L_b, \quad R_{ab} = R_b R_a, \quad L_{a+b} = L_a + L_b, \quad R_{a+b} = R_a + R_b.$$

Definition 61. We denote by $\mathcal{O}$ the associative algebra generated by all operators of left and right multiplication by elements (1.99). This algebra is called algebra of local operators.

Let us extend the set of generators (1.99) by additional symbols $V_0, V_1, \ldots$ and define $D(V_i) = V_{i+1}$.

Given $H(U, U_1, U_2, \ldots, U_k) \in \mathscr{F}$ we find

$$L_H = \frac{\partial}{\partial \varepsilon} H(U + \varepsilon V_0,\ U_1 + \varepsilon V_1,\ U_2 + \varepsilon V_2, \ldots) \Big|_{\varepsilon=0}$$

and represent this expression as $H_*(v)$, where H_* is a linear differential operator of order k, whose coefficients belong to $\mathcal{O}$. For example, $(U_2 + U_1)_* = D^2 + L_U D + R_{U_1}$.

In contrast to the definition of the symmetry, which is a straightforward generalization of the corresponding scalar notion, the definition of conserved density has to be essentially modified.

Recall (see Sect. 3) that in the scalar case the conserved density is a function $\rho \in \mathscr{F}$ such that $D_t(\rho) = D(\sigma)$ for some $\sigma \in \mathscr{F}$. It is supposed that the equivalent densities define the same conservation law. Here the equivalence relation is defined as follows: $\rho_1 \sim \rho_2$ iff $\rho_1 - \rho_2 = D(s)$, $s \in \mathscr{F}$. In others, the density is an equivalence class in $\mathscr{F}$ such that D_t takes it to zero equivalence class.

This definition is motivated by the fact that if ρ is a polynomial such that $\rho(0) = 0$, then the functional $\int_{-\infty}^{\infty} \rho(u, u_x, \dots) dx$, where $u(x)$ is a rapidly decreasing function, does not depend on the choice of a representative from the equivalence class. If ρ is a conserved density then the functional applying to a solution $u(x,t)$ of our evolution equation does not depend on t.

In the nonabelian case we hold the same line. The following elementary operations define an equivalence relation:

1. addition of elements of the form $D(s)$, $s \in \mathscr{F}$ to the polynomial $\rho \in \mathscr{F}$;
2. the cyclic permutation of factors in any monomial of the polynomial ρ.

Two polynomials ρ_1 and ρ_2 related to each other through a finite sequence of the elementary operations are called *equivalent*. It is clear that in abelian case this definition coincides with the standard one.

A motivation of the definition is that in the matrix case the functional

$$\int_{-\infty}^{\infty} \mathrm{trace}(\rho(u, u_x, \dots)) \, dx$$

is correctly defined on the equivalence classes.

At least for nonabelian equations of the form

$$U_t = U_n + f(U, U_1, \dots, U_{n-1}) \tag{1.102}$$

all definitions and results concerning the formal recursion operator (see Sect. 3.4) can be easily generalized.

Definition 62. A formal series

$$\Lambda = D + l_0 + l_{-1}D^{-1} + \cdots, \qquad l_k \in \mathscr{O}, \tag{1.103}$$

is called a formal recursion operator for Eq. (1.102) if it satisfies the equation

$$D_t(\Lambda) - [F_*, \Lambda] = 0. \tag{1.104}$$

Notice that now coefficients of both F_* and Λ belong to the associative algebra $\mathscr{O}$ of local operators (see Definition 61 above).

For example, in the case of nonabelian Korteweg–de Vries equation (1.88) one can take $\Lambda = R^{1/2}$, where R is the recursion operator for (1.88):

$$R = D^2 + 2(L_U + R_U) + (L_{U_x} + R_{U_x})D^{-1} + (L_U - R_U)D^{-1}(L_U - R_U)D^{-1}.$$

In the abelian case this recursion operator coincides with the standard one (see Sect. 3.4).

The analogs of Theorems 24, 31, 32 can be proved by similar reasoning as the original statements.

1.4.3.2 Vector Equations

Let us consider equations of the form

$$U_t = f_n U_n + f_{n-1} U_{n-1} + \cdots + f_1 U_1 + f_0 U, \tag{1.105}$$

where $U(x,t)$ is unknown vector, and f_i are scalar functions[4] of variables

$$u_{[i,j]} = <U_i, U_j>, \qquad i \le j, \tag{1.106}$$

where $0 \le i, j \le n$. Here and in the sequel, $<\cdot,\cdot>$ stands for the standard scalar product in a vector space V. We denote the set of all such functions by $\mathscr{F}$.

It is clear that any equation (1.105) is invariant with respect to arbitrary orthogonal transformations of the vector U. Equations of the form (1.105) are called *isotropic vector equations*.

Variables (1.106) are regarded as *independent*. The algebraic independence of $u_{[i,j]}$ is a crucial requirement in all computations. Note that if V is finite dimensional and the dimension N is fixed, we cannot suppose that. For instance, if $N = 3$, then the determinant of the matrix with entries $a_{ij} = u_{[i,j]}, i, j = 1, 2, 3, 4$ identically equals to zero.

In other respects our considerations are formal. The signature of the scalar product is inessential for us. Furthermore, the assumptions that the space V is finite dimensional and the constant field is $\mathbb{R}$ are also unimportant. For instance, U could be a function of t, x and y and the scalar product be

$$<U, V> = \int_{-\infty}^{\infty} U(t,x,y) V(t,x,y) \, dy.$$

In such a way, our formulas and statements are valid also for this particular sort of $1 + 2$-dimensional nonlocal equations.

The vector-modified Korteweg–de Vries equations (1.97) and (1.98) give us examples of integrable isotropic equation.

In this section we establish an infinite consequence of necessary conditions for the existence of higher symmetries and/or conserved densities for Eq. (1.105). These conditions have the following form:

$$D_t \rho_i = D \theta_i, \qquad i = 0, 1, 2, \ldots, \qquad \rho_i, \theta_i \in \mathscr{F}.$$

Here ρ_i, θ_i can be recursively found in terms of the coefficients f_i of Eq. (1.105).

These conditions are very close to the canonical conservation laws from Sect. 3.3 by spirit but do not coincide with them. Our componentless conditions are more convenient for classification problems related to Eq. (1.105) since they are much simpler than the standard canonical densities for multi-component systems.

---

[4] In contrast with the nonabelian case we do not assume that functions under consideration are polynomials.

Theorem 63. *If Eq. (1.105) possesses an infinite series of commuting flows of the form*

$$U_\tau = g_m U_m + g_{m-1} U_{m-1} + \cdots + g_1 U_1 + g_0 U, \qquad g_i \in \mathscr{F},$$

then

(i) *there exists a formal series*

$$L = a_1 D + a_0 + a_{-1} D^{-1} + a_{-2} D^{-2} + \cdots, \qquad a_i \in \mathscr{F},$$

satisfying the operator relation

$$L_t = [A, L], \qquad A = \sum_0^n f_i D^i. \tag{1.107}$$

Here f_i are the coefficients of Eq. (1.105).

(ii) *The following functions*

$$\rho_{-1} = \frac{1}{a_1}, \qquad \rho_0 = \frac{a_0}{a_1}, \qquad \rho_i = \operatorname{res} L^i, \qquad i \in \mathbb{N}, \tag{1.108}$$

are conserved densities for Eq. (1.105).

(iii) *If Eq. (1.105) possesses an infinite series of conserved densities depending on variables (1.106), then there exists a series L satisfying (1.107), and a series S of the form*

$$S = s_1 D + s_0 + s_{-1} D^{-1} + s_{-2} D^{-2} + \cdots, \qquad s_i \in \mathscr{F},$$

such that

$$S_t + A^T S + SA = 0, \qquad S^T = -S,$$

where the superscript T stands for a formal conjugation.

(iv) *Under the conditions of item (iii) densities (1.108) with $i = 2k$ are of the form $\rho_{2k} = D_x(\sigma_k)$ for some functions $\sigma_k \in \mathscr{F}$.*

Comment. In Sect. 4.1 the notion of the formal symmetry has been generalized to the case of systems of evolution equations. However, in these papers the formal symmetry is a series with *matrix* coefficients that satisfies (1.101). In Theorem 63 both the operators A and L are scalar objects and of course A does not coincide with the Frechét derivative F_* of the right-hand side of the system.

Equation (see [22])

$$U_t = \left(U_{xx} + \frac{3}{2} < U_x, U_x > U \right)_x + \frac{3}{2} < U, R(U) > U_x, \qquad < U, U >= 1, \tag{1.109}$$

give us an example of integrable anisotropic vector equation. Here R is an arbitrary symmetric operator. Equations of such type can also be classified in the framework of our componentless approach. To do that we assume that the coefficients f_i of Eq. (1.105), besides (1.106), depend on additional variables

$$v_{[i,j]} = < U_i, R(U_j) >, \qquad i \leq j, \qquad (1.110)$$

$0 \leq i, j \leq n$. We assume variables (1.110) to be independent both with each other and in respect of variables (1.106). Theorem 63 remains to be valid for such anisotropic vector equations.

1.4.4 Nonlocal Integrable Equations and the Symmetry Approach in the Symbolic Representation

1.4.4.1 Symbolic Representation

The aim of this section is to formulate a perturbative version of the symmetry approach in the symbolic representation and to generalize it in order to make it suitable for study of nonlocal and nonevolution equations. We illustrate this theory on the example of Camassa–Holm type equations.

In what follows we shall consider equations for which the right-hand side is a differential polynomial or can be represented in the form of a series

$$F(u_n, \ldots, u_1, u_0) = F_1[u] + F_2[u] + F_3[u] + \cdots, \qquad (1.111)$$

where $F_k[u]$ is a homogeneous differential polynomial, i.e. a polynomial of variables $u_n, \ldots, u_1, u_0$ with complex constant coefficients satisfying the condition $F_k[\lambda u] = \lambda^k F_k[u], \lambda \in \mathbb{C}$, linear part $F_1[u] = L(u_0)$ and L is a linear operator ($\mathrm{ord}(L) \geq 2$)

$$L = \sum_{k=0}^n r_k D^k, \qquad r_k \in \mathbb{C}.$$

For such equations we develop here a perturbative method to construct formal recursion operator and test for integrability. For simplicity we shall consider the case when function F is a differential polynomial, i.e. the series (1.111) contains a finite number of terms. The generalization to the case of infinite series will be obvious.

Differential polynomials over $\mathbb{C}$ form a differential ring $\mathscr{R}(u, D)$ which has a natural gradation

$$\mathscr{R}(u, D) = \bigoplus_{n \geq 1} \mathscr{R}_n(u, D), \qquad (1.112)$$

where $\mathscr{R}_n(u, D)$ is a set of homogeneous differential polynomials of degree n. The condition $n \geq 1$ in (1.112) means that $1 \notin \mathscr{R}(u, D)$. In order to develop a perturbation theory and for further generalization of the approach to nonlocal cases it is convenient to introduce a symbolic representation of this ring.

Symbolic representation (or symbolic method) was used in mathematics since the middle of the nineteenth century. It was successfully applied to the theory of integrable equations by I.M. Gel'fand and L.A. Dikii [20] in 1975 and also by V.E.

Zakharov and E.I. Schulman [77]. Recently the power of this method has been demonstrated again in the series of works of J. Sanders and Jing Ping Wang (see for example [53, 76]) where they have given ultimate description of integrable hierarchies of polynomial homogeneous evolution equations.

Actually the symbolic representation is a simplified form of notations and rules for formal Fourier images of dynamical variables u_n, differential polynomials and formal series (1.43) with coefficients from the ring $\mathscr{R}(u,D) \oplus \mathbb{C}$.

Let $\hat{u}(\kappa,t)$ denote a Fourier image of $u(x,t)$

$$u(x,t) = \int_{-\infty}^{\infty} \hat{u}(\kappa,t) \exp(i\kappa x)\, d\kappa,$$

then we have the following correspondences: $u_0 \to \hat{u}$, $u_1 \to i\kappa\hat{u}$, ..., $u_m \to (i\kappa)^m \hat{u}$, ... The Fourier image of a monomial $u_n u_m$ can obviously be represented as

$$u_n u_m = \iiint \delta(\kappa_1 + \kappa_2 - \kappa)(i\kappa_1)^n (i\kappa_2)^m \hat{u}(\kappa_1,t)\hat{u}(\kappa_2,t) \exp(i\kappa x)\, d\kappa_1\, d\kappa_2\, d\kappa$$

and can be rewritten in a symmetrized form

$$u_n u_m = \iiint \delta(\kappa_1 + \kappa_2 - \kappa) \cdot$$

$$\frac{[(i\kappa_1)^n (i\kappa_2)^m + (i\kappa_2)^n (i\kappa_1)^m]}{2} \hat{u}(\kappa_1,t)\hat{u}(\kappa_2,t) \exp(i\kappa x)\, d\kappa_1\, d\kappa_2\, d\kappa,$$

therefore $u_n u_m \to$

$$\iint \delta(\kappa_1 + \kappa_2 - \kappa) \frac{[(i\kappa_1)^n (i\kappa_2)^m + (i\kappa_2)^n (i\kappa_1)^m]}{2} \hat{u}(\kappa_1,t)\hat{u}(\kappa_2,t)\, d\kappa_1\, d\kappa_2.$$

We shall simplify notations further omitting the integration, the delta function, replacing $i\kappa_n$ by ξ_n and $\hat{u}(\kappa_1,t)\hat{u}(\kappa_2,t)$ by u^2. Thus we shall represent the monomial $u_n u_m$ by a symbol

$$u_n u_m \to u^2 a(\xi_1,\xi_2), \quad \text{where} \quad a(\xi_1,\xi_2) = \frac{[\xi_1^n \xi_2^m + \xi_2^n \xi_1^m]}{2}$$

is a symmetric polynomial of its arguments. Following this rule we shall represent any differential monomial $u_0^{n_0} u_1^{n_1} \cdots u_q^{n_q}$ by the symbol

$$u_0^{n_0} u_1^{n_1} \cdots u_q^{n_q} \to u^m \left\langle \xi_1^0 \cdots \xi_{n_0}^0 \xi_{n_0+1}^1 \cdots \xi_{n_0+n_1}^1 \xi_{n_0+n_1+1}^2 \cdots \xi_{n_0+n_1+n_2}^2 \cdots \xi_m^q \right\rangle$$

where $m = n_0 + n_1 + \cdots + n_q$ and the brackets $\langle \rangle$ mean the symmetrization over the group of permutation of m elements (i.e. permutation of all arguments ξ_j)

$$\langle f(\xi_1,\xi_2,\dots,\xi_m) \rangle = \frac{1}{m!} \sum_{\sigma \in \Sigma_m} f(\sigma(\xi_1), \sigma(\xi_2), \dots, \sigma(\xi_m)).$$

For example

$$u_n \to u\xi_1^n, \quad u_3^2 \to u^2\xi_1^3\xi_2^3, \quad u^3u_2 \to u^4\frac{\xi_1^2 + \xi_2^2 + \xi_3^2 + \xi_4^2}{4}.$$

We want to emphasize that the symmetrization over the permutation group is important and it is the symmetrization that makes the symbol defined uniquely. Equality of symbols implies the equality of the corresponding differential polynomials.

The symbolic representation $\hat{\mathscr{R}}(u,\eta)$ of the differential ring $\mathscr{R}(u,D)$ can be defined as follows. The sum of differential monomials is represented by the sum of the corresponding symbols. To the multiplication of monomials f and g with symbols $f \to u^p a(\xi_1, \dots, \xi_p)$ and $g \to u^q b(\xi_1, \dots, \xi_q)$ corresponds the symbol

$$fg \to u^{p+q}\langle a(\xi_1, \dots, \xi_p)b(\xi_{p+1}, \dots, \xi_{p+q})\rangle.$$

Here the symmetrization is taken over the group of permutation of all $p + q$ arguments $\xi_1, \dots, \xi_{p+q}$. The derivative $D(f)$ of a monomial f with the symbol $u^p a(\xi_1, \dots, \xi_s)$ is represented by

$$D(f) \to u^s(\xi_1 + \xi_2 + \cdots + \xi_p)a(\xi_1, \dots, \xi_s).$$

The following rules are motivated by the theory of linear pseudo-differential operators in Fourier representation and are nothing but abbreviated notations. To the operator D (1.34) we shall assign a special symbol η and the following rules of action on symbols:

$$\eta(u^n a(\xi_1, \dots, \xi_n)) = u^n a(\xi_1, \dots, \xi_n)\sum_{j=1}^{n}\xi_j$$

and the composition rule

$$\eta \circ u^n a(\xi_1, \dots, \xi_n) = u^n a(\xi_1, \dots, \xi_n)\left(\sum_{j=1}^{n}\xi_j + \eta\right).$$

The latter corresponds to the Leibnitz rule $D \circ f = D(f) + fD$. Now it can be shown that the composition rule (1.44) can be represented as follows. Let we have two operators fD^q and gD^s such that f and g have symbols $u^i a(\xi_1, \dots, \xi_i)$ and $u^j b(\xi_1, \dots, \xi_j)$, respectively. Then $fD^q \to u^i a(\xi_1, \dots, \xi_i)\eta^q, gD^s \to u^j b(\xi_1, \dots, \xi_j)\eta^s$ and

$$fD^q \circ gD^s \to u^{i+j}\left\langle a(\xi_1, \dots, \xi_i)\left(\eta + \sum_{m=i+1}^{i+j}\xi_m\right)^q b(\xi_{i+1}, \dots, \xi_{i+j})\eta^s\right\rangle. \quad (1.113)$$

Here the symmetrization is taken over the group of permutation of all $i + j$ arguments $\xi_1, \dots, \xi_{i+j}$ and the symbol η is not included in this set. In particular it follows from (1.113) that $D^q \circ D^s \to \eta^{q+s}$. The composition rule (1.113) is valid for positive and negative exponents q, s. In the case of positive exponents it is a polynomial in η and the result is a Fourier image of a differential operator. In the case of negative exponents one can expand the result on η at $\eta \to \infty$ in order to identify it with (1.44). In the symbolic representation instead of formal series (1.43) it is natural to consider formal series of the form

$$B = b(\eta) + ub_1(\xi_1, \eta) + u^2 b_2(\xi_1, \xi_2, \eta) + u^3 b_3(\xi_1, \xi_2, \xi_3, \eta) + \cdots, \quad b(\eta) \neq 0.$$
(1.114)

Using the composition rule (1.113) one can compute the square of the series B

$$\begin{aligned}
B^2 = {} & b^2(\eta) + u(b(\eta + \xi_1)b_1(\xi_1, \eta) + b_1(\xi_1, \eta)b(\eta)) \\
& + u^2 \left(\frac{1}{2}b_1(\xi_1, \eta + \xi_2)b_1(\xi_2, \eta) + \frac{1}{2}b_1(\xi_2, \eta + \xi_1)b_1(\xi_1, \eta) \right. \\
& \left. + b(\eta + \xi_1 + \xi_2)b_2(\xi_1, \xi_2, \eta) + b_2(\xi_1, \xi_2, \eta)b(\eta) \right) + \cdots,
\end{aligned}$$

any integer power B^k, the inverse series B^{-1}, etc.

Let $fD^q \to u^i a(\xi_1, \ldots, \xi_i)\eta^q$ then the symbolic representation for the formally conjugated operator is

$$(-1)^q D^q \circ f \to u^i a(\xi_1, \ldots, \xi_i) \left(-\eta - \overset{i}{\underset{n=1}{\sum}} \xi_n \right)^q.$$

The symbolic representation of the Fréchet derivative of the element $f \to u^n a(\xi_1, \ldots, \xi_n)$ is

$$f_* \to n u^{n-1} a(\xi_1, \ldots, \xi_{n-1}, \eta).$$

For example, let $F = u_3 + 6uu_1$, then $F \to u\xi_1^3 + 3u^2(\xi_1 + \xi_2)$ and

$$F_* \to \eta^3 + 6u(\xi_1 + \eta).$$

It is interesting to notice that the symbol of the Fréchet derivative is always symmetric with respect to all permutations of arguments, including the argument η. Moreover, the following obvious, but useful proposition holds:

Proposition 64. *A differential operator is a Fréchet derivative of an element of $\mathcal{R}(u, D)$ if and only if its symbol is invariant with respect to all permutations of its argument, including the argument η.*

The variational derivative $\delta f / \delta u$ of $f \to u^m a(\xi_1, \ldots, \xi_m)$ can be represented as

$$\frac{\delta f}{\delta u} \to m u^{m-1} a \left(\xi_1, \ldots, \xi_{m-1}, - \overset{m-1}{\underset{i=1}{\sum}} \xi_i \right).$$

The symbolic representation has been extended and proved to be very useful in the case of noncommutative differential rings [52]. It can be easily generalized to the case of many dependent variables [7], suitable for study of system of equations. Here we are going to extend it further to the case of nonlocal and multi-dimensional equations.

Let the right-hand side of Eq. (1.53) be a differential polynomial or can be represented in the form of a series (1.111). In the symbolic representation it can be written as

$$u_t = u\omega(\xi_1) + \frac{u^2}{2}a_1(\xi_1,\xi_2) + \frac{u^3}{3}a_2(\xi_1,\xi_2,\xi_3) + \frac{u^4}{4}a_3(\xi_1,\xi_2,\xi_3,\xi_4) + \cdots = F,$$

(1.115)

where $\omega(\xi_1), a_n(\xi_1,\ldots,\xi_{n+1})$ are symmetrical polynomials and deg $\omega(\xi_1) \geq 2$. According to the previous section the Fréchet derivative of the right-hand side is of the form

$$F_* = \omega(\eta) + ua_1(\xi_1,\eta) + u^2 a_2(\xi_1,\xi_2,\eta) + u^3 a_3(\xi_1,\xi_2,\xi_3,\eta) + \cdots .$$

Symmetries of Eq. (1.115), if they exist, can be found recursively:

Proposition 65. *Suppose Eq.* (1.115) *has a symmetry*

$$u_\tau = u\Omega(\xi_1) + \sum_{j \geq 1} \frac{u^{j+1}}{j+1} A_j(\xi_1,\ldots,\xi_{j+1}) = G,$$

then functions $A_j(\xi_1,\ldots,\xi_{j+1})$ *of the symmetry are related to functions* $a_i(\xi_1, \ldots,\xi_{i+1})$ *of the equation by the following formulae:*

$$A_1(\xi_1,\xi_2) = \frac{N^\omega(\xi_1,\xi_2)}{N^\Omega(\xi_1,\xi_2)}a_1(\xi_1,\xi_2),$$

$$A_m(\xi_1,\ldots,\xi_{m+1}) = \frac{N^\omega(\xi_1,\ldots,\xi_{m+1})}{N^\Omega(\xi_1,\ldots,\xi_{m+1})}a_m(\xi_1,\ldots,\xi_{m+1}) + N^\omega(\xi_1,\ldots,\xi_{m+1})$$

$$\times \left\langle \sum_{j=1}^{m-1} \frac{m+1}{m-j+1}A_j(\xi_1,\ldots,\xi_j,\xi_{j+1}+\cdots+\xi_{m+1})a_{m-j}(\xi_{j+1},\ldots,\xi_{m+1}) \right.$$

$$\left. - \sum_{j=1}^{m-1} \frac{m+1}{j+1}a_{m-j}(\xi_1,\ldots,\xi_{m-j}\xi_{m-j+1}+\cdots+\xi_{m+1})A_j(\xi_{m-j+1},\ldots,\xi_{m+1}) \right\rangle,$$

where

$$N^\omega(\xi_1,\ldots,\xi_m) = \left(\omega\left(\sum_{n=1}^m \xi_n\right) - \sum_{n=1}^m \omega(\xi_n)\right)^{-1},$$

$$N^\Omega(\xi_1,\ldots,\xi_m) = \left(\Omega\left(\sum_{n=1}^m \xi_n\right) - \sum_{n=1}^m \Omega(\xi_n)\right)^{-1}.$$

For any function F of the form (1.115) we can solve the linear operator equation (1.55) to find a formal recursion operator Λ.

Proposition 66. *Operator Λ is a solution of Eq. (1.55) if its symbol is of the form*

$$\Lambda = \phi(\eta) + u\phi_1(\xi_1,\eta) + u^2\phi_2(\xi_1,\xi_2,\eta) + u^3\phi_3(\xi_1,\xi_2,\xi_3,\eta) + \cdots ,$$

where $\phi(\eta)$ is an arbitrary function and $\phi_m(\xi_1,\ldots,\xi_m,\eta)$ are determined recursively:

$$\phi_1(\xi_1,\eta) = N^\omega(\xi_1,\eta)a_1(\xi_1,\eta)(\phi(\eta+\xi_1) - \phi(\eta)),$$

$$\phi_m(\xi_1,\ldots,\xi_m,\eta) = \bigg\{ (\phi(\eta+\xi_1+\cdots+\xi_m) - \phi(\eta))a_m(\xi_1,\ldots,\xi_m,\eta) +$$

$$\sum_{n=1}^{m-1}\bigg\langle \frac{n}{m-n+1}\phi_n(\xi_1,\ldots,\xi_{n-1},\xi_n+\cdots+\xi_m,\eta)a_{m-n}(\xi_n,\ldots,\xi_m) +$$

$$\phi_n(\xi_1,\ldots,\xi_n,\eta+\xi_{n+1}+\cdots+\xi_m)a_{m-n}(\xi_{n+1},\ldots,\xi_m,\eta) -$$

$$a_{m-n}(\xi_{n+1},\ldots,\xi_m,\eta+\xi_1+\cdots+\xi_n)\phi_n(\xi_1,\ldots,\xi_n,\eta) \bigg\rangle \bigg\} N^\omega(\xi_1,\ldots,\xi_m,\eta).$$

We immediately see the advantage of the perturbative approach. Now we are able to obtain explicit recursion relations for determining the coefficients of a symmetry and a formal recursion operator while in the standard Symmetry Approach the corresponding problem was quite difficult.

Existence of a symmetry means that all coefficients $A_m(\xi_1,\ldots,\xi_{m+1})$ are polynomials (not rational functions). In other words the symbols

$$u^{m+1}A_m(\xi_1,\ldots,\xi_{m+1}) \in \hat{\mathscr{R}}(u,\eta)$$

and correspond to differential polynomials in the standard representation. This requirement can be used for testing for integrability and even for complete classification of integrable equations (see [52, 53, 76]).

In the standard Symmetry Approach the integrability, i.e. the existence of infinite hierarchies, of local symmetries or conservation laws implies (Theorems 24,32) that all coefficients l_n are local and belong to the corresponding differential field or ring. In the symbolic representation it suggests the following definition.

Definition 67. We say that the function $b_m(\xi_1,\ldots,\xi_m,\eta)$, $m \geq 1$, is *k*-local if the first k coefficients $\beta_{mn}(\xi_1,\ldots,\xi_m)$, $n = n_s,\ldots,n_s+k$ of its expansion at $\eta \to \infty$

$$b_m(\xi_1,\ldots,\xi_m,\eta) = \sum_{n=s_n}^{\infty} \beta_{mn}(\xi_1,\ldots,\xi_m)\eta^{-n}$$

are symmetric polynomials. We say that the coefficient $b_m(\xi_1,\ldots,\xi_m,\eta)$ of a formal series (1.114) is local if it is *k*-local for any *k*.

Theorem 68. *Suppose Eq. (1.115) has an infinite hierarchy of symmetries*

$$u_{t_i} = u\Omega_i(\xi_1) + \sum_{j\geq 1}\frac{u^{j+1}}{j+1}A_{ij}(\xi_1,\ldots,\xi_{j+1}) = G_i, \quad i = 1,2,\ldots,$$

where $\Omega_i(\xi_1)$ are polynomials of degree $m_i = \deg(\Omega_i(\xi_1))$ and $m_1 < m_2 < \cdots < m_i < \cdots$. Then the coefficients $\phi_m(\xi_1,\ldots,\xi_m,\eta)$ of the formal recursion operator

$$\Lambda = \eta + u\phi_1(\xi_1,\eta) + u^2\phi_2(\xi_1,\xi_2,\eta) + \cdots$$

are local.

The symmetry approach in symbolic representation suggests the following test for integrability of equations of the form (1.115):

- Find a first few coefficients $\phi_n(\xi_1,\ldots,\xi_n,\eta)$.
- Expand these coefficients in series of $1/\eta$

$$\phi_n(\xi_1,\ldots,\xi_n,\eta) = \sum_{s=s_n} \Phi_{ns}(\xi_1,\ldots,\xi_n)\eta^{-s} \tag{1.116}$$

and find the corresponding functions $\Phi_{ns}(\xi_1,\ldots,\xi_n)$.
- Check that functions $\Phi_{ns}(\xi_1,\ldots,\xi_n)$ are polynomials (not rational functions).

As an example of application we consider equations of the form

$$u_t = u\omega(\xi_1) + \sum_{i\geq1} \frac{u^{i+1}}{i+1}a_i(\xi_1,\ldots,\xi_{i+1}), \tag{1.117}$$

where $\omega(\xi_1)$ is a polynomial on ξ_1 of the degree $deg(\omega(\xi_1)) = n \geq 2$ and $a_i(\xi_1,\ldots,\xi_{i+1})$ are symmetric polynomials on its arguments of degree $deg(a_i(\xi_1,\ldots,\xi_{i+1})) \leq n-2$. The following propositions are valid.

Proposition 69. *If Eq. (1.117) is integrable, then $\omega(0) = 0$, i.e. the polynomial $\omega(\xi_1)$ can be factorized $\omega(\xi_1) = \xi_1 f(\xi_1)$, where $f(\xi_1)$ is a polynomial.*

Proposition 70. *Suppose n is even and $a_1(\xi_1,\xi_2) \equiv 0$ or n is odd and $a_1(\xi_1,\xi_2) \equiv 0, a_2(\xi_1,\xi_2,\xi_3) \equiv 0$. Then Eq. (1.117) is not integrable.*

The statement of proposition 70 in the homogeneous case was proved in the works of J. Sanders and J.P. Wang [53].

1.4.4.2 Nonlocal and Nonevolutionary Integrable Equations

In order to deal with nonlocal or nonevolutionary equations we have to extend the differential ring properly. Here we illustrate the construction of the extension and the corresponding generalization of the symmetry approach on two examples.

The first example is the generalized Camassa–Holm–Degasperis–Procesi equation, which can be written as a scalar nonlocal evolution equation[5]

$$u_t = (1-D^2)^{-1}(u_3 - (c+1)u_1 - uu_3 + (c+1)uu_1 - cu_1u_2), \quad c \neq 0, \tag{1.118}$$

---

[5] If we apply operator $1-D^2$ to Eq. (1.118) we obtain local, but non-evolutionary equation.

where $\Delta = (1 - D^2)^{-1}$. It is well known that Eq. (1.118) is integrable if $c = 2, 3$ (see [8, 11, 12]). For these values of c Eq. (1.118) has an infinite hierarchy of higher symmetries. The higher symmetries contain nested operator Δ. In order to be able to consider such symmetries we need to extend the differential ring $\mathscr{R}(u, D)$ (c.f.[46]). We shall build the following sequence of ring extensions:

$$\mathscr{R}_\Delta^0 \subset \mathscr{R}_\Delta^1 \subset \mathscr{R}_\Delta^2 \subset \mathscr{R}_\Delta^3 \subset \cdots \subset \mathscr{R}_\Delta,$$

where

$$\mathscr{R}_\Delta^0 = \mathscr{R}(u, D), \quad \mathscr{R}_\Delta^1 = \overline{\mathscr{R}_\Delta^0 \bigcup \Delta(\mathscr{R}_\Delta^0)}, \quad \mathscr{R}_\Delta^{n+1} = \overline{\mathscr{R}_\Delta^n \bigcup \Delta(\mathscr{R}_\Delta^n)}.$$

Here the over-line denotes the ring closure, index n in $\mathscr{R}_\Delta^n$ denotes the depth of the nesting for the operator Δ and $\mathscr{R}_\Delta = \lim_{n \to \infty} \mathscr{R}_\Delta^n$. Symbolic representation of operator Δ is $\Delta \to \frac{1}{1 - \eta^2}$. For example if A is an element from $\mathscr{R}_\Delta^0$ with corresponding symbol $u^n a(\xi_1, \dots, \xi_n)$ then $\Delta(A)$ has a symbol $u^n \frac{a(\xi_1, \dots, \xi_n)}{1 - (\xi_1 + \cdots + \xi_n)^2}$.

The way of testing for integrability of a given equation is the one described in the previous section. The only difference is that we have to replace the requirement of locality for the coefficients of the formal recursion operator by quasi-locality, i.e. we have to require that the coefficients Φ_{ns} in the expansion (1.116) correspond to the symbolic representation of elements from $\mathscr{R}_\Delta$.

Let us illustrate the application of this test to the Camassa–Holm–Degasperis–Procesi type equations.

Theorem 71. *Equation* (1.118) *is integrable only if* $c = 2$ *or* $c = 3$.

Proof. In the symbolic representation Eq. (1.118) has the form

$$u_t = u\omega(\xi_1) + \frac{u^2}{2} a(\xi_1, \xi_2) = F,$$

where

$$\omega(k) = \frac{k^3 - (c+1)k}{1 - k^2},$$

$$a(\xi_1, \xi_2) = \frac{(c+1)(\xi_1 + \xi_2) - (\xi_1^3 + \xi_2^3) - c\xi_1\xi_2(\xi_1 + \xi_2)}{1 - (\xi_1 + \xi_2)^2}.$$

Calculating first two coefficients of the corresponding formal recursion operator

$$\Lambda = \eta + u\phi_1(\xi_1, \eta) + u^2\phi_2(\xi_1, \xi_2, \eta) + \cdots,$$

we find that the first coefficient

$$\phi_1(\xi_1, \eta) = \frac{(\xi_1^2 - 1)(\eta^2 - 1)(\xi_1^2 + \eta^2 - \xi_1\eta - 1 + c(\xi_1\eta - 1))}{c\eta(\eta^2 + \xi_1^2 + \xi_1\eta - 3)}$$

is local because all the coefficients of its expansion in $1/\eta$ are polynomials on ξ_1. For the second coefficient $\phi_2(\xi_1,\xi_2,\eta)$ we have the following expansion:

$$\phi_2(\xi_1,\xi_2,\eta) = \Phi_{21}(\xi_1,\xi_2)\eta + \Phi_{20}(\xi_1,\xi_2) + \Phi_{2,-1}(\xi_1,\xi_2)\eta^{-1}+$$

$$+\Phi_{2,-2}(\xi_1,\xi_2)\eta^{-2} + \Phi_{2,-3}(\xi_1,\xi_2)\eta^{-3} + \cdots,$$

where coefficients $\Phi_{21}(\xi_1,\xi_2), \dots, \Phi_{2,-2}(\xi_1,\xi_2)$ are polynomials on their arguments (we do not present here the explicit expressions for $\Phi_{21}(\xi_1,\xi_2), \dots, \Phi_{2,-2}(\xi_1,\xi_2)$ – they are quite large and complex), while the coefficient $\Phi_{2,-3}$ has the form

$$\Phi_{2,-3}(\xi_1,\xi_2) = \frac{f(\xi_1,\xi_2)}{1-\xi_1\xi_2}$$

and $f(\xi_1,\xi_2)$ is a polynomial. If the numerator $f(\xi_1,\xi_2)$ does not have $1-\xi_1\xi_2$ as a factor, then the symbol $u^2\Phi_{2,-3}$ does not correspond to any element of our extended ring $\mathscr{R}_\Delta$ and hence it is not quasi-local. It is easy to check that the polynomial $f(\xi_1,\xi_2)$ can be divided by $1-\xi_1\xi_2$ if and only if the condition

$$(c-2)(c-3)=0$$

is satisfied and in these cases the coefficient $\Phi_{2,-3}(\xi_1,\xi_2)$ is a polynomial. Therefore conditions $c=2$ or $c=3$ are necessary for the integrability (i.e. existence of higher quasi-local symmetries) of Eq. (1.118).

1.5 Short Description of Solved Classification Problems and References

1.5.1 Hyperbolic Equations

The first classification result [78] in the frame of the symmetry approach was as follows:

Theorem 72. *Nonlinear hyperbolic equation of the form*

$$u_{xy} = F(u)$$

possesses higher symmetries iff (up to scalings and shifts)

$$F(u) = e^u, \quad F(u) = e^u + e^{-u} \quad or \quad F(u) = e^u + e^{-2u}.$$

In [79] all integrable hyperbolic systems of the form

$$u_x = p(u,v), \qquad v_y = q(u,v)$$

have been described. Notice that in the nondegenerate case $\frac{\partial p}{\partial v} \neq 0$, such a system is equivalent to a second-order hyperbolic equation of the form

$$u_{xy} = A_1(u, u_x) u_y + A_2(u, u_x).$$

The complete classification of integrable hyperbolic equations of the form

$$u_{xy} = F(x, y, u, u_x, u_y) \tag{1.119}$$

is an open problem till now. The following examples show that the dependence of the right-hand side F on the derivatives u_x and u_y can be rather complicated.

Example:

$$u_{xy} = S(u) \sqrt{1 - u_x^2} \sqrt{1 - u_y^2}, \qquad \text{where} \qquad S'' - 2S^3 + \lambda S = 0,$$

$$u_{xy} = S(u) b(u_x) \bar{b}(u_y), \qquad \text{where}$$

$$S'' - 2S' - 4S^3 = 0, \qquad (u_x - b)(b + 2u_x)^2 = 1, \qquad (u_y - \bar{b})(\bar{b} + 2u_y)^2 = 1.$$

In [80] all Darboux integrable equations (1.119) have been listed.

1.5.2 One Component Evolution Equations

1.5.2.1 Second-Order Equations

All nonlinear integrable equations of the form

$$u_t = F(u_2, u_1, u, x, t)$$

were listed in [64] and [59]. The answer is

$$u_t = u_2 + 2uu_x + h(x),$$
$$u_t = u^2 u_2 - \lambda x u_1 + \lambda u,$$
$$u_t = u^2 u_2 + \lambda u^2,$$
$$u_t = u^2 u_2 - \lambda x^2 u_1 + 3\lambda x u.$$

This list is complete up to contact transformations of the form

$$\hat{t} = \chi(t), \qquad \hat{x} = \varphi(x, u, u_1), \qquad \hat{u} = \psi(x, u, u_1),$$

$$\hat{u}_i = \left(\frac{1}{D(\varphi)} D \right)^i (\psi),$$

where the contact condition

$$D(\varphi)\frac{\partial \psi}{\partial u_1} = D(\psi)\frac{\partial \varphi}{\partial u_1}$$

is satisfied.

Three first equations of the list possess local symmetries and form a list obtained in [64]. The latter equation has so-called weakly nonlocal symmetries (see [59, 71]).

1.5.2.2 Third-Order Equations

All equations of the form

$$u_t = u_3 + F(u_1, u)$$

possessing higher symmetries have been obtained in [16, 17, 27] (see Sect. 3.7).

All equations of the form

$$u_t = u_3 + F(u_2, u_1, u, x)$$

possessing higher conservation laws or higher symmetries were found in [69, 70]. In order to derive all integrable equations the following four necessary integrability conditions have been used:

$$D_t\left(\frac{\partial F}{\partial u_2}\right) = D(\sigma_1),$$

$$D_t\left(3\frac{\partial F}{\partial u_1} - \left(\frac{\partial F}{\partial u_2}\right)^2\right) = D(\sigma_2),$$

$$D_t\left(9\sigma_1 + 2\left(\frac{\partial F}{\partial u_2}\right)^3 - 9\left(\frac{\partial F}{\partial u_2}\right)\left(\frac{\partial F}{\partial u_1}\right) + 27\frac{\partial F}{\partial u}\right) = D(\sigma_3),$$

$$D_t(\sigma_2) = D(\sigma_4).$$

An analysis of the answer of the classification problem shows that any equation that satisfies these conditions is really integrable. Thus to verify the integrability of a given equation, it suffices to check the four conditions presented above.

Using transformations of different types, one can reduce any integrable equation to one of the canonical forms contained in [45, 69, 70].

Theorem 73. *A complete list (up to "almost invertible" transformations) of nonlinear equations with infinite hierarchy of conservation laws can be written as*

$$u_t = u_{xxx} + u u_x,$$
$$u_t = u_{xxx} + u^2 u_x,$$
$$u_t = u_{xxx} - \frac{1}{2}u_x^3 + (\alpha e^{2u} + \beta e^{-2u})u_x,$$

$$u_t = u_{xxx} - \frac{1}{2}Q'' u_x + \frac{3}{8}\frac{\left(Q - u_x^2\right)_x^2}{u_x\left(Q - u_x^2\right)},$$

$$u_t = u_{xxx} - \frac{3}{2}\frac{u_{xx}^2 + Q}{u_x},$$

where $Q = c_4 u^4 + c_3 u^3 + c_2 u^2 + c_1 u + c_0$.

The first and the latter equations (i.e. the Korteweg–de Vries and the Krichever–Novikov equations) form a complete list of integrable equations (see [73]) up to differential substitutions.

Third-order equations of more general form have been considered in [25, 45].

1.5.2.3 Fifth-Order Equations

All equations of the form

$$u_t = u_5 + F(u_4, u_3, u_2, u_1, u)$$

possessing higher conservation laws were found in [15] (see also [45]).

Example: Well-known equations:

$$u_t = u_5 + 5 u u_3 + 5 u_1 u_2 + 5 u^2 u_1,$$

$$u_t = u_5 + 5 u u_3 + a frac252 u_1 u_2 + 5 u^2 u_1,$$

$$u_t = u_5 + 5(u_1 - u^2)u_3 + 5u_2^2 - 20 u u_1 u_2 - 5 u_1^3 + 5 u^4 u_1.$$

A new equation:

$$u_t = u_5 + 5\left(u_2 - u_1^2 + \lambda_1 e^{2u} - \lambda_2^2 e^{-4u}\right) u_3$$
$$- 5 u_1 u_2^2 + 15\left(\lambda_1 e^{2u} + 4\lambda_2^2 e^{-4u}\right) u_1 u_2 + u_1^5$$
$$- 90 \lambda_2^2 e^{-4u} u_1^3 + 5\left(\lambda_1 e^{2u} - \lambda_2^2 e^{-4u}\right)^2 u_1.$$

1.5.3 Two-Component Systems of Evolution Equations

The most significant work has been done in [40–42, 54] where all systems of the form

$$u_t = u_2 + F(u, v, u_1, v_1), \qquad u_t = -v_2 + G(u, v, u_1, v_1) \tag{1.120}$$

possessing higher conservation laws were listed and studied.

Example: Well-known NLS equation is in the form (1.120)

$$u_t = u_2 + u^2 v, \qquad v_t = -v_2 - v^2 u.$$

The Boussinesq equation can be written in this form (1.120):

$$u_t = u_2 + (u+v)^2, \qquad v_t = -v_2 - (u+v)^2.$$

The Landau–Lifshitz equation

$$\dot{\mathbf{S}} = \mathbf{S} \times \mathbf{S}_{xx} + \mathbf{S} \times \hat{J}\mathbf{S}, \quad \hat{J} = \mathrm{diag}(J_1, J_2, J_3), \quad (\mathbf{S} \cdot \mathbf{S}) = 1$$

in proper coordinates can be rewritten in the form (1.120)

$$u_t = u_2 - \frac{2u_1^2}{u+v} - \frac{4\,(p(u,v)\,u_1 + r(u)\,v_1)}{(u+v)^2},$$

$$v_t = -v_2 + \frac{2v_1^2}{u+v} - \frac{4\,(p(u,v)\,v_1 + r(-v)\,u_1)}{(u+v)^2},$$

where $r(y) = c_4 y^4 + c_3 y^3 + c_2 y^2 + c_1 y + c_0$ and

$$p(u,v) = 2c_4 u^2 v^2 + c_3 (uv^2 - vu^2) - 2c_2 uv + c_1 (u-v) + 2c_0.$$

That are examples of equations in a very long list of integrable systems given in [42, 43, 45].

Quasilinear systems of the form

$$u_t = \lambda_1 u_2 + A_1(u,v)u_1 + A_2(u,v)v_1 + A_3(u,v),$$
$$v_t = \lambda_2 v_2 + B_1(u,v)u_1 + B_2(u,v)v_1 + B_3(u,v),$$

where $\lambda_1 \neq -\lambda_2, \lambda_1 \neq \lambda_2, \lambda_i \neq 0$, that have higher symmetries were considered in [5]. All quasilinear systems of the above form with $\lambda_1 = -\lambda_2$ having higher symmetries were found in [60]. In the case $\lambda_1 \neq -\lambda_2, \lambda_1 \lambda_2 \neq 0$ all homogeneous differential polynomial systems with higher symmetries have been found in [7].

In a recent paper [31] the following very interesting example of integrable system of the form

$$u_t = u_2 + A(u,v)v_2 + F(u,v,u_1,v_1), \qquad u_t = v_2 + G(u,v,u_1,v_1)$$

has been found:

$$u_t = D(u_1 - 2v_1 + uv^2 - u^2), \qquad v_t = D(v_x - 2uv + v^3).$$

Third-order integrable systems of the form

$$u_t = \lambda_1 u_3 + f(u,v,u_1,v_1,u_2,v_2),$$
$$v_t = \lambda_2 v_3 + g(u,v,u_1,v_1,u_2,v_2)$$

were studied by A. Meshkov [32, 33]. Almost all known integrable systems of this kind are related by differential substitutions to examples of such systems found in [14].

Nonevolutionary equations of Boussinesq type

$$u_{tt} = K(u, u_x, u_{xx}, \cdots, \partial_x^n u, u_t, u_{tx}, u_{txx}, \cdots, \partial_x^m u_t) \qquad (1.121)$$

can always be replaced by a system of two evolutionary equations

$$\begin{cases} u_t = v, \\ v_t = K(u, u_x, u_{xx}, \dots, \partial_x^n u, v, v_x, v_{xx}, \dots, \partial_x^m v). \end{cases}$$

If $K = D_x^r(G(u, u_x, u_{xx}, \cdots, \partial_x^{n-r} u, u_t, u_{tx}, u_{txx}, \cdots, \partial_x^{m-r} u_t))$ then Eq. (1.121) has also representation

$$\begin{cases} u_t = \partial_x^r v, \\ v_t = G(u, u_x, u_{xx}, \dots, \partial_x^{n-r} u, v, v_x, v_{xx}, \dots, \partial_x^{m-r} v). \end{cases}$$

Integrable systems of these types have been studied in [38, 48].

In the case of systems of two (or more) evolutionary equations

$$\begin{aligned} u_t &= \lambda_1(n)u_n + F(u, v, \dots u_{n-1}, v_{n-1}), \qquad (1.122) \\ u_t &= \lambda_2(n)v_n + G(u, v, \dots u_{n-1}, v_{n-1}), \end{aligned}$$

the ratio $\lambda_1(n)/\lambda_2(n)$ is invariant with respect to point transformations, which we call the spectrum of the system (or dispersion law). The spectrum of integrable systems and higher symmetries is not arbitrary. Here we present two new and rather nontrivial examples of integrable systems which have been found recently [39] using the symbolic method. System

$$\begin{cases} \begin{aligned} u_t &= \left(9 - 5\sqrt{3}\right) u_5 + D_x \left\{ 2\left(9 - 5\sqrt{3}\right) u u_2 + \left(-12 + 7\sqrt{3}\right) u_1^2 \right\} \\ &\quad + 2\left(3 - \sqrt{3}\right) u_3 v + 2\left(6 - \sqrt{3}\right) u_2 v_1 + 2\left(3 - 2\sqrt{3}\right) u_1 v_2 \\ &\quad - 6\left(1 + \sqrt{3}\right) u v_3 + D_x \left\{ 2\left(33 + 19\sqrt{3}\right) v v_2 + \left(21 + 12\sqrt{3}\right) v_1^2 \right\} \\ &\quad + \frac{4}{5}\left(-12 + 7\sqrt{3}\right) u^2 u_1 + \frac{8}{5}\left(3 - 2\sqrt{3}\right)\left(v u u_1 + u^2 v_1\right) \\ &\quad + \frac{4}{5}\left(24 + 13\sqrt{3}\right) v^2 u_1 + \frac{8}{5}\left(36 + 20\sqrt{3}\right) u v v_1 - \frac{8}{5}\left(45 + 26\sqrt{3}\right) v^2 v_1, \\[8pt] v_t &= \left(9 + 5\sqrt{3}\right) v_5 + D_x \left\{ 2\left(33 - 19\sqrt{3}\right) u u_2 + \left(21 - 12\sqrt{3}\right) u_1^2 \right\} \\ &\quad - 6\left(1 - \sqrt{3}\right) u_3 v + 2\left(3 + 2\sqrt{3}\right) u_2 v_1 + 2\left(6 + \sqrt{3}\right) u_1 v_2 \\ &\quad + 2\left(3 + \sqrt{3}\right) u v_3 + D_x \left\{ 2\left(9 + 5\sqrt{3}\right) v v_2 - \left(12 + 7\sqrt{3}\right) v_1^2 \right\} \\ &\quad - \frac{8}{5}\left(45 - 26\sqrt{3}\right) u^2 u_1 + \frac{8}{5}\left(36 - 20\sqrt{3}\right) v u u_1 + \frac{4}{5}\left(24 - 13\sqrt{3}\right) u^2 v_1 \\ &\quad + \frac{8}{5}\left(3 + 2\sqrt{3}\right)\left(v^2 u_1 + u v v_1\right) - \frac{4}{5}\left(12 + 7\sqrt{3}\right) v^2 v_1 \end{aligned} \end{cases}$$

possesses an infinite-dimensional algebra of higher symmetries with

$$\frac{\lambda_2(m)}{\lambda_1(m)} = \frac{\left(1+\exp\left(\frac{\pi i}{6}\right)\right)^m}{1+\exp\left(\frac{m\pi i}{6}\right)}, \qquad m \equiv 1,5,7,11 \bmod 12.$$

System

$$\begin{cases} u_t = -\dfrac{5}{3}u_5 - 10vv_3 - 15v_1v_2 + 10uu_3 + 25u_1u_2 - 6v^2v_1 \\ \qquad +6v^2u_1 + 12uvv_1 - 12u^2u_1, \\ v_t = 15v_5 + 30v_1v_2 - 30v_3u - 45v_2u_1 - 35v_1u_2 - 10vu_3 \\ \qquad -6v^2v_1 + 6v^2u_1 + 12u^2v_1 + 12vuu_1 \end{cases}$$

possesses symmetries of orders $m \equiv 1,5 \bmod 6$ with

$$\frac{\lambda_2(m)}{\lambda_1(m)} = \frac{\left(1+\exp\left(\frac{\pi i}{3}\right)\right)^m}{1+\exp\left(\frac{m\pi i}{3}\right)}.$$

There is a reduction $v = 0$ to the Kaup–Kupershmidt equation.

1.5.4 Nonpolynomial Multi-component Systems

Several classes of nonpolynomial integrable systems are related to deformations of nonassociative structures [24, 61, 72]. Let $\{X,Y,Z\}$ be a Jordan triple system, $\phi(u)$ be a solution of the following overdetermined consistent system of PDEs:

$$\frac{\partial \phi}{\partial u^k} = -\{\phi,\, e_k,\, \phi\},$$

$k = 1,\ldots,N$. Denote

$$\alpha_u(X,Y) = \{X,\phi(u),Y\}, \qquad \sigma_u(X,Y,Z) = \{X,\{\phi(u),Y,\phi(u)\},Z\}.$$

For any given u, α_u and σ_u define a Jordan algebra and a Jordan triple system, correspondingly.

It turns out the following systems possess higher symmetries:

$$u_{xy} = \alpha_u(u_x,u_y),$$

$$u_t = u_{xxx} - 3\alpha_u(u_x,u_{xx}) + \frac{3}{2}\sigma_u(u_x,u_x,u_x),$$

$$v_t = v_{xxx} - \frac{3}{2}\alpha_{v_x}(v_{xx},v_{xx}),$$

$$u_t = u_{xx} - 2\alpha_{u+v}(u_x,u_x), \qquad v_t = -v_{xx} + 2\alpha_{u+v}(v_x,v_x).$$

The explicit formulas for $\phi(u)$ for triple Jordan systems (1.94), (1.95), and (1.96) (see [72]) provide examples of matrix- and vector-integrable nonpolynomial systems. For instance, in the matrix case (1.94) we have $\phi(U) = U^{-1}$ and the above formulas give rise to the following matrix-integrable systems

$$U_{xy} = \frac{1}{2}(U_x U^{-1} U_y + U_y U^{-1} U_x),$$

$$U_t = U_{xxx} - \frac{3}{2}U_x U^{-1} U_{xx} - \frac{3}{2}U_{xx} U^{-1} U_x + \frac{3}{2}U_x U^{-1} U_x U^{-1} U_x,$$

$$U_t = U_{xxx} - \frac{3}{2}U_{xx} U_x^{-1} U_{xx},$$

$$U_t = U_{xx} - 2U_x(U+V)^{-1}U_x, \qquad V_t = -V_{xx} + 2V_x(U+V)^{-1}V_x.$$

1.5.5 Nonabelian Evolution Equations

Integrable nonabelian polynomial homogeneous evolution equations having higher symmetries have been considered in [50, 51].

In many interesting examples, the right-hand side of an integrable evolution equation turns out to be a homogeneous differential polynomial with respect to some weighting of its constituent monomials. We introduce a weighting scheme by assigning a weight $m = \deg u$ to the dependent variable and $n = \deg x$ to the independent variable, so that the kth order derivative of u with respect to x has weight $m + kn$. Without loss of generality, we assume that $n = 1$. It was proved in [52, 53] that for any integrable homogeneous equation with $m > 0$ the number m belongs to the set $\{1/2, 1, 2\}$.

For example, the weighting for the mKdV equation $u_t = u_3 + 6u_1$ is: $\deg u = 1$, $\deg x = 1$. All integrable nonabelian equations $U_t = U_3 + P(U_2, U_1, U)$ of the mKdV weighting belong to the following list:

$$U_t = U_3 + 3UU_2 + 3U_1^2 + 3U^2U_1,$$

$$U_t = U_3 + 3U_2U + 3U_1^2 + 3U_1U^2,$$

$$U_t = U_3 + 3U^2U_1 + 3U_1U^2,$$

$$U_t = U_3 + 3UU_2 - 3U_2U - 6UU_1U,$$

$$U_t = U_3 + 3U_1^2.$$

Second-order nonabelian homogeneous systems of NLS and DNLS types (see also [74]) were also listed and several new integrable models were found.

Example:

$$U_t = U_2 + 2(U+V)U_1, \qquad V_t = -V_2 + 2V_1(U+V);$$
$$U_t = U_2 + 2U_1VU, \qquad V_t = -V_2 + 2VUV_1.$$

1.5.6 Nonabelian Ordinary Differential Equations

Polynomial nonabelian ODEs have been considered in [44]. Some partial classification results have been obtained.

For example the following system

$$U_t = V^2, \qquad V_t = U^2$$

possesses infinitely many symmetries of the form

$$U_{\tau_i} = P_i(U, V), \qquad V_{\tau_i} = Q_i(U, V)$$

and first integrals (the definition of the first integral for nonabelian ODEs is similar to the definition of the conserved density from Sect. 4.2.1).

There exists two basic integrable nonabelian equations containing arbitrary constant element C:

$$U_t = CU^2 - U^2C \tag{1.123}$$

and

$$U_t = UCU^2 - U^2CU.$$

Different reductions of these equations give rise to known integrable multi-component ODEs. For example, let M and C in (1.123) be represented by matrices of the form

$$M = \begin{pmatrix} 0 & u_1 & 0 & 0 & \cdot & 0 \\ 0 & 0 & u_2 & 0 & \cdot & 0 \\ \cdot & \cdot & \cdot & \cdot & \cdot & \cdot \\ 0 & 0 & 0 & 0 & \cdot & u_{N-1} \\ u_N & 0 & 0 & 0 & \cdot & 0 \end{pmatrix}, \qquad C = \begin{pmatrix} 0 & 0 & 0 & \cdot & 0 & 1 \\ 1 & 0 & 0 & \cdot & 0 & 0 \\ 0 & 1 & 0 & \cdot & 0 & 0 \\ \cdot & \cdot & \cdot & \cdot & \cdot & \cdot \\ 0 & 0 & 0 & \cdot & 1 & 0 \end{pmatrix}.$$

Then it follows from Eq. (1.123) that u_k, $k = 1, \ldots, N$, satisfy the Volterra chain

$$\frac{d}{dt} u_k = u_k (u_{k+1} - u_{k-1}), \qquad \text{where} \quad u_{N+1} = u_1, \quad u_0 = u_N.$$

The following nonabelian Painlevé equations [4]

$$U_{xx} + 3U^2 = xE + C,$$
$$U_{xx} + 2U^3 + xU = \lambda E,$$
$$U_{xx} + \frac{1}{x}U_x = U_x U^{-1} U_x,$$

where E is the unity and λ is arbitrary constant, can be derived from integrable nonabelian PDEs by means of the symmetry reductions.

1.5.7 Integrable Isotropic Evolution Equations on the N-Dimensional Sphere

The following class of vector equations

$$U_t = U_3 + f_2 U_2 + f_1 U_1 + f_0 U, \tag{1.124}$$

where functions f_i depend on

$$u_{[i,j]} = <U_i, U_j>,$$

$0 \le i \le j \le 2$, was considered in [34]. Under additional assumption $< U, U >= 1$ the following complete list of integrable equations

$$U_t = U_{xxx} - 3\frac{u_{[1,2]}}{u_{[1,1]}} U_{xx} + \frac{3}{2}\left(\frac{u_{[2,2]}}{u_{[1,1]}} + \frac{u_{[1,2]}^2}{u_{[1,1]}^2(1+au_{[1,1]})}\right) U_x, \tag{1.125}$$

$$U_t = U_{xxx} + \frac{3}{2}\left(\frac{a^2 u_{[1,2]}^2}{1+au_{[1,1]}} - a(u_{[2,2]} - u_{[1,1]}^2) + u_{[1,1]}\right) U_x + 3u_{[1,2]} U, \tag{1.126}$$

$$U_t = U_{xxx} - 3\frac{u_{[1,2]}}{u_{[1,1]}} U_{xx} + \left(\frac{3}{2}\frac{u_{[2,2]}}{u_{[1,1]}}\right) U_x, \tag{1.127}$$

$$U_t = U_{xxx} - 3\frac{(q+1) u_{[1,2]}}{2q u_{[1,1]}} U_{xx} + 3\frac{(q-1) u_{[1,2]}}{2q} U$$
$$+ \frac{3}{2}\left(\frac{(q+1) u_{[2,2]}}{u_{[1,1]}} - \frac{(q+1) a u_{[1,2]}^2}{q^2 u_{[1,1]}} + u_{[1,1]}(1-q)\right) U_x, \tag{1.128}$$

where a is arbitrary constant, $q = \varepsilon\sqrt{1+au_{[1,1]}}$, $\varepsilon^2 = 1$, has been obtained.

Remark 74. The constant a can be reduced to $a = 0$ or to $a = 1$ by an appropriate scaling of x and t. Thus in fact, the list contains many non-equivalent equations. In particular, Eq. (1.128) with $a = 0$ and $\varepsilon = -1$ reads as

$$U_t = U_{xxx} + 3u_{[1,1]} U_x + 3u_{[1,2]} U.$$

Equation (1.126) with $a = 0$ coincides with (1.109), where $R = 0$. If $a = 0$ and $\varepsilon = 1$, then Eq. (1.128) becomes

$$U_t = U_{xxx} - 3\frac{u_{[1,2]}}{u_{[1,1]}} U_{xx} + 3\frac{u_{[2,2]}}{u_{[1,1]}} U_x.$$

Remark 75. Equation (1.127) on $\mathbb{R}^N$ has arisen in the papers [24, 58, 69]. This equation is related to vector triple Jordan systems. It is a vector generalization of well-known Schwartz–KdV equation

$$v_t = v_{xxx} - \frac{3}{2}\frac{v_{xx}^2}{v_x}.$$

Remark 76. In the case $N = 2$ Eqs. (1.125) and (1.126) with $a = 0$ can be reduced to the potential KdV equation

$$v_t = v_{xxx} + v_x^3$$

by the stereographic projection and some point-wise transformations. Equations (1.125) with $a = -1$ and (1.126) with $a = -1$ come to

$$v_t = v_{xxx} - \frac{1}{8}Q'' v_x + \frac{3}{32}\frac{((Q - 4v_x^2)_x)^2}{v_x (Q - 4v_x^2)}, \tag{1.129}$$

where $Q(v) = (v^2 + 1)^2$. The last equation is a special case of the generic Calogero–Degasperis equation (see [10]), which can also be written in the form (1.129) but with $Q(v)$ being arbitrary polynomial of fourth degree. Our particular case corresponds to a trigonometric degeneration of Jacobi's elliptic sine implicitly involved in (1.129).

1.5.8 Integrable Anisotropic Evolution Equations on the N-Dimensional Sphere

The anisotropic equations (1.124) with

$$f_i = f_i(u_{[1,1]}, u_{[1,2]}, u_{[2,2]}, v_{[0,0]}, v_{[0,1]}, v_{[1,1]})$$

were also considered in [34]. Here the variables $v_{[i,j]}$ are given by (1.110). The following complete list of integrable equations was obtained:

$$U_t = U_3 + \left(\frac{3}{2}u_{[1,1]} + c\,v_{[0,0]}\right) U_1 + 3\,u_{[1,2]}\, U_0, \tag{1.130}$$

$$U_t = U_3 - 3\frac{u_{[1,2]}}{u_{[1,1]}} U_2 + \frac{3}{2}\left(\frac{u_{[2,2]}}{u_{[1,1]}} + \frac{u_{[1,2]}^2}{u_{[1,1]}^2} + \frac{c\,v_{[1,1]}}{u_{[1,1]}}\right) U_1, \tag{1.131}$$

$$U_t = U_3 - 3\frac{u_{[1,2]}}{u_{[1,1]}} U_2 + \tag{1.132}$$

$$\frac{3}{2}\left(\frac{u_{[2,2]}}{u_{[1,1]}} + \frac{u_{[1,2]}^2}{u_{[1,1]}^2} - \frac{(v_{[0,1]} + u_{[1,2]})^2}{(u_{[1,1]} + v_{[0,0]} + a)\,u_{[1,1]}} + \frac{v_{[1,1]}}{u_{[1,1]}}\right) U_1.$$

Equation (1.130) coincides with (1.109). In the case $N = 2$, after the trigonometric parametrization of the circle

$$u^1 = \frac{\tan^2(s) - 1}{\tan^2(s) + 1}, \qquad u^2 = \frac{2\tan(s)}{\tan^2(s) + 1},$$

both Eqs. (1.131) and (1.130) become

$$s_t = s_{xxx} + 2\, s_x^3 + \frac{3}{4}\left(c_1 + c_2\cos(4s)\right) s_x.$$

The last equation is well known in the theory of integrable PDEs [10, 17].

The parametrization

$$u^1 = \frac{v^2 - 1}{v^2 + 1}, \qquad u^2 = \frac{2v}{v^2 + 1}$$

brings Eq. (1.132) with $N = 2$ to the form (1.129), where Q has the form $Q = \alpha v^4 + \beta v^2 + \alpha$ with arbitrary parameters α and β. Thus (1.132) is an integrable vector generalization of generic Calogero–Degasperis equation (1.129).

Some lists of both isotropic and anisotropic equations on $\mathbb{R}^N$ can be found in [6, 9, 35, 62, 75].

1.5.9 Nonlocal Integrable PDEs

Recently the first results on classification of integrable nonlinear integro-differential equations have been obtained [36, 37]. Using the perturbative symmetry approach in symbolic representation, described in Sect. 1.4.4, the complete classification of integrable Benjamin–Ono type equations have been obtained. The Benjamin–Ono equation, which is known to be integrable, can be written in the form

$$u_t = iH(u_2) + 2uu_1, \qquad (1.133)$$

where $H(f)$ denotes the Hilbert transform

$$H(f) = \frac{1}{\pi i} \int_{-\infty}^{\infty} \frac{f(y)}{y - x}\, dy. \qquad (1.134)$$

Its higher symmetries and conservation laws are even more nonlocal, i.e. have nesting Hilbert transform and we have to define the adequate extension $\mathscr{R}_H$ of the ring of differential polynomials. The construction of the extension is similar to $\mathscr{R}_\Delta$ in Sect. 1.4.4.2, the only difference is that we have to replace the operator Δ by the Hilbert transform operator H (1.134).

We call equations of the form

$$u_t = \hat{H}(u_2) + c_1 uu_1 + c_2 \hat{H}(uu_1) + c_3 u\hat{H}(u_1) \qquad (1.135)$$
$$+ c_4 u_1 \hat{H}(u) + c_5 \hat{H}(u\hat{H}(u_1)) + c_6 \hat{H}(u)\hat{H}(u_1),$$

where c_j are complex constants, the Benjamin–Ono type equations. That is a natural generalization of Eq. (1.133) and has the same scaling properties. We have applied the approach formulated in Sect. 4.4 and isolated all cases with quasi-local (i.e. from $\mathscr{R}_H$) higher symmetries. The result can be formulated as follows:

Theorem 77. *Equation of the form* (1.135) *is integrable if and only if it is up to the point transformations of the form* $u \to au + b\hat{H}(u)$, $a^2 - b^2 \neq 0$ *coincides with one of the list*

$$u_t = \hat{H}(u_2) + D\left(\frac{1}{2}c_1 u^2 + c_2 u\hat{H}(u) + \frac{1}{2}c_1\hat{H}(u)^2\right), \tag{1.136}$$

$$u_t = \hat{H}(u_2) + D\left(\frac{1}{2}c_1 u^2 + \frac{1}{2}c_2\hat{H}(u^2) - c_2 u\hat{H}(u)\right), \tag{1.137}$$

$$u_t = \hat{H}(u_2) + uu_1 \pm \hat{H}(uu_1) \mp u\hat{H}(u_1) \mp 2u_1\hat{H}(u) + \hat{H}(u\hat{H}(u_1)), \tag{1.138}$$

$$u_t = \hat{H}(u_2) + \hat{H}(uu_1) + u_1\hat{H}(u) \pm \hat{H}(u\hat{H}(u_1)) \pm \hat{H}(u)\hat{H}(u_1). \tag{1.139}$$

References

1. V.E. Adler, A.B. Shabat, and R.I. Yamilov, The symmetry approach to the problem of integrability, Theor. Math. Phys. 125(3), 355–424, 2000.
2. M. Adler, On the trace functional for formal pseudodifferential operators and the symplectic structure of the KdV type equations, Inventiones Math. 50, 219–248, 1979.
3. C. Athorne and A. Fordy, Generalized KdV and MKdV equations associated with symmetric spaces, J. Phys. A. 20, 1377–1386, 1987.
4. S.P. Balandin and V.V. Sokolov, On the Painlevé test for non-Abelian equations, Phys. Lett. A 246(3–4), 267–272, 1998.
5. I.M. Bakirov and Popkov V.Yu.,Completely integrable systems of brusselator type, Phys. Lett. A 141(5), 275–277, 1989.
6. M.Ju. Balakhnev, A class of integrable evolutionary vector equations. Theor. Math. Phys. 142(1), 8–14, 2005.
7. F. Beukers, J. Sanders, and Jing Ping Wang On Integrability of Systems of Evolution Equations, J. Differ. Equations 172, 396–408, 2001.
8. R. Camassa, D.D.D. Holm, An integrable shallow water equation with peaked solutions, Phys. Rev. Lett. 71, 1661–1664, 1993.
9. F. Calogero, Why Are Certain Nonlinear PDE's Both Widely Applicable and Integrable?, in bookWhat is integrability?, Springer-Verlag (Springer Series in Nonlinear Dynamics), 1–62, 1991.
10. F. Calogero and A. Degasperis, Spectral transforms and solitons,North-Holland Publ. Co., Amsterdam-New York-Oxford, 1982.
11. A. Degasperis and M. Procesi, Asymptotic integrability, inSymmetry and Perturbation Theory, A. Degasperis and G. Gaeta (eds.), World Scientific, 23–37, 1999.
12. A. Degasperis, D.D. Holm, and A.N.W. Hone, A New Integrable Equation with Peakon Solutions, to appear in NEEDS 2001 Proceedings, Theoretical and Mathematical Physics, 2002.
13. I.Ya. Dorfman, Dirac Structures and Integrability of Nonlinear Evolution Equations, John Wiley&Sons, Chichester, 1993.
14. V.G. Drinfeld and V.V. Sokolov, Lie algebras and equations of Korteweg de Vries type. J. Sov. Math. 30, 1975–2036, 1985.

15. V.G. Drinfeld, S.I. Svinolupov, and Sokolov, V.V., Classification of fifth order evolution equations with infinite series of conservation laws, Doklady of Ukrainian Akademy, Section A 10, 7–10, 1985.

16. A.S. Fokas, Symmetries and integrability, Stud. Appl. Math. 77, 253–299, 1987.

17. A.S. Fokas, A symmetry approach to exactly solvable evolution equations, J. Math. Phys. 21(6), 1318–1325, 1980.

18. A.P. Fordy and P. Kulish, Nonlinear Schrödinger equations and simple Lie algebras, Commun. Math. Phys. 89, 427–443, 1983.

19. A.P. Fordy, Derivative nonlinear Schrödinger equations and Hermitian symmetric spaces, J. Phys. A.: Math. Gen. 17, 1235–1245, 1984.

20. I.M. Gel'fand and L.A. Dickii, Asymptotic properties of the resolvent of Sturm-Lioville equations, and the algebra of Korteweg de Vries equations. Russian Math. Surveys 30, 77–113, 1975.

21. I.M. Gel'fand, Yu. I.Manin, and M.A. Shubin Poisson brackets and kernel of variational derivative in formal variational calculus. Funct. Anal. Appl. 10(4), 30–34, 1976.

22. I.Z. Golubchik and V.V. Sokolov Multicomponent generalization of the hierarchy of the Landau-Lifshitz equation, Theor. Math. Phys. 124(1), 909–917, 2000.

23. I.T. Habibullin, Phys. Lett. A 369, 1993.

24. I.T. Habibullin, V.V. Sokolov, and R.I. Yamilov, Multi-component integrable systems and non-associative structures, in Nonlinear Physics: theory and experiment, E. Alfinito, M. Boiti, L. Martina, F. Pempinelli (eds.), World Scientific Publisher: Singapore, 139–168, 1996.

25. R.H. Heredero, V.V. Sokolov, and S.I. Svinolupov Toward the classification of third order integrable evolution equations, J. Phys. A: Math. General 13, 4557–4568, 1994.

26. E. Husson, Sur un thereme de H.Poincaré, relativement d'un solide pesant, Acta Math. 31, 71–88, 1908.

27. N.Kh. Ibragimov and A.B. Shabat, Evolution equation with non-trivial Lie-Bäcklund group, Funct. Anal. Appl. 14(1), 25–36, 1980. [in Russian]

28. N.Kh. Ibragimov and A.B. Shabat, Infinite Lie-Bäcklund algebras, Funct. Anal. Appl. 14(4), 79–80, 1980. [in Russian]

29. N. Jacobson, Structure and representations of Jordan algebras, Amer. Math. Soc. Colloq. Publ., Providence R.I. 39, 1968.

30. I. Kaplansky, An Introduction to Differential Algebra, Hermann, Paris, 1957.

31. L. Martínez Alonso and A.B. Shabat, Towards a theory of differential constraints of a hydrodynamic hierarchy, J. Nonlin. Math. Phys. 10, 229–242, 2003.

32. A.G. Meshkov, On symmetry classification of third order evolutionary systems of divergent type, Fund. Appl. Math. 12(7), 141–161, 2006. [in Russian]

33. A.G. Meshkov and M.Ju. Balakhnev, Two-field integrable evolutionary systems of the third order and their differential substitutions. Symmetry, Integrability and Geometry: Methods and Applications. 4, 018, 29, 2008.

34. A.G. Meshkov and V.V. Sokolov, Integrable evolution equations on the N-dimensional sphere, Comm. Math. Phys. 232(1), 1–18, 2002.

35. A.G. Meshkov and V.V. Sokolov, Classification of integrable divergent N-component evolution systems, Theoret. Math. Phys. 139(2), 609–622, 2004.

36. A.V. Mikhailov and V.S. Novikov Perturbative Symmetry Approach, J. Phys. A 35, 4775–4790, 2002.

37. A.V. Mikhailov and V.S. Novikov Classification of Integrable Benjamin-Ono type equations, Moscow Math. J. 3(4), 1293–1305, 2003.

38. A.V. Mikhailov, V.S. Novikov, and J.P. Wang. On classification of integrable non-evolutionary equations. Stud. Appl. Math. 118, 419–457, 2007.

39. A.V. Mikhailov, V.S. Novikov, and J.P. Wang., Symbolic representation and classification of integrable systems, in Algebraic Theory of Differential Equations, M.A.H. MacCallum and A.V. Mikhailov (eds.), CUP, 2008 (to appear)

40. A.V. Mikhailov and A.B. Shabat, Integrability conditions for systems of two equations $u_t = A(u)u_{xx} + B(u, u_x)$. I, Theor. Math. Phys. 62(2), 163–185, 1985.

41. A.V. Mikhailov and A.B. Shabat, Integrability conditions for systems of two equations $u_t = A(u)u_{xx} + B(u, u_x)$. II, Theor. Math. Phys. 66(1), 47–65, 1986

42. A.V. Mikhailov, A.B. Shabat, and R.I. Yamilov, The symmetry approach to the classification of non-linear equations. Complete lists of integrable systems, Russian Math. Surveys 42(4), 1–63, 1987.

43. A.V. Mikhailov, A.B. Shabat, and R.I. Yamilov, Extension of the module of invertible transformations. Classification of integrable systems, Commun. Math. Phys. 115, 1–19, 1988.

44. A.V. Mikhailov and V.V. Sokolov, Integrable ODEs on Associative Algebras, Comm. Math. Phys. 211(1), 231–251, 2000.

45. A.V. Mikhailov, V.V. Sokolov, A.B. Shabat, The symmetry approach to classification of integrable equations, in What is Integrability? V.E. Zakharov (ed.), Springer series in Nonlinear Dynamics, 115–184, 1991.

46. A.V. Mikhailov, R.I. Yamilov, Towards classification of $(2 + 1)-$ dimensional integrable equations. Integrability conditions I., J. Phys. A: Math. Gen. 31, 6707–6715, 1998.

47. R.M. Miura, Korteweg-de Vries equation and generalization. I. A remarkable explicit nonlinear transformation, J. Math. Phys. 9, 1202–1204, 1968.

48. V.S. Novikov and J.P. Wang. Symmetry structure of integrable nonevolutionary equations. Stud. Appl. Math. 119(4):393–428, 2007.

49. P.J. Olver, Applications of Lie groups to differential equations, Volume 107 of Graduate texts in Mathematics, Springer Verlag, New York, 1993.

50. P.J. Olver and V.V. Sokolov, Integrable evolution equations on associative algebras, Comm. Math. Phys. 193(2), 245–268, 1998.

51. P.J. Olver and V.V. Sokolov, Non-abelian integrable systems of the derivative nonlinear Schrödinger type, Inverse Problems 14(6), L5–L8, 1998.

52. P. Olver, J.P. Wang, Classification of integrable one-component systems on associative algebras, Proc. London Math. Soc. 81(3), 566–586, 2000.

53. J. Sanders, J.P. Wang, On the Integrability of homogeneous scalar evolution equations, J. Differ. Equations 147, 410–434, 1998.

54. A.B. Shabat and R.I. Yamilov On a complete list of integrable systems of the form $iu_t = u_{xx} + f(u, v, u_x, v_x)$, $-iv_t = v_{xx} + g(u, v, u_x, v_x)$, *Preprint BFAN*, Ufa, 28 pages, 1985.

55. V.V. Sokolov and A.B. Shabat, Classification of Integrable Evolution Equations, Soviet Sci. Rev., Section C 4, 221–280, 1984.

56. V.V. Sokolov, On the symmetries of evolution equations, Russian Math. Surveys 43(5), 165–204, 1988.

57. V.V. Sokolov, A new integrable case for the Kirchhoff equation, Theoret. Math. Phys. 129(1), 1335–1340, 2001.

58. V.V. Sokolov and S.I. Svinolupov, Vector-matrix generalizations of classical integrable equations, Theor. Math. Phys. 100(2), 959–962, 1994.

59. V.V. Sokolov and S.I. Svinolupov, Weak nonlocalities in evolution equations, Math. Notes 48(5–6), 1234–1239, 1991.

60. V.V. Sokolov and T. Wolf, A symmetry test for quasilinear coupled systems, Inverse Problems 15, L5–L11, 1999

61. V.V. Sokolov and S.I. Svinolupov, Deformation of nonassociative algebras and integrable differential equations, Acta Applicandae Mathematica, 41(1–2), 323–339, 1995.

62. V.V. Sokolov, T. Wolf, Classification of integrable polynomial vector evolution equations, J. Phys. A 2001, 34, 11139–11148.

63. V.A. Steklov On the motion of a rigid body in a fluid, Kharkov, 234 pages, 1893.

64. S.I. Svinolupov, Second-order evolution equations with symmetries, Uspehi Mat. Nauk 40(5), 263, 1985.

65. S.I. Svinolupov, On the analogues of the Burgers equation, Phys. Lett. A 135(1), 32–36, 1989.

66. S.I. Svinolupov, Generalized Schrödinger equations and Jordan pairs, Comm. Math. Phys. 143(1), 559–575, 1992.

67. S.I. Svinolupov, Jordan algebras and generalized Korteweg-de Vries equations, Theor. Math. Phys. 87(3), 391–403, 1991.

68. S.I. Svinolupov and V.V. Sokolov, Deformations of Jordan triple systems and integrable equations, Theor. Math. Phys. 108(3), 1160–1163, 1996.

69. S.I. Svinolupov and V.V. Sokolov, Evolution equations with nontrivial conservation laws, Func. analiz i pril. 16(4), 86–87, 1982. [in Russian],

70. S.I. Svinolupov and V.V. Sokolov, On conservation laws for equations with nontrivial Lie-Bäcklund algebra, in Integrable systems, A.B. Shabat (ed.), Ufa, BFAN SSSR 53–67, 1982. [in Russian].

71. S.I. Svinolupov and V.V. Sokolov, Factorization of evolution equations, Russian Math. Surveys 47(3), 127–162, 1992.

72. S.I. Svinolupov and V.V. Sokolov, Deformations of Jordan triple systems and integrable equations, Theoret. Math. Phys. 1996, 108(3), 1160–1163, 1997.

73. S.I. Svinolupov, V.V. Sokolov, and R.I. Yamilov, Bäcklund transformations for integrable evolution equations, Dokl. Akad. Nauk SSSR 271(4), 802–805, 1983.

74. T. Tsuchida, M. Wadati, New integrable systems of derivative nonlinear Schrödinger equations with multiple components, Phys. Lett. A 257, 53–64, 1999.

75. T. Tsuchida, T. Wolf, Classification of polynomial integrable systems of mixed scalar and vector evolution equations, J. Phys. A: Math. Gen. 38, 7691–7733, 2005.

76. Jing Ping Wang, Symmetries and Conservation Laws of Evolution Equations, PhD thesis, published by Thomas Stieltjes Institute for Mathematics, Amsterdam, 1998.

77. V.E. Zakharov, E.I. Schulman, Integrability of Nonlinear Systems and Perturbation Theory, in What is Integrability? V.E. Zakharov (ed.), Springer series in Nonlinear Dynamics, 185–250, 1991.

78. A.V. Zhiber and A.B. Shabat, Klein-Gordon equations with a nontrivial group, Sov. Phys. Dokl. 247(5), 1103–1107, 1979.

79. A.V. Zhiber and A.B. Shabat, Systems of equations $u_x = p(u, v)$, $v_y = q(u, v)$ possessing symmetries, Sov. Math. Dokl. 30, 23–26, 1984.

80. A.V. Zhiber and V.V. Sokolov Exactly integrable hyperbolic equations of Liouville type, Russian Math. Surveys 56(1), 63–106, 2001.

81. A.V. Zhiber, V.V. Sokolov, and Startsev S. Ya, On nonlinear Darbouxintegrable hyperbolic equations, Doklady RAN 343(6), 746–748, 1995.

82. S.L. Ziglin The branching of solutions and non-existing of first integrals in Hamiltonian mechanics. I, II, Funct. Anal. Appl. 16(3), 30–41, 1982; 17(1), 8–23, 1983.

Chapter 2
Number Theory and the Symmetry Classification of Integrable Systems

J.A. Sanders and J.P. Wang

2.1 Introduction

The theory of integrable systems has developed in many directions, and although the interconnections between the different subjects are clearly suggested by the similarity of the results, they are not always so easy to prove or even formulate. Of the various methods used to characterize integrable differential equations, including existence of infinitely many symmetries and/or conservation laws, soliton solutions, linearization by inverse scattering or differential substitution, Bäcklund transformation, Painlevé property, bi-Hamiltonian structure, recursion operator, formal symmetry of infinite rank, etc. [35], the most fruitful for systematic classification and discovery of new systems has been the characterization of integrable systems by the existence of a sufficient number of higher order symmetries. The main questions in this respect are the following:

- Can we decide, given an equation, whether there exists a generalized symmetry (the recognition problem)?
- And if so, can we answer the question whether this leads to infinitely many symmetries (the symmetry-integrability problem)?
- Given a class of equations with arbitrary parameters, possibly functions of given type, can we completely classify this class with respect to the existence of symmetries (The classification problem)?

As it turns out, these three questions are strongly related. In certain cases, they can be effectively and completely analyzed by an adaptation of the symbolic method of classical invariant theory [22], after which powerful number-theoretic results on factorizability of polynomials based on Diophantine approximation theory [2] are applied to complete the classification.

The history of the subject experienced two developmental periods. In the first, following the discovery of the Korteweg–de Vries (KdV) equation, a surprisingly large number of other integrable hierarchies, including mKdV, Sawada–Kotera,

J.A. Sanders (✉)
Department of Mathematics, Faculty of Sciences, Vrije Universiteit, Amsterdam, The Netherlands,
jansa@cs.vu.nl

Sanders, J.A., Wang, J.P.: *Number Theory and the Symmetry Classification of Integrable Systems.* Lect. Notes Phys. **767**, 89–118 (2009)
DOI 10.1007/978-3-540-88111-7_2

Kaup–Kupershmidt, were soon found. However, the second period was more disappointing in this respect, as the integrable well quickly dried up, at least in the most basic case that scalar, polynomial evolution equations are linear in the highest order derivative. This led to the conjecture that all integrable systems of this particular form had been found. In this chapter, we describe rigorous classification results for both commutative and noncommutative systems [24, 27, 28, 33], including a proof of this particular conjecture and a discussion of the general methods by which such complete classification results are established, cf. Sect. 2.4.

To do so, we prove that symmetry-integrability of an equation of the form

$$u_t = u_n + f(u, \cdots, u_{n-1}), \quad where \quad u_n = D_x^n u \tag{2.1}$$

with f a formal power series starting with terms that are at least quadratic, is determined by

- the existence of one generalized symmetry,
- the existence of approximate symmetries.

This led to the proof of the remark made in [7]

> Another interesting fact regarding the symmetry structure of evolution equations is that in all known cases the existence of one generalized symmetry implies the existence of infinitely many. (However, this has not been proved in general.)

under fairly relaxed conditions. In particular, for homogeneous scalar evolution equations, to prove the integrability of an equation of order 2 we need a symmetry of order 3; for an equation of order 3 we need a symmetry of order 5; for an equation of order 5 we need a symmetry of order 7; and for an equation of order 7 we need a symmetry of order 13; this enables us to give the complete list of integrable homogeneous equations. The result also confirms the remark made in [10]:

> It turns out from practice that if the first integrability conditions [...] are fulfilled, then often all the others are fulfilled as well.

However, the conjecture

> the existence of one symmetry implies the existence of (infinitely many) others

has been disproved using the example in [1]. This example does not contradict our theorem, since it proves the nonexistence of certain quadratic terms, the existence of which is one of the conditions in our theorem. In this chapter, we give the strict proof that Bakirov's example has only one symmetry using p-adic analysis, cf. Sect. 2.6.

We should remark that the modified conjecture made in [8]

> ... Similarly for n-component equations one needs n symmetries

has also been disproved in [11, 12], where the authors found a two-component system that has only two symmetries.

This theory was soon successfully applied to noncommutative evolution equations of the form (2.1) in which the field variable u takes its values in an associative, non-commutative algebra [24]. In this manner, it was rigorously proved that the list of integrable evolution equation in [23] is complete. These equations can be regarded as quantizing classical integrable systems; see [6], where the authors treated the Korteweg–de Vries equation.

The classification problem has been noticed and studied since the 1960s. The group consisting of A.B. Shabat, A.V. Mikhailov, V.V. Sokolov, S.I. Svinolupov, R.I. Yamilov and co-workers, cf. [19, 25], was successful in giving the complete classification for equations of fixed order, allowing for much bigger equivalence classes. We only work with homogeneous equations and transformations that do not change the weight of the dependent variables, but this restriction enables us, at least in the scalar case, to obtain results for all orders of the evolution equation.

2.2 The Symbolic Method

2.2.1 Basic Definitions

The symbolic method was first introduced by Gel'fand-Dikiĭ [9] and used in [32] to show (as an example) that the symmetries of the Sawada–Kotera equation have to be of order 1 or 5 (mod 6). The basic idea of the symbolic method is simply to replace u_i, where i is an index – in our case counting the number of derivatives – by $u\xi^i$, where ξ is now a symbol. We see that the basic operation of differentiation, i.e. replacing u_i by u_{i+1}, is now replaced by multiplication with ξ, as is the case in Fourier transform theory. For higher degree terms with multiple us, one uses different symbols to denote differentiation; for example, the noncommutative binomial $u_i u_j$ has symbolic form $u^2 \xi_1^i \xi_2^j$. In the commutative case, one needs to average over permutations of the differentiation symbols so that $u_i u_j$ and $u_j u_i$ have the same symbolic form. However, in the noncommutative case, this is no longer necessary. In other words, the noncommutative symbolic method works with general tensors, while in the commutative case one restricts to (multi)-symmetric tensors, or polynomials for short.

Usually one replaces u_i by ξ^i, but this leads to confusion for the expressions like u^n since the distinction between the powers disappears.

With this method one can readily translate solvability questions into divisibility questions and we can use generating functions to handle infinitely many orders at once. While this does not mean that the questions are much easier to answer, we do now have the whole machinery which has been developed in number theory available, and this makes a crucial difference.

For simplicity, we restrict our attention to the case of a single independent variable x and a single dependent field variable u. Extensions of the basic ideas to several (noncommutative) dependent variables are immediate, see Sect. 2.5, and to several commutative independent variables can be found in [34].

A differential monomial takes the form $u_I = u_{i_1} u_{i_2} \cdots u_{i_k}$. We call k the *degree* of the monomial, $\#I = i_1 + \cdots + i_k$ the *index*, and $\max(i_j, \ j = 1, \cdots, k)$ the *order*. For brevity, $[u]$ is used to denote the set of arguments $u, u_1, u_2, \ldots$.

We denote by $\mathcal{U}_n^k$ the set of differential polynomials in $[u]$ of degree $k+1$ and index n. Let $\mathcal{U}^k = \bigoplus_n \mathcal{U}_n^k$, and $\mathcal{U} = \bigoplus_{k \geq 0} \mathcal{U}^k$, the algebra of all differential polynomials. Notice that we consider $k \geq 0$ that excludes the constant case, i.e. $1 \notin \mathcal{U}$. The *order* of a differential polynomial is the maximum of the orders of its constituent monomials.

The *symbolic transform* defines a linear isomorphism between the space $\mathcal{U}^k$ of (non)-commutative differential polynomials of degree $k+1$ and the space $\mathscr{A}^k = R[\xi_1, \ldots, \xi_{k+1}]$ of algebraic polynomials in $k+1$ variables. It is uniquely defined by its action on monomials.

Definition 1. The symbolic form of a differential monomial is defined as

$$u_{i_1} u_{i_2} \cdots u_{i_k} \quad \longmapsto \quad u^k < \xi_1^{i_1} \xi_2^{i_2} \cdots \xi_k^{i_k} >$$

$$= \begin{cases} u^k \xi_1^{i_1} \xi_2^{i_2} \cdots \xi_k^{i_k} & \text{(noncommutative)}; \\ \dfrac{u^k}{k!} \displaystyle\sum_{\pi \in \mathbb{S}^k} \xi_{\pi(1)}^{i_1} \xi_{\pi(2)}^{i_2} \cdots \xi_{\pi(k)}^{i_k} & \text{(commutative)}, \end{cases}$$

where $\mathbb{S}^k$ is the permutation group of k elements.

In general, in analogy with Fourier transforms, we denote the symbolic form of $P \in \mathcal{U}^k$, whether it is commutative or not, by $\widehat{P}$. The transform has two basic properties:

$$\widehat{D_x P}(\xi_1, \ldots, \xi_{k+1}) = (\xi_1 + \cdots + \xi_{k+1}) \, \widehat{P}(\xi_1, \ldots, \xi_{k+1}),$$

$$\widehat{\frac{\partial P}{\partial u_i}}(\xi_1, \cdots, \xi_k) = \frac{1}{i!} \frac{1}{k+1} \sum_{j=1}^{k+1} \frac{\partial^{i+1} \widehat{P}}{\partial u (\partial \xi_j)^i}(\xi_1, \cdots, \xi_{j-1}, 0, \xi_j, \cdots, \xi_k). \tag{2.2}$$

The following key result is a consequence of these formulae.

Proposition 2. *Let $K \in \mathcal{U}^m$ and $Q \in \mathcal{U}^n$. Then $D_K(Q) \in \mathcal{U}^{m+n}$, where D_K is the Fréchet derivative of K, and*

$$\widehat{D_K[Q]} = \frac{1}{m+1} \sum_{\tau=1}^{m+1}$$

$$\times \left\langle \frac{\partial \widehat{K}}{\partial u} \left(\xi_1, \ldots, \xi_{\tau-1}, \sum_{\kappa=0}^n \xi_{\tau+\kappa}, \xi_{\tau+n+1}, \ldots, \xi_{m+n+1} \right) \widehat{Q}(\xi_\tau, \ldots, \xi_{\tau+n}) \right\rangle.$$

Proof. Using (2.2), we compute

$$\widehat{D_K(Q)} = \left\langle \sum_i \frac{\widehat{\partial K}}{\partial u_i} \widehat{D_x^i Q} \right\rangle$$

$$= \left\langle \sum_i \frac{1}{i!} \frac{1}{m+1} \sum_{\tau=1}^{m+1} \frac{\partial^{i+1} \widehat{K}}{\partial u (\partial \xi_\tau)^i}(\xi_1, \cdots, \xi_{\tau-1}, 0, \xi_\tau, \cdots, \xi_m) \right.$$

$$\left. \times (\zeta_1 + \cdots + \zeta_{n+1})^i \widehat{Q}(\zeta_1, \cdots, \zeta_{n+1}) \right\rangle$$

$$= \frac{1}{m+1} \sum_{\tau=1}^{m+1} \left\langle \frac{\partial \widehat{K}}{\partial u} \left(\xi_1, \cdots, \xi_{\tau-1}, \sum_{\kappa=1}^{n+1} \zeta_\kappa, \xi_\tau, \cdots, \xi_m \right) \widehat{Q}(\zeta_1, \cdots, \zeta_{n+1}) \right\rangle$$

and the conclusion follows.   □

For any $K, Q \in \mathcal{U}$, we define $[K, Q] = D_Q(K) - D_K(Q)$. This bracket makes $\mathcal{U}$ into a graded Lie algebra.

The following polynomials play a critical role in the analysis.

Definition 3. The G-functions are the (commutative) polynomials

$$G_k^{(m)} = \xi_1^k + \cdots + \xi_{m+1}^k - \left(\xi_1 + \cdots + \xi_{m+1} \right)^k.$$

The key fact is the following formula for the bracket of a differential polynomial with a linear differential polynomial:

$$\widehat{[u_k, Q]} = G_k^{(m)} \widehat{Q}, \qquad \text{whenever} \qquad Q \in \mathcal{U}^m. \tag{2.3}$$

This follows directly from Proposition 2 and the fact that u_k has symbolic form $\widehat{u_k} = u\,\xi_1^k$. An immediate application is the known result that the space of the symmetries of linear evolution equations $u_t = u_n$ with $n > 1$ is $\mathcal{U}^0$, as shown in the following:

Proposition 4. *Consider the linear evolution equation* $u_t = \sum_{j=1}^{p} \lambda_j u_j$, *where the* λ_j *are constants and* $\lambda_p \neq 0$. *The space of its symmetries is*

- $\mathcal{U}$ *iff* $p = 1$;
- $\mathcal{U}^0$ *iff* $p > 1$.

Proof. Let $Q \in \mathcal{U}$ and $Q = \sum Q^i$, where $Q^i \in \mathcal{U}^i$. Since $\mathcal{U}$ is a graded Lie algebra, Q is a symmetry of this equation iff $[\sum_{j=1}^{p} \lambda_j u_j, Q^i] = 0$ for any $i \geq 0$. Formula (2.3) leads to

$$\sum_{j=1}^{p} \lambda_j G_j^{(i)} = 0.$$

Under the assumption, this holds iff either $p = 1$ or $p \neq 1$ and $i = 0$.   □

The crucial step is the following result [2] on the divisibility properties of the G-functions. The proof relies on sophisticated techniques from diophantine analysis.

Proposition 5. *The symmetric polynomials* $G_n^{(k)}$ *can be factorized as*

$$G_n^{(k)} = t_n^k g_n^{(k)}, \text{where} \quad (g_n^{(k)}, g_m^{(k)}) = 1, \quad \text{for all } n < m,$$

and t_n^k *is one of the following polynomials:*

- $k = 1$:

  - $m = 0 \pmod{2}$: $\xi_1 \xi_2$
  - $m = 3 \pmod{6}$: $\xi_1 \xi_2 (\xi_1 + \xi_2)$

$$- m = 5 \ (\text{mod } 6): \quad \xi_1\xi_2(\xi_1+\xi_2)(\xi_1^2+\xi_1\xi_2+\xi_2^2)$$
$$- m = 1 \ (\text{mod } 6): \quad \xi_1\xi_2(\xi_1+\xi_2)(\xi_1^2+\xi_1\xi_2+\xi_2^2)^2$$

- $k = 2$:

  $$- m = 0 \ (\text{mod } 2): \quad 1$$
  $$- m = 1 \ (\text{mod } 2): \quad (\xi_1+\xi_2)(\xi_1+\xi_3)(\xi_2+\xi_3)$$

- $k > 2 : 1$

Proof. For $k = 1$, this was proved by F. Beukers using diophantine approximation theory [2]; for $k = 2$, see Appendix 2.8; and $k > 2$ is a special case of Theorem 16. □

Despite the innocent look of the polynomials involved, we have not been able to find a simpler proof for $k = 1$. It is conjectured that the $g_m^{(1)}$ are $\mathbb{Q}[\xi]$-irreducible.

2.2.2 Computational Example: Fifth-Order Symmetry of KdV

To illustrate how the symbolic method works, we give the symbolic calculation for the fifth-order symmetry of the Korteweg–de Vries equation. When one computes a symmetry, the natural approach is to do this degree by degree. So for instance, if we have the equation

$$u_t = K = K_3^0 + K_1^1 = u_3 + uu_1 \qquad \text{(KdV)}$$

then we try a symmetry

$$S_5 = S_5^0 + S_3^1 + \cdots = u_5 + a_1 uu_3 + a_2 u_1 u_2 + \cdots,$$

where $K_i^j, S_i^j \in \mathcal{U}_i^j$. We have to solve $[K_3^0, S_3^1] + [K_1^1, S_5^0] = 0$, i.e.

$$D_x^3 S_3^1 + u D_x S_5^0 + u_1 S_5^0$$
$$= D_x^5 K_1^1 + a_1 u D_x^3 K_3^0 + a_1 u_3 K_3^0 + a_2 u_1 D_x^2 K_3^0 + a_2 u_2 D_x K_3^0.$$

Translating this to the symbols, we have

$$(\xi_1+\xi_2)^3 \hat{S}_3^1 + \left(\xi_1^5+\xi_2^5\right) \hat{K}_1^1$$
$$= (\xi_1+\xi_2)^5 \hat{K}_1^1 + \left(\xi_1^3+\xi_2^3\right) \hat{S}_3^1,$$

where $\hat{K}_1^1 = \frac{u^2}{2}(\xi_1+\xi_2)$. We can now (formally) express $\hat{S}_3^1$ in terms of $\hat{K}_1^1$:

$$\hat{S}_3^1 = \frac{(\xi_1+\xi_2)^5 - \xi_1^5 - \xi_2^5}{(\xi_1+\xi_2)^3 - \xi_1^3 - \xi_2^3} \hat{K}_1^1.$$

By Definition 3, this can be rewritten as $\hat{S}_3^1 = \dfrac{G_5^{(1)}}{G_3^{(1)}}\hat{K}_1^1$. This is a real solution if $\hat{S}_3^1$ turns out to be a polynomial. Thus we have translated our problem into the question whether the polynomials $G_5^{(1)}$ and $G_3^{(1)}$ have common factors.

The symbolic method brings the possibility to apply the invariant theory of the permutation group to attack the classification problem.

Let us introduce ξ_0 by requiring that $\xi_0 + \xi_1 + \xi_2 = 0$. For odd n, we have

$$G_n^{(1)} = \sum_{i=0}^{2} \xi_i^n,$$

that is, the $G_n^{(1)}$ are $\mathbb{S}^3$-invariants, where $\mathbb{S}^3$ permutes the ξ-indices. Let

$$c_n = \sum_{i=0}^{2} \xi_i^n, \qquad n = 2,3. \tag{2.4}$$

The invariants of $\mathbb{S}^3$ are generated by c_2 and c_3. This implies that $G_3^{(1)} \equiv c_3$ and $G_5^{(1)} \equiv c_2 c_3$ up to multiplication by constants, since there is only one way in which we can write 5 as an additive combination of 2 and 3. Therefore $\hat{S}_3^1 \equiv c_2 \hat{K}_1^1$. To be explicit,

$$\hat{S}_3^1 = \frac{5}{3}\left(\xi_1^2 + \xi_1\xi_2 + \xi_2^2\right)\hat{K}_1^1 = \frac{5}{6}\left(\xi_1^3 + 2\xi_1^2\xi_2 + 2\xi_1\xi_2^2 + \xi_2^3\right)u^2.$$

Let us compute S_1^2 by solving $[S_3^1, K_1^1] + [S_1^2, K_3^0] = 0$. By Proposition 2, this leads to

$$\hat{S}_1^2 = \frac{5}{6}\frac{(\xi_1 + \xi_2)(\xi_2 + \xi_3)(\xi_1 + \xi_3)(\xi_1 + \xi_2 + \xi_3)}{(\xi_1 + \xi_2 + \xi_3)^3 - \xi_1^3 - \xi_2^3 - \xi_3^3}u^3 = \frac{5}{18}(\xi_1 + \xi_2 + \xi_3)u^3.$$

Note that $[S_1^2, K_1^1] = 0$ in the next degree. Thus, the fifth-order symmetry is

$$S_5 = S_5^0 + S_3^1 + S_1^2 = u_5 + \frac{5}{3}uu_3 + \frac{10}{3}u_1u_2 + \frac{5}{6}u^2u_1,$$

the well-known Lax equation.

This illustrates both the simplification induced by the symbolic method as well as the role of the G-functions in the whole analysis. The fact that the fifth-order integrable equations like Kaup–Kupershmidt and Sawada–Kotera have hierarchies with period 6 can now be explained by the invariant group $\mathbb{S}^3$.

2.2.3 The Higher Order Symmetries of KdV

What do we need in order to show that there exists a symmetry at every odd order for the Korteweg–de Vries equation? Let us sketch the computation for a higher order symmetry

$$S_{2k+1} = S_{2k+1}^0 + S_{2k-1}^1 + \cdots = u_{2k+1} + a_1 u u_{2k-1} + a_2 u_1 u_{2k-2} + \cdots.$$

First we have to solve $[K_3^0, S_{2k-1}^1] + [K_1^1, S_{2k+1}^0] = 0$. If we translate this to the symbols, by Definition 3 we obtain

$$G_3^{(1)} \hat{S}_{2k-1}^1 - G_{2k+1}^{(1)} \hat{K}_1^1 = 0.$$

We can now (formally) express $\hat{S}_{2k-1}^1$ in terms of $\hat{K}_1^1$ as

$$\hat{S}_{2k-1}^1 = \frac{G_{2k+1}^{(1)}}{G_3^{(1)}} \hat{K}_1^1,$$

and this is a real solution if $\hat{S}_{2k-1}^1$ turns out to be a polynomial. Since the invariants of $\mathbb{S}^3$ are generated by c_2 and c_3, cf. (2.4), that is, $G_{2k+1}^{(1)}$ is a polynomial in these two, we must have $c_3 | G_{2k+1}^{(1)}$. Therefore, $\hat{S}_{2k-1}^1$ is polynomial. Note that the whole argument is completely independent from the fact that we started with the Korteweg–de Vries equation; it only depends on the equation being third order. This means in general that there are no obstructions to be expected in computing the quadratic terms of an odd-order symmetry for third-order equation. The first obstructions do occur in the computation of the cubic terms.

2.3 An Implicit Function Theorem

In this section we formulate a theorem that leads to the proof that the existence of one generalized symmetry implies infinitely many under fairly relaxed condition. The theorem itself, stated in the context of graded (or filtered) Lie algebras, is not difficult to prove. Its difficulty lies in formulating and checking some technical conditions, which derive immediately from the symbolic formulation. Here we give the theorem in graded Lie algebra version so that the reader can understand it better. The filtered Lie algebra version is put in Appendix 2.9.

Consider a graded Lie algebra $\mathfrak{g} = \prod_{i=0}^{\infty} \mathfrak{g}^i$ and let V be a graded $\mathfrak{g}$-module $\prod_{i=0}^{\infty} V^i$, where the action of $\mathfrak{g}$ on V is such that if $X^i \in \mathfrak{g}^i$ and $v^j \in V^j$, then $X^i \cdot v^j \in V^{i+j}$.

Example 6. A typical example is $\mathfrak{g}^j = \mathcal{V}^j = \mathcal{U}^j$, the set of differential polynomials of degree $j + 1$, and the action of $\mathfrak{g}$ on $\mathcal{V}$ is the usual adjoint action given by the Lie bracket.

Definition 7. We call $K^0 \in \mathfrak{g}^0$ **nonlinear injective** if $\operatorname{Ker} K^0 \cdot \subset V^0$.

For Example 6, any element in $\mathfrak{g}^0$ is nonlinear injective unless it is a multiple of u_1, cf. Proposition 4.

Definition 8. We call $S^0 \in \mathfrak{g}^0$ **relatively l-prime** with respect to $K^0 \in \mathfrak{g}^0$ if $S^0 \cdot X^j \in \operatorname{Im} K^0 \Rightarrow X^j \in \operatorname{Im} K^0$ for all $j \geq l$.

We know from formula (2.3) that the Lie bracket of a differential polynomial with an element in $\mathfrak{g}^0$ equals the multiplication with a G-function, cf. Definition 3. In this case, this definition can be checked by answering whether the corresponding G-functions of S^0 and K^0 have common factors.

Theorem 9. *Let* $K = \sum_{i=0}^{k} K^i$ *and* $S = \sum_{i=0}^{s} S^i$, *where* $K^i, S^i \in \mathfrak{g}^i$ *and* $0 < k, s \in \mathbb{N}$. *Suppose there exists* $Q^j \in V^j$, $j = 0, \cdots, l-1$ *such that*

- $[K,S] = 0$,
- K^0 *is nonlinear injective,*
- S^0 *is relatively l-prime with respect to* K^0,
- $\sum_{i=0}^{p} K^i \cdot Q^{p-i} = 0$ *for* $p = 0, \cdots, l-1$ *and* $S^0 \cdot Q^0 = 0$.

Then there exists a unique $Q = \sum_{i=0}^{\infty} Q^i$, $Q^i \in V^i$, *such that* $K \cdot Q = S \cdot Q = 0$.

Proof. First we prove that $\sum_{i=0}^{j} S^{j-i} \cdot Q^i = 0$ for all $0 \leq p < l$ by induction.

For $p = 0$ this is true by assumption. Suppose it is true for all $j \leq p < l-1$. Now we show it is also true for $p+1$. We know that the action of a Lie algebra on a module is $[K,S] \cdot = K \cdot S \cdot - S \cdot K \cdot$ and that the assumption that $[K,S] = 0$ implies that $\sum_{j=0}^{q} [K^j, S^{q-j}] = 0$, for any $q \in \mathbb{N}$. It follows

$$K^0 \cdot \sum_{i=0}^{p+1} S^{p+1-i} \cdot Q^i = \sum_{i=0}^{p+1} [K^0, S^{p+1-i}] \cdot Q^i + \sum_{i=0}^{p+1} S^{p+1-i} \cdot K^0 \cdot Q^i$$

$$= -\sum_{i=0}^{p+1} \sum_{j=1}^{p+1-i} [K^j, S^{p+1-i-j}] \cdot Q^i - \sum_{i=0}^{p+1} S^{p+1-i} \cdot \sum_{j=1}^{i} K^j \cdot Q^{i-j}$$

$$= -\sum_{j=1}^{p+1} \sum_{i=0}^{p+1-j} [K^j, S^{p+1-i-j}] \cdot Q^i - \sum_{j=1}^{p+1} \sum_{i=j}^{p+1} S^{p+1-i} \cdot K^j \cdot Q^{i-j}$$

$$= -\sum_{j=1}^{p+1} \sum_{i=0}^{p+1-j} \left([K^j, S^{p+1-i-j}] \cdot Q^i + S^{p+1-i-j} \cdot K^j \cdot Q^i \right)$$

$$= -\sum_{j=1}^{p+1} K^j \cdot \sum_{i=0}^{p+1-j} S^{p+1-i-j} \cdot Q^i = -\sum_{j=0}^{p} K^{p-j+1} \cdot \sum_{i=0}^{j} S^{j-i} \cdot Q^i = 0.$$

By the nonlinear injectiveness of K^0, we obtain that $\sum_{i=0}^{p+1} S^{p+1-i} \cdot Q^i = 0$.

Next we suppose that there exists $\sum_{j=0}^{p-1} Q^j$ satisfying $\sum_{i=0}^{j} K^{j-i} \cdot Q^i = 0$ and $\sum_{i=0}^{j} S^{j-i} \cdot Q^i = 0$ for $j = 0, \ldots, p-1$.

For $p = l$, this follows from the previous.

$$K^0 \cdot \sum_{i=0}^{p-1} S^{p-i} \cdot Q^i = \sum_{i=0}^{p-1} [K^0, S^{p-i}] \cdot Q^i + \sum_{i=0}^{p-1} S^{p-i} \cdot K^0 \cdot Q^i$$

$$= -\sum_{i=0}^{p-1} \sum_{j=1}^{p-i} [K^j, S^{p-i-j}] \cdot Q^i - \sum_{i=0}^{p-1} S^{p-i} \cdot \sum_{j=1}^{i} K^j \cdot Q^{i-j}$$

$$= -\sum_{j=1}^{p}\sum_{i=0}^{p-j}\left[K^j,S^{p-i-j}\right]\cdot Q^i - \sum_{j=1}^{p}\sum_{i=j}^{p-1}S^{p-i}\cdot K^j\cdot Q^{i-j}$$

$$= -\sum_{j=1}^{p}\left(\sum_{i=0}^{p-j}\left[K^j,S^{p-i-j}\right]\cdot Q^i + \sum_{i=0}^{p-1-j}S^{p-i-j}\cdot K^j\cdot Q^i\right)$$

$$= -\sum_{j=1}^{p}K^j\cdot\sum_{i=0}^{p-j}S^{p-i-j}\cdot Q^i + S^0\cdot\sum_{j=1}^{p}K^j\cdot Q^{p-j} = S^0\cdot\sum_{j=1}^{p}K^j\cdot Q^{p-j}.$$

We have $\sum_{j=1}^{p}K^j\cdot Q^{p-j}\in \operatorname{Im}K^0$ since S^0 is relatively l-prime with respect to K^0. So we can uniquely define Q^p by

$$K^0\cdot Q^p = -\sum_{j=1}^{p}K^j\cdot Q^{p-j}\,.$$

We then automatically have $\sum_{i=0}^{p}K^{p-i}Q^i = 0$. That $\sum_{i=0}^{p}S^{p-i}Q^i = 0$ follows from the first part of the proof. Again by induction on p, we prove that Q can always be extended such that all graded parts of $K\cdot Q$ and $S\cdot Q$ vanish. $\square$

If one thinks of the application of this theorem to the computation of symmetries of evolution equations, cf. Example 6, then this proves (at least up till the existence of $\sum_{i=0}^{l}Q^i$) the long-held belief that one nontrivial symmetry S of the equation K is enough for integrability. With such a strong result one has to inspect the conditions. The strangest of them seems to be the relative prime condition. In the next sections, however, we show that for scalar equations with linear part $u_t = u_k$ any symmetry S starting with $u_s, s \notin \{1,k\}$ satisfies the conditions of the theorem with $l = 2$ when $K^1 \neq 0$ and $l = 3$ when $K^1 = 0$ and $K^2 \neq 0$.

2.4 Symmetry-Integrable Evolution Equations

2.4.1 Symmetries of λ-Homogeneous Equations

In this section we give the complete classification for homogeneous scalar commutative and noncommutative evolution equations. A key result is that it suffices to compute the linear and quadratic terms, or cubic if the quadratic terms are zero, of a nontrivial odd-order symmetry in order to guarantee its existence. This speeds up the classification process, since any obstructions to the existence of symmetries have to show up early in the computation.

The differential equation (2.1) is said to be λ-**homogeneous** of **weight** μ if it admits the one-parameter group of scaling symmetries

$$(x,t,u) \longmapsto (a^{-1}x, a^{-\mu}t, a^{\lambda}u), \qquad a \in \mathbb{R}^+.$$

For example, the Korteweg–de Vries equation $u_t = u_{xxx} + uu_x$ is homogeneous of weight 3 for $\lambda = 2$.

Two evolution equations $u_t = K$ and $u_t = Q$ are symmetries of each other if and only if [21]

$$[K, Q] = 0. \tag{2.5}$$

An equation is called (symmetry-)**integrable** if it has infinitely many linearly independent higher order symmetries.

Any λ-homogeneous evolution equation of order n can be broken up into its homogeneous components, and so it takes the form

$$u_t = K = \sum_{i \geq 0} K^i_{n-\lambda i}, \quad \left(K^i_{n-\lambda i} \in \mathscr{U}^i_{n-\lambda i} \right). \tag{2.6}$$

We assume that $K^0_n = u_n$, $n \geq 2$, and $0 < \lambda \in \mathbb{Q}$. When $i\lambda \notin \mathbb{N}$, $K^i_{n-i\lambda} = 0$. This reduces the number of relevant λ to a finite set.

For $\lambda = 1$, this describes the family of Burgers-like equations and for $\lambda = 2$ the family of KdV-like equations.

Let $S \in \mathscr{U}$ be a symmetry of order m of the evolution equation (2.6). We break up the bracket condition $[S, K] = 0$ into its homogeneous summands, leading to the series of successive symmetry equations

$$\sum_{i+j=r} \left[S^j_{m-\lambda j}, K^i_{n-\lambda i} \right] = 0, \qquad \text{for} \qquad r = 0, 1, 2, \dots . \tag{2.7}$$

According to Proposition 4, S must have nontrivial linear term, $S^0_m \neq 0$, and we can set $S^0_m = u_m$ without loss of generality. Clearly we have $[S^0_m, K^0_n] = 0$. The next equation to be solved is

$$\left[S^0_m, K^1_{n-\lambda} \right] + \left[S^1_{m-\lambda}, K^0_n \right] = 0. \tag{2.8}$$

Condition (2.8) is trivially satisfied if K has no quadratic terms: $K^1_{n-\lambda} = 0$. Let us concentrate on the case $K^1_{n-\lambda} \neq 0$. We use (2.3) and Proposition 5 to rewrite (2.8) in symbolic form:

$$\widehat{K}^1_{n-\lambda} = \frac{\widehat{S}^1_{m-\lambda}}{G^{(1)}_m} G^{(1)}_n = u^2 \frac{p(\xi_1, \xi_2)}{\xi_1 \xi_2 (\xi_1 + \xi_2)} G^{(1)}_n, \tag{2.9}$$

where $\lim_{\xi_1 + \xi_2 \to 0} p(\xi_1, \xi_2)$ exists. We next set $r = 2$ in (2.7) and find

$$\widehat{S}^2_{m-2\lambda} = \frac{\widehat{K}^2_{n-2\lambda} G^{(2)}_m + \widehat{M}}{G^{(2)}_n}, \tag{2.10}$$

where $\widehat{M}$ is the symbolic form of the commutator

$$M = \left[S^1_{m-\lambda}, K^1_{n-\lambda} \right] \tag{2.11}$$

between the quadratic terms.

We use the notation $q|p$ to indicate that the polynomial q divides the polynomial p. Consider the set

$$\mathscr{I} = \{\, p(\xi_1, \xi_2) \,:\, (\xi_1 + \xi_2)|p(\xi_1, \xi_2) \ \text{ or } \ \xi_1 \xi_2 | p(\xi_1, \xi_2) \,\}$$

consisting of bivariate polynomials $p(\xi_1, \xi_2)$ that have either $\xi_1 + \xi_2$ or $\xi_1 \xi_2$ as a factor.

Proposition 10. *Suppose m and n are both odd. Let $\hat{M}$ and p be given by (2.11) and (2.9), respectively. Then $(\xi_1 + \xi_2)(\xi_2 + \xi_3)(\xi_1 + \xi_3)$ divides $\hat{M}$ iff $p \in \mathscr{I}$.*

Proof. Using formula (2.9), we compute $\hat{M}$ to be

$$\hat{M} = u^3 \left\langle \frac{p(\xi_1 + \xi_2, \xi_3)p(\xi_1, \xi_2)F_{\xi_2, \xi_3}(\xi_1 + \xi_2)}{\xi_1 \xi_2 \xi_3 (\xi_1 + \xi_2)^2(\xi_1 + \xi_2 + \xi_3)} \right\rangle$$
$$+ u^3 \left\langle \frac{p(\xi_1, \xi_2 + \xi_3)p(\xi_2, \xi_3)F_{\xi_2, \xi_1}(\xi_2 + \xi_3)}{\xi_1 \xi_2 \xi_3 (\xi_2 + \xi_3)^2(\xi_1 + \xi_2 + \xi_3)} \right\rangle,$$

where

$$F_{\xi_i, \xi_j}(\eta) = G_n^{(1)}(\eta, \xi_j)G_m^{(1)}(\eta - \xi_i, \xi_i) - G_m^{(1)}(\eta, \xi_j)G_n^{(1)}(\eta - \xi_i, \xi_i).$$

Here we only write out the analysis for noncommutative case. For the commutative case, the expression of $\hat{M}$ needs to be symmetrized. However, the proof is quite similar, cf. [27].

Notice that $\xi_1 + \xi_3$ is a factor of $\hat{M}$. We now prove that $\lim_{\xi_1 + \xi_2 \to 0} \hat{M} = 0$. The second summand has

$$\lim_{\xi_1 + \xi_2 \to 0} F_{\xi_2, \xi_1}(\xi_2 + \xi_3)$$
$$= G_n^{(1)}(-\xi_2, \xi_2 + \xi_3)G_m^{(1)}(\xi_2, \xi_3) - G_m^{(1)}(-\xi_2, \xi_2 + \xi_3)G_n^{(1)}(\xi_2, \xi_3)$$
$$= -G_n^{(1)}(\xi_2, \xi_3)G_m^{(1)}(\xi_2, \xi_3) + G_m^{(1)}(\xi_2, \xi_3)G_n^{(1)}(\xi_2, \xi_3) = 0.$$

As for the first part, a straightforward computation shows that

$$F_{\xi_2, \xi_3}(0) = 0 = \frac{d}{d\eta}F_{\xi_2, \xi_3}(0).$$

Moreover,

$$\frac{d^2}{d\eta^2}F_{\xi_2, \xi_3}(0) = 2\left(\frac{d}{d\eta}G_n^{(1)}(\xi_3, \eta)\frac{d}{d\eta}G_m^{(1)}(\eta - \xi_2, \xi_2) \right.$$
$$\left. - \frac{d}{d\eta}G_n^{(1)}(\eta - \xi_2, \xi_2)\frac{d}{d\eta}G_m^{(1)}(\xi_3, \eta) \right) \Big|_{\eta = 0}$$
$$= 2nm\left(\xi_3^{m-1}\xi_2^{n-1} - \xi_3^{n-1}\xi_2^{m-1} \right) \neq 0.$$

This implies that

$$\lim_{\xi_1 + \xi_2 \to 0} \frac{F_{\xi_2, \xi_3}(\xi_1 + \xi_2)}{(\xi_1 + \xi_2)^2} \neq 0$$

and therefore $(\xi_1 + \xi_2) \nmid \widehat{M}$ unless $(\xi_1 + \xi_2) | p(\xi_1 + \xi_2, \xi_3) p(\xi_1, \xi_2)$ or, equivalently, $(\xi_1 + \xi_2) | p(\xi_1, \xi_2)$ or $\xi_1 | p(\xi_1, \xi_2)$. Similarly, when we deal with factor $\xi_2 + \xi_3$, we obtain $(\xi_2 + \xi_3) \nmid \widehat{M}$ unless $(\xi_1 + \xi_2) | p(\xi_1, \xi_2)$ or $\xi_2 | p(\xi_1, \xi_2)$. Therefore, the statement of the proposition follows.   □

Corollary 11. *Assume m and n are odd. Then* $(\xi_1 + \xi_2)(\xi_2 + \xi_3)(\xi_1 + \xi_3)$ *divides* $\widehat{K}^2_{n-2\lambda} G_m^{(2)} + \widehat{M}$ *if and only if* $\widehat{K}^1_{n-\lambda}(\xi_1, \xi_2) \in \mathscr{I}$.

We next state a result that says the symmetry algebra of a commutative or non-commutative polynomial evolution equation is commutative. Moreover, every symmetry is uniquely determined by its quadratic terms.

Theorem 12. *Suppose the evolution equation* (2.6) *has a nonzero symmetry S of order* $m \geq 2$. *Suppose* $Q^1_{q-\lambda}$ *is a nonzero quadratic differential polynomial* $(q \geq \lambda)$, *where* $q \notin \{m, n\}$, *and q is odd if n is odd, which satisfies the leading order symmetry condition* $[K^0_n, Q^1_{q-\lambda}] + [K^1_{n-\lambda}, Q^0_q] = 0$, *cf.* (2.8). *Then there exists a unique symmetry of the form* $Q = \sum_{i \geq 0} Q^i_{q-i\lambda}$. *Moreover, the symmetries Q and S commute.*

Proof. For even n or m, this follows from Theorem 9, since S^0_m is relatively 2-prime with respect to K^0_n.

We conclude from the existence of S that $(\xi_1 + \xi_2)(\xi_2 + \xi_3)(\xi_1 + \xi_3)$ divides

$$\widehat{K}^2_{n-2\lambda} G_m^{(2)} + \left[\widehat{S}^1_{m-\lambda}, \widehat{K}^1_{n-\lambda} \right] \tag{2.12}$$

for odd n and m. In other words, $\widehat{K}^1_{n-\lambda}(\xi_1, \xi_2) \in \mathscr{I}$.

Since S is a symmetry, i.e. $[K, S] = 0$, we have

$$[K, [S, Q]] = [S, [K, Q]]$$

from Jacobi identity. We break it up into its homogeneous summands leading to

$$g_n^{(2)} \left(\left[\widehat{S}^1, \widehat{Q}^1 \right] + \left[\widehat{S}^2, \widehat{Q}^0 \right] \right) = g_m^{(2)} \left(\left[\widehat{K}^1, \widehat{Q}^1 \right] + \left[\widehat{K}^2, \widehat{Q}^0 \right] \right).$$

We know that $(g_m^{(2)}, g_n^{(2)}) = 1$, and (by exactly the same argument as for S)

$$(\xi_1 + \xi_2)(\xi_2 + \xi_3)(\xi_1 + \xi_3) | \left(\left[\widehat{K}^1, \widehat{Q}^1 \right] + \left[\widehat{K}^2, \widehat{Q}^0 \right] \right).$$

This implies that $G_n^{(2)}$ divides $[\widehat{K}^1, \widehat{Q}^1] + [\widehat{K}^2, \widehat{Q}^0]$. Therefore,

$$\widehat{Q}^2_{q-2\lambda} = \frac{[\widehat{Q}^1, \widehat{K}^1] + [\widehat{Q}^0, \widehat{K}^2]}{G_n^{(2)}}$$

is well defined. Since the $G_n^{(k)}$ are relative prime for $k > 2$, this means that K_m^0 is relatively 2-prime and we can apply Theorem 9 to draw the conclusion that there indeed exists a symmetry Q commuting with S.   □

We make a very interesting observation. Suppose Q is a nontrivial qth odd-order symmetry of (2.6) with odd n, whose quadratic terms, cf. (2.9), have the following symbolic expression:

$$\widehat{Q}^1_{q-\lambda} = \frac{\widehat{K}^1_{n-\lambda} \, (\xi_1^2 + \xi_1\xi_2 + \xi_2^2)^{s-s'} \, g_q^{(1)}}{g_n^{(1)}}.$$

Proposition 5 implies that $\lambda \le 3 + 2\min(s, s')$, where $s' = \frac{n+3}{2}$ (mod 3) and $s = \frac{q+3}{2}$ (mod 3). Then Theorem 12 implies that

$$\widehat{Q}^1_{2s+3-\lambda} = \frac{\widehat{K}^1_{n-\lambda} \, (\xi_1^2 + \xi_1\xi_2 + \xi_2^2)^{s-s'} \, g_{2s+3}^{(1)}}{g_n^{(1)}}$$

gives rise to a symmetry $Q = Q^0_{2s+3} + Q^1_{2s+3-\lambda} + \cdots$ of the original equation. (Of course, one can use this argument to generate an entire hierarchy of symmetries.) This implies that the evolution equations defined by Q and K have the same symmetries, so instead of considering K we may consider the equation given by Q, which is of order $q = 2s + 3$ for $s = 0, 1, 2$. It follows that we only need to find the symmetries of λ-homogeneous equations (with $\lambda \le 7$) of order ≤ 7 in order to obtain the complete classification of symmetries of λ-homogeneous scalar polynomial equations starting with linear terms.

A similar observation can be made for even $n > 2$. Suppose we have found a nontrivial symmetry with quadratic term

$$\widehat{Q}^1_{q-\lambda} = \frac{\widehat{K}^1_{n-\lambda} \, G_q^{(1)}}{\xi_1\xi_2 \, g_n^{(1)}}.$$

This immediately implies $\lambda \le 2$. Then $\widehat{Q}^1_{2-\lambda} = 2\widehat{K}^1_{n-\lambda}/g_n^{(1)}$ gives rise to a symmetry $Q = Q^0_2 + Q^1_{2-\lambda} + \cdots$ of the original equation. Therefore, we only need to find the symmetries of equations of order 2 to get the complete classification of symmetries of λ-homogeneous scalar polynomial equations (with $\lambda \le 2$) starting with an even linear term.

Finally, we must analyze the case when K has no quadratic terms. Assume that $K^i_{n-\lambda i} = 0$ for $i = 1, \ldots, j - 1$, and $K^j_{n-\lambda j} \ne 0$ for some $j > 1$. In place of (2.8), we now need to solve the leading order equation

$$[S^0_m, K^j_{n-j\lambda}] + [S^j_{m-j\lambda}, K^0_n] = 0.$$

Using (2.3), the symbolic form of this condition is

$$\widehat{S}^j_{m-j\lambda} = \frac{\widehat{K}^j_{n-j\lambda} G^{(j)}_m}{G^{(j)}_n}. \tag{2.13}$$

Proposition 5 implies that this polynomial identity has no solutions when $j \geq 3$, or when $j = 2$ and n is even, since $G^{(j)}_m$ and $G^{(j)}_n$ have no common factors, and the degree of $K^j_{n-j\lambda}$ is $n - j\lambda < n$, which is the degree of $G^{(j)}_n$. Thus there are no symmetries for such equations. When $j = 2$ and n is odd, the equation can only have odd-order symmetries. If Eq. (2.13) can be solved for any m, it can also be solved for $m = 3$.

By now, we have proved the following

Theorem 13. *A nontrivial symmetry of a λ-homogeneous equation with $\lambda > 0$ is part of a hierarchy starting at order $3, 5$ or 7 in the odd case, and at order 2 in the even case.*

2.4.2 The List of Symmetry-Integrable Equations

Only an equation with nonzero quadratic or cubic terms can have a nontrivial symmetry. For each possible $\lambda > 0$, we must find a third-order symmetry for a second-order equation, a fifth-order symmetry for a third-order equation, a seventh-order symmetry for a fifth-order equation with quadratic terms and the thirteenth-order symmetry for a seventh-order equation with quadratic terms. The last case can be easily reduced to the case of fifth-order equations by determining the quadratic terms of the equation. The details of this final symbolic computation are completed as in the commutative case described in [26].

2.4.2.1 Commutative Case

We list all integrable hierarchies which are λ-homogeneous, with $\lambda \geq 0$. For $\lambda = 0$, details can be found in [28]. For $\lambda > 0$ the equivalence transformations are just scalings $u \mapsto \alpha u$, while for $\lambda = 0$ we allow arbitrary change of variables $u \mapsto h(u)$. The classification theorem states that every λ-homogeneous evolution equation with linear leading term is equivalent, modulo homogeneous transformations in u, to an equation lying in one of the following hierarchies.

Korteweg–de Vries

$$u_t = u_3 + uu_1$$

Kaup–Kupershmidt

$$u_t = u_5 + 10uu_3 + 25u_1u_2 + 20u^2u_1$$

Sawada–Kotera
$$u_t = u_5 + 10uu_3 + 10u_1u_2 + 20u^2u_1$$

Burgers
$$u_t = u_2 + uu_1$$

Potential Korteweg–de Vries
$$u_t = u_3 + u_1^2$$

Modified Korteweg–de Vries
$$u_t = u_3 + u^2u_1$$

Potential Kaup–Kupershmidt
$$u_t = u_5 + 10u_1u_3 + \frac{15}{2}u_2^2 + \frac{20}{3}u_1^3$$

Potential Sawada–Kotera
$$u_t = u_5 + 10u_1u_3 + \frac{20}{3}u_1^3$$

Kupershmidt Equation [19, 4.2.6]
$$u_t = u_5 + 5u_1u_3 + 5u_2^2 - 5u^2u_3 - 20uu_1u_2 - 5u_1^3 + 5u^4u_1$$

Ibragimov–Shabat [5]
$$u_t = u_3 + 3u^2u_2 + 9uu_1^2 + 3u^4u_1$$

Potential Burgers/Heat Equation
$$u_t = u_2 \quad \sim \quad u_t = u_2 + u_1^2$$

Potential modified Korteweg–de Vries
$$u_t = u_3 + u_1^3$$

Potential Kupershmidt Equation
$$u_t = u_5 + 5u_2u_3 - 5u_1^2u_3 - 5u_1u_2^2 + u_1^5$$

2.4.2.2 Noncommutative Case

Recently, the analysis of integrable evolution equations in which the field variable u takes its values in an associative, noncommutative algebra, such as matrix, operator, Clifford and group algebras, has attracted attention. A complete classification for $\lambda > 0$ homogeneous equations with linear leading term was established in

[24]. (The case $\lambda = 0$ poses considerable technical difficulties.) There are only five noncommutative hierarchies, each generalizing one of the preceding commutative hierarchies.

Korteweg–de Vries
$$u_t = u_3 + uu_1 + u_1 u$$

Burgers
$$u_t = u_2 + uu_1, \qquad u_t = u_2 + u_1 u$$

Potential Korteweg–de Vries
$$u_t = u_3 + u_1^2$$

Modified Korteweg–de Vries I
$$u_t = u_3 + u^2 u_1 + u_1 u^2$$

Modified Korteweg–de Vries II
$$u_t = u_3 + uu_2 - u_2 u - \frac{2}{3} uu_1 u$$

Interestingly, whereas the mKdV has two inequivalent noncommutative versions, there is no noncommutative generalization of the Sawada–Kotera, Kaup-Kupershmidt, Kupershmidt, or Ibragimov–Shabat hierarchies.

2.5 Evolution Systems with k Components

In this section, we use a simple geometric fact to prove that homogeneous evolution systems with positive weights of order larger than 1 and their linear parts with distinct nonzero eigenvalues are not symmetry-integrable without quadratic and cubic terms.

As we mentioned in Sect. 2.2, the generalization of the symbolic method to more dependent variables is straightforward. We introduce a symbol for each of dependent variables, like u and v, for instance ξ and η. Thus the symbolic expression for $u_1 u_2 v_3$ is $\frac{1}{2}\xi_1\xi_2\eta_1^3(\xi_1 + \xi_2)u^2 v$, symmetric with respect to ξ_1 and ξ_2, the symbols from us, and with respect to η_1, the symbol from v. If we would not carry along the u's and v's, information would be lost: consider the expressions uv and u^2. The alternative would be to keep the zeroth power of any symbol, so that uv would go to $\xi^0\eta^0$, but this is very awkward in actual polynomial computations.

Consider evolutionary vectorfields with two components u and v. Let $\mathscr{U}_m^{(i,j)}$ denote a set of differential polynomial vectorfields with index m, total number of x-derivatives, and degree i in u and j in v. This degree can be -1: $\frac{\partial}{\partial u} \in \mathscr{U}_0^{(-1,0)}$.

Assume the weights of u and v are λ_1 and λ_2, respectively, and $\lambda_2 \geq \lambda_1 > 0$. So any nth-order homogeneous system can be written:

$$u_t \frac{\partial}{\partial u} + v_t \frac{\partial}{\partial v} = K = \sum_{i,j} K^{(i,j)}_{n-i\lambda_1-j\lambda_2}, \quad K^{(i,j)}_l \in \mathcal{U}^{(i,j)}_l, \quad i,j \geq -1. \quad (2.14)$$

Only when $n - i\lambda_1 - j\lambda_2 < n \in \mathbb{N}$ does the term $K^{(i,j)}$ make sense and can appear in the system. The linear part of the system can be written as $K^{(0,0)}_n + K^{(-1,1)}_{n-\lambda_2+\lambda_1}$, where $K^{(0,0)}_n = a_1 u_n \frac{\partial}{\partial u} + a_2 v_n \frac{\partial}{\partial v}$, $K^{(-1,1)}_{n-\lambda_2+\lambda_1} = a_3 v_{n-\lambda_2+\lambda_1} \frac{\partial}{\partial u}$, and $a_i \in \mathbb{C}$.

Assumption 14. *We assume that the linear part of the system equals*

$$K^{(0,0)}_n = a_1 u_n \frac{\partial}{\partial u} + a_2 v_n \frac{\partial}{\partial v}, \quad a_1 a_2 \neq 0, \quad a_1 \neq a_2, \quad n \geq 2. \quad (2.15)$$

Since the linear part is diagonal, it will act semisimply on polynomial vectorfields. This simplifies the analysis considerably. Let us compute the action of the diagonal linear part on vectorfields of $Q^{(i,j)}$ using the symbolic method:

$$\left[Q^{(i,j)}, \widehat{\begin{pmatrix} a_1 u_n \\ a_2 v_n \end{pmatrix}} \right] = \begin{pmatrix} f^{(i,j)}_{u;n}(a_1,a_2;\xi;\eta) & 0 \\ 0 & f^{(i,j)}_{v;n}(a_1,a_2;\xi;\eta) \end{pmatrix} \hat{Q}^{(i,j)}(\xi;\eta),$$

where $\hat{Q}^{(i,j)}(\xi;\eta)$ is the symbolic expression of $Q^{(i,j)}$ and

$$f^{(i,j)}_{u;n}(a_1,a_2;\xi;\eta) = a_1 \left(\sum_{l=1}^{i+1} \xi_l + \sum_{k=1}^{j} \eta_k \right)^n - a_1 \sum_{l=1}^{i+1} \xi_l^n - a_2 \sum_{k=1}^{j} \eta_k^n; \quad (2.16)$$

$$f^{(i,j)}_{v;n}(a_1,a_2;\xi;\eta) = a_2 \left(\sum_{l=1}^{i} \xi_l + \sum_{k=1}^{j+1} \eta_k \right)^n - a_1 \sum_{l=1}^{i} \xi_l^n - a_2 \sum_{k=1}^{j+1} \eta_k^n.$$

These are two important polynomials corresponding to the G-functions in scalar case, cf. Definition 3, and related by

$$f^{(i,j)}_{u;n}(a_1,a_2;\xi;\eta) = f^{(j,i)}_{v;n}(a_2,a_1;\eta;\xi). \quad (2.17)$$

This calculation immediately leads to the following result (cf. Proposition 4):

Proposition 15. *The space of the symmetries of a linear system of the form of* (2.15) *is* $\mathcal{U}^{(0,0)} = \bigoplus_m \mathcal{U}^{(0,0)}_m$.

We are now in the position to do the same analysis as in Sect. 2.4. However, since we do not have the neat results on functions (2.16) as in Proposition 5 for the G-functions, the analysis is more complicated and difficult, for details see [31], where we did complete classification for second-order evolution equations with two components.

Let S be a symmetry of order m of system (2.14). Its linear part is in $\mathcal{U}^{(0,0)}$. Without loss of generality, we set $S^{(0,0)}_m = b_1 u_m \frac{\partial}{\partial u} + b_2 v_m \frac{\partial}{\partial v}$. The next equation to be solved is

$$\left[S_m^{(0,0)}, K_{n-i\lambda_1-j\lambda_2}^{(i,j)} \right] = \left[K_n^{(0,0)}, S_{m-i\lambda_1-j\lambda_2}^{(i,j)} \right], \quad i+j=1. \tag{2.18}$$

Assume that system (2.14) has no quadratic and cubic terms, that is,

$$K_{n-i\lambda_1-j\lambda_2}^{(i,j)} = 0, \quad 1 \le i+j \le 2.$$

We then need to solve (2.18) for $i+j=3$. Translating this to the symbolic language, we need to study

$$\left(f_{u;n}^{(i,j)}(a_1,a_2;\xi;\eta), f_{u;m}^{(i,j)}(b_1,b_2;\xi;\eta) \right), \quad i+j=3.$$

If they have no common factors, system (2.14) has no such symmetry.

The following theorem is due to Frits Beukers.

Theorem 16. *For any positive integer m the polynomial*

$$h_{c,m} = (\xi_1 + \xi_2 + \xi_3 + \xi_4)^m - c_1^{m-1}\xi_1^m - c_2^{m-1}\xi_2^m - c_3^{m-1}\xi_3^m - c_4^{m-1}\xi_4^m,$$

where $\Pi_{i=1}^4 c_i \ne 0$, is irreducible over $\mathbb{C}$.

Proof. Suppose that $h_{a,m} = A \cdot B$ with A, B polynomial of positive degree. Then the projective hypersurface Σ given by $h_{a,m} = 0$ consists of two components Σ_A, Σ_B given by $A = 0$ and $B = 0$, respectively. $\Sigma_A \cap \Sigma_B$ consists of an infinite number of points, which should be singularities of Σ since

$$\frac{dh_{a,m}}{d\xi_i} = \frac{dA}{d\xi_i} \cdot B + A \cdot \frac{dB}{d\xi_i} \Big|_{\Sigma_A \cap \Sigma_B} = 0.$$

Thus it suffices to show that Σ has finitely many singular points.

We compute the singular points by setting the partial derivatives of $h_{c,m}$ equal to zero, i.e.

$$(\xi_1 + \xi_2 + \xi_3 + \xi_4)^{m-1} - (c_1\xi_1)^{m-1} = 0,$$
$$(\xi_1 + \xi_2 + \xi_3 + \xi_4)^{m-1} - (c_2\xi_2)^{m-1} = 0,$$
$$(\xi_1 + \xi_2 + \xi_3 + \xi_4)^{m-1} - (c_3\xi_3)^{m-1} = 0,$$
$$(\xi_1 + \xi_2 + \xi_3 + \xi_4)^{m-1} - (c_4\xi_4)^{m-1} = 0.$$

From these equations it follows in particular that

$$\xi_1 = \zeta_1/c_1, \ \xi_2 = \zeta_2/c_2, \ \xi_3 = \zeta_3/c_3, \ \xi_4 = \zeta_4/c_4,$$

where $\zeta_i^{m-1} = 1$ and $\zeta_1/c_1 + \zeta_2/c_2 + \zeta_3/c_3 + \zeta_4/c_4 = 1$. For given $c_i, i = 1, \cdots, 4$, we get finitely many singular points. $\square$

In two-component case, the c_i are determined by a_1 and a_2. The condition $\Pi_{i=1}^4 c_i \ne 0$ is automatically satisfied due to the assumption that $a_1 a_2 \ne 0$. This implies that when system (2.14) has no quadratic and cubic terms, i.e. $K^{(i,j)} = 0$

$(1 \leq i + j \leq 2)$, it is not integrable. One can even make the stronger statement that it has no nontrivial generalized symmetries at all!

We can draw the similar conclusion to k-component systems from this theorem that homogeneous evolution systems with positive weights of order large than 1 and their linear parts with distinct nonzero eigenvalues cannot have nontrivial generalized symmetries without quadratic and cubic terms.

2.6 One Symmetry Does not Imply Integrability

As we proved in Sect. 2.4, scalar evolution equations are integrable once one nontrivial generalized symmetry exists. However, this cannot be generalized to multicomponent systems. The first example was found by Bakirov [1] (see also [21, p. 381], exercise 5.15 and [3]) that the system

$$\begin{cases} u_t = u_4 + v^2 \\ v_t = \frac{1}{5} v_4 \end{cases} \tag{2.19}$$

has one symmetry of order 6, but no others were found up till order 53. In this section, we prove that indeed no other symmetries exist for this system. Further classification and recognition of integrable such type of equations can be found in [13].

2.6.1 The Symbolic Interpretation of Bakirov's Example

We rewrite system (2.19) as $(u_4 + v^2)\frac{\partial}{\partial u} + \frac{1}{5} v_4 \frac{\partial}{\partial v}$. Its symbolic form is

$$(\xi_1^4 u + v^2) \frac{\partial}{\partial u} + \frac{1}{5} \eta_1^4 v \frac{\partial}{\partial v}$$

Since the system satisfies Assumption 14, from Proposition 15, its symmetry of a given order m has to start with $a u_m \frac{\partial}{\partial u} + b v_m \frac{\partial}{\partial v}$, i.e.

$$a \xi_1^m u \frac{\partial}{\partial u} + b \eta_1^m v \frac{\partial}{\partial v}.$$

At first sight we are losing some candidates (for being a symmetry) here, since we implicitly assume the vectorfield to be polynomial. As is shown in [1], however, this is not a restriction.

Computing the commutator of the quadratic part of system (2.19) with this linear part of the (potential) symmetry, we have

$$\left[\frac{\eta_1^0 + \eta_2^0}{2} v^2 \frac{\partial}{\partial u}, a \xi_1^m u \frac{\partial}{\partial u} + b \eta_1^m v \frac{\partial}{\partial v} \right] = (a(\eta_1 + \eta_2)^m - b(\eta_1^m + \eta_2^m)) v^2 \frac{\partial}{\partial u}.$$

Notation 17. Let $F_a^{(n)} = a(\eta_1 + \eta_2)^n - (\eta_1^n + \eta_2^n)$ and $\bar{F}_a^{(n)} = a(x+1)^n - (x^n + 1)$.

We now construct the quadratic terms of the symmetry. Provided $b \neq 0$, we compute

$$\left[(\xi_1^4 u + v^2) \frac{\partial}{\partial u} + \frac{1}{5} \eta_1^4 v \frac{\partial}{\partial v}, (a\xi_1^m u + \hat{A}v^2) \frac{\partial}{\partial u} + b\eta_1^m v \frac{\partial}{\partial v} \right]$$
$$= \left(bF_{a/b}^{(m)} - \frac{1}{5}\hat{A}F_5^{(4)} \right) v^2 \frac{\partial}{\partial u}.$$

Let $\hat{A} = 5\, b\, F_{a/b}^{(m)}/F_5^{(4)}$. If $\hat{A}$ is polynomial in η_1, η_2, then

$$(a\xi_1^p u + \hat{A}v^2) \frac{\partial}{\partial u} + b\eta_1^p v \frac{\partial}{\partial v}$$

is a symmetry of system (2.19).

Therefore, the question about the existence of symmetries of an evolution system of the form (2.19) is translated into:

Question 18. Given a, n, for which $b \in \mathbb{C}$ and $m \in \mathbb{N}$ does $F_a^{(n)}$ divide $F_b^{(m)}$?

This can be answered by the following results.

Theorem 19. *Let $a \in \mathbb{C}\setminus\{0,1\}$ and $n \in \mathbb{N}_{\geq 2}$. We consider $\bar{F}_a^{(n)}$. Suppose that at least one of the following conditions holds:*

1. $n \geq 6$,
2. $n = 4, 5$ and $\bar{F}_a^{(n)}$ has two zeros $\alpha, \beta \neq 0, -1$ such that $\alpha/\beta, (1+\alpha)/(1+\beta)$ or $\alpha\beta, (1+\alpha)/(1+1/\beta)$ are not simultaneously roots of unity.

Then there exist at most finitely many pairs $b \in \mathbb{C}, m \in \mathbb{N}$ such that $\bar{F}_a^{(n)}$ divides $\bar{F}_b^{(m)}$.

This theorem will be proved in Sect. 2.6.2.

Remark 20. • For $n = 2$ or 3, it is easy to check that there are infinitely many such pairs. Condition 2 in the theorem is violated only in seven cases including $a = 1$, see [4] for details.
• Since there is one-to-one correspondence between $F_a^{(n)}$ and $\bar{F}_a^{(n)}$, we can translate the results on $\bar{F}_a^{(n)}$ to those on $F_a^{(n)}$, and further back on symmetries of the evolution systems.

In particular given a, n it is often possible to compute the complete set of b, m explicitly. This will be done for the example $a = 5, n = 4$ in Sect. 2.6.3, which is precisely Bakirov's example. Here we only give the result.

Theorem 21. *Suppose $F_5^{(4)}$ divides $F_b^{(m)}$. Then (b, m) equals $(5, 4)$ or $(11, 6)$.*

In the first case, it leads to the system itself. For $(b, n) = (11, 6)$, we find $\hat{A} = \frac{25}{22} \eta_2^2 + \frac{20}{11} \eta_1 \eta_2 + \frac{25}{22} \eta_1^2$. We now translate these results back to results on symmetries of system (2.19).

Corollary 22. *The system*

$$\begin{cases} u_t = u_4 + v^2, \\ v_t = \frac{1}{5} v_4 \end{cases}$$

has one and only one nontrivial symmetry:

$$\left(u_6 + \frac{5}{11} \left(5vv_2 + 4v_1^2 \right) \right) \frac{\partial}{\partial u} + \frac{1}{11} v_6 \frac{\partial}{\partial v}.$$

2.6.2 The Lech–Mahler Theorem

In this section we prove Theorem 19 by using the Lech–Mahler theorem from number theory.

First we realize that $\bar{F}_a^{(n)}$ has double zeros for some values of a, which is important for our analysis later on.

Lemma 23. *Suppose that $\bar{F}_a^{(n)}$ has a multiple zero. Then this is given by an $(n-1)$th root of unity ζ and $a = 1/(\zeta+1)^{n-1}$. Together with $1/\zeta$ these are the only multiple zeros and they have multiplicity two.*

Proof. We solve the simultaneous equations $\bar{F}_a^{(n)} = d\bar{F}_a^{(n)}/dx = 0$. Explicitly, $a(x+1)^n = x^n + 1$ and $a(x+1)^{n-1} = x^{n-1}$. Multiply the second by $x+1$ and subtract the equations. We obtain $0 = 1 - x^{n-1}$. Hence the roots of $\bar{F}_a^{(n)}$, denoted by X, are an $n-1$th root of unity and from the second equation we get $a = 1/(1+X)^{n-1}$. Since

$$\frac{d^2 \bar{F}_a^{(n)}}{dx^2}\Big|_X = n(n-1)\left(a(x+1)^{n-2} - x^{n-2} \right)\Big|_X = n(n+1)\left(\frac{1}{X+1} - \frac{1}{X} \right) \neq 0,$$

the root X is a double zero. Suppose we have a second $(n-1)$th root of unity Y such that $a(1+Y)^{n-1} = 1$. In particular we find that $|1+Y| = |1+X|$ and $|X| = |Y|$. This implies that either $X = Y$ or $X = \bar{Y} = 1/Y$. This proves our lemma. $\qquad\square$

For the proof of Theorem 19 we shall use the following theorem from number theory [14].

Theorem 24 (Lech, Mahler). *Let $A_1, A_2, \ldots, A_n \in \mathbb{C}$ be nonzero complex numbers and similarly for $a_1, a_2, \ldots, a_n$. Suppose that none of the ratios A_i/A_j with $i \neq j$ is a root of unity. Then the equation*

$$a_1 A_1^k + a_2 A_2^k + \cdots + a_n A_n^k = 0$$

in the unknown integer k has finitely many solutions.

Repeatedly applying this theorem, we obtain the following corollary:

Corollary 25. *Let $A,B,C,D \in \mathbb{C}$ be nonzero complex numbers. Suppose that the equation*

$$A^k + B^k = C^k + D^k$$

has infinitely many integers k with $A^k + B^k \neq 0$ as solution. Then at least one of the pairs $A/C, B/D$ or $A/D, B/C$ consists of roots of unity.

Proof of Theorem 19. Let α, β be complex zeros of $\bar{F}_a^{(n)}$ not equal to $0, -1$ such that condition (2) of Theorem 19 is satisfied.

For $n = 4$ or 5 such zeros exist by assumption. For $n \geq 6$ we shall prove that such zeros also exist.

Suppose that $\alpha/\beta, (1+\alpha)/(1+\beta)$ are roots of unity. Then we have $|\alpha| = |\beta|$ and $|1+\alpha| = |1+\beta|$. Hence β lies on the intersection of the circles $|z| = |\alpha|$ and $|z+1| = |1+\alpha|$ which implies $\beta = \alpha$ or $\beta = \bar{\alpha}$. Similarly if $\alpha\beta$ and $(1+\alpha)/(1+1/\beta)$ are roots of unity then $\beta = 1/\alpha$ or $\beta = 1/\bar{\alpha}$. As a consequence of the statement, we need to prove there exists a root of $\bar{F}_a^{(n)}$ such that it is not in a set of the form $V_\alpha = \{0, -1, \alpha, 1/\alpha, \bar{\alpha}, 1/\bar{\alpha}\}$. If $\bar{F}_a^{(n)}$ has multiple zeros then, according to Lemma 23, the multiple zero is an $(n-1)$th root of unity, which we may assume to be equal to α. Together with $1/\alpha$ these are the only multiple zeros and they have multiplicity two. Whether $G_a^{(m)}$ has multiple zeros or not, it is clear that if $a \neq 1$ and $m \geq 6$, $\bar{F}_a^{(m)}$ has a zero outside V_α.

Note that α, β being zeros of $\bar{F}_a^{(n)}$ implies

$$(\alpha^n + 1)/(\alpha+1)^n = (\beta^n + 1)/(\beta+1)^n = a,$$

that is,

$$\left(\frac{1}{1+1/\alpha}\right)^n + \left(\frac{1}{\alpha+1}\right)^n = \left(\frac{1}{1+1/\beta}\right)^n + \left(\frac{1}{\beta+1}\right)^n.$$

Suppose $\bar{F}_a^{(n)}$ divides $\bar{F}_b^{(m)}$ for some $b \in \mathbb{C}, m \in \mathbb{N}$. Then we also have

$$\left(\frac{1}{1+1/\alpha}\right)^m + \left(\frac{1}{\alpha+1}\right)^m = \left(\frac{1}{1+1/\beta}\right)^m + \left(\frac{1}{\beta+1}\right)^m.$$

Suppose there are infinitely many such pairs (b, m). Then, according to Corollary 25, the ratios

$$\frac{1+1/\alpha}{1+1/\beta}, \frac{1+\alpha}{1+\beta} \quad \text{or} \quad \frac{1+1/\alpha}{1+\beta}, \frac{1+\alpha}{1+1/\beta}$$

are roots of unity. Let us assume the first. Then we see that the ratios α/β and $(1+\alpha)/(1+\beta)$ are roots of unity. This was excluded by our assumptions. We deal similarly with the second case. $\qquad\square$

2.6.3 Skolem's Method

In this section we prove Theorem 21. We assume that the reader is familiar with the concept of p-adic numbers. The set of p-adic numbers is denoted by $\mathbb{Q}_p$ and the set of p-adic integers by $\mathbb{Z}_p$.

Lemma 26. *(Skolem's method) Suppose p is an odd prime. Let $A, B, C, D \in \mathbb{Z}_p$ and suppose they are not zero modulo p. Write*

$$A^{p-1} = 1 + p\alpha, \ B^{p-1} = 1 + p\beta, \ C^{p-1} = 1 + p\gamma, \ D^{p-1} = 1 + p\delta,$$

where $\alpha, \beta, \gamma, \delta \in \mathbb{Z}_p$. Denote for every $k \in \mathbb{Z}$, $H_k = A^k + B^k - C^k - D^k$.
   Suppose that $H_k \not\equiv 0 \pmod{p}$. Then $H_{k+r(p-1)} \neq 0$ for all $r \in \mathbb{Z}$.
   Suppose $H_k = 0$ and $\alpha A^k + \beta B^k - \gamma C^k - \delta D^k \not\equiv 0 \pmod{p}$. Then, for $r \in \mathbb{Z}$, $H_{k+r(p-1)} = 0$ implies $r = 0$.

Proof. Note that by Fermat's little theorem,

$$H_{k+r(p-1)} = A^{k+r(p-1)} + B^{k+r(p-1)} - C^{k+r(p-1)} - D^{k+r(p-1)}$$
$$\equiv A^k + B^k - C^k - D^k \equiv H_k \pmod{p}.$$

Since $H_k \not\equiv 0 \pmod{p}$ we conclude that $H_{k+r(p-1)} \not\equiv 0 \pmod{p}$ for all $r \in \mathbb{Z}$ and our first statement follows.
   Suppose $H_{k+r(p-1)} = 0$ and assume $r \geq 0$. Then

$$0 = A^{k+r(p-1)} + B^{k+r(p-1)} - C^{k+r(p-1)} - D^{k+r(p-1)}$$
$$= A^k(1 + p\alpha)^r + B^k(1 + p\beta)^r - C^k(1 + p\gamma)^r - D^k(1 + p\delta)^r$$
$$= \sum_{i=1}^{r} \binom{r}{i} p^i \left(A^k \alpha^i + B^k \beta^i - C^k \gamma^i - D^k \delta^i \right).$$

Suppose that $r \neq 0$. Dividing by pr and using the fact that

$$\frac{1}{r}\binom{r}{i} = \frac{1}{i}\binom{r-1}{i-1},$$

we obtain

$$0 = A^k \alpha + B^k \beta - C^k \gamma - D^k \delta + \sum_{i=2}^{r} \binom{r-1}{i-1} \frac{p^{i-1}}{i} \left(A^k \alpha^i + + B^k \beta^i - C^k \gamma^i - D^k \delta^i \right).$$

The summation is of course empty when $r = 1$. Since $p \geq 3$ the number $\frac{p^{i-1}}{i}$ has p-adic valuation less than $1/p$. So after reduction modulo p we obtain

$$0 \equiv A^k \alpha + B^k \beta - C^k \gamma - D^k \delta \pmod{p}$$

which contradicts our assumption. Hence we conclude $r = 0$. When $r < 0$ we can repeat the above proof with $A^{-1}, B^{-1}, C^{-1}, D^{-1}$ instead of A, B, C, D. □

Proof of Theorem 21. When $F_5^{(4)}$ divides $F_b^{(m)}$ this means in particular that the zeros of $f = \bar{F}_5^{(4)}$ are a subset of the zeros of $\bar{F}_b^{(m)}$. This holds true in any field, also p-adic fields. Let r, s be two zeros of f. Then clearly, $\frac{(r+1)^4}{r^4+1} = \frac{(s+1)^4}{s^4+1}$. Suppose f divides $\bar{F}_b^{(m)}$ for some b, m. Then we also have $\frac{(r+1)^m}{r^m+1} = \frac{(s+1)^m}{s^m+1}$ and hence

$$((r+1)s)^m + (r+1)^m - ((s+1)r)^m - (s+1)^m = 0.$$

Note that when modulo 181 we have the factorisation

$$f \equiv 4(x-66)(x-139)(x-96)(x-56) \pmod{181}.$$

Since 181 does not divide the discriminant of f, this implies that f has four roots in $\mathbb{Q}_{181}$. They are

$$66 + 13 \cdot 181, \ 139 + 29 \cdot 181, \ 96 + 93 \cdot 181, \ 56 + 44 \cdot 181 \pmod{181^2}.$$

We now apply Lemma 26 with $p = 181$ and $A = (r+1)s, B = r+1, C = r(s+1), D = s+1$. We take r, s to be the first two roots. Then, using modulo 181^2, we get

$$A \equiv 67 + 13 \cdot 181, \ B \equiv 82, \ C \equiv 140 + 29 \cdot 181, \ D \equiv 9 + 165 \cdot 181 \pmod{181^2}.$$

We also compute modulo 181,

$$\alpha \equiv 33, \ \beta \equiv 46, \ \gamma \equiv 40, \ \delta \equiv 140 \pmod{181}.$$

A straightforward computation shows that $H_k \equiv 0 \pmod{181}$ and $0 \le k < 180$ yields $k = 0, 1, 4, 6$. Lemma 26 now implies that $H_{k+180r} \ne 0$ for all r when $k \ne 0, 1, 4, 6$. When $k = 0, 1, 4$ or 6 we easily check that $H_k = 0$ and

$$\alpha A^k + \beta B^k - \gamma C^k - \delta D^k \ne 0 \pmod{181}.$$

Again, application of Lemma 26 shows that $H_k = 0 \Rightarrow k = 0, 1, 4, 6$. When $k = 6$ we check that $b = \frac{r^6+1}{(r+1)^6} = 11$ and f divides indeed $11(x+1)^6 - x^6 - 1$. □

We finally remark that the method sketched in this section works also for other cases. When $(a, b, n, m) = (29, 3599, 4, 10)$ we can take $p = 491$. When $(a, b, n, m) = (11, 14867171, 4, 28)$ or $(a, b, n, m) = (17/3, 78719/81, 4, 16)$ we can take $p = 101$.

2.7 Concluding Remarks, Open Problems and Further Development

We have shown in this chapter that the symbolic method, combined with the implicit function theorem for filtered Lie algebras, gives us a powerful technique, which translates our classification questions into questions about divisibility. To attack these, we have at our disposal the results of centuries of mathematics, ranging from number theoretical methods as diophantine approximation theory and p-adic methods, to algebraic geometry. Still not all problems have been solved, and the two- and three-variable version of Theorem 16 would be very welcome, even in some restricted form with relations between the parameters. Nevertheless, all this seems to be within range, and we may hope that further results along these lines will enable us to completely classify evolution systems under certain conditions.

We have not discussed here the application of these methods to for instance the classification of co-symmetries. In principle the same techniques apply, but there are two difficulties. First of all, the G-functions do not belong to the same class now, and we have to look at the quotient of a regular G-function and a dual G-function. This complicates the analysis and makes the results less regular than for symmetries. The second problem arises when the system does not have a symmetry. In this case we cannot apply the implicit function theorem for filtered Lie algebras and we have to go back to ad hoc techniques. These issues are discussed in [26, 29]. Similar remarks apply to the classification of other objects like recursion operators or formal symmetries.

One can also start, once partial classification results are available, to apply larger transformation 'groups' to the integrable equations, to see which can be transformed into one another. The introduction of canonical densities as new coordinates can lead to remarkable simplification of the results, and smaller lists, as was pointed out to us by Prof. V.V. Sokolov and A. Meshkov.

Further development has shown that symbolic representation can be extended to differential [30] and pseudo-differential operators [15]. It has been a suitable tool to study integrability of nonevolutionary [15, 17, 18, 20], nonlocal (integro-differential) [16] and multi-dimensional equations [34].

2.8 Some Irreducibility Results by F. Beukers

The results in this appendix are obtained by F. Beukers, Mathematical Department, University of Utrecht and are published here with his kind permission.

Theorem 27. *Consider the polynomial* $G_k^{(2)} = \xi_1^k + \xi_2^k + \xi_3^k + (-\xi_1 - \xi_2 - \xi_3)^k$. *Then* $G_k^{(2)}$ *is absolutely irreducible if k is even. When k is odd it factors as* $(\xi_1 + \xi_2)(\xi_1 + \xi_3)(\xi_2 + \xi_3)g_k^{(2)}$, *where* $g_k^{(2)}$ *is absolutely irreducible.*

Proof. Consider the projective curve $\mathscr{C}$ defined by $G_k^{(2)} = 0$. Suppose that $G_k^{(2)} = A \cdot B$, where A and B are two polynomials of positive degree. Geometrically the curve $\mathscr{C}$ now consists of two components $\mathscr{C}_1, \mathscr{C}_2$ given by $A = 0, B = 0$, respectively. The curves $\mathscr{C}_1$ and $\mathscr{C}_2$ intersect in at least one point, which implies that the curve $\mathscr{C}$ has a singularity.

Let us now determine the singularities of $\mathscr{C}$, i.e. the projective points (ξ_1, ξ_2, ξ_3) where all partial derivatives of $G_k^{(2)}$ vanish. Hence

$$k\xi_1^{k-1} - k(-\xi_1 - \xi_2 - \xi_3)^{k-1} = 0,$$
$$k\xi_2^{k-1} - k(-\xi_1 - \xi_2 - \xi_3)^{k-1} = 0,$$
$$k\xi_3^{k-1} - k(-\xi_1 - \xi_2 - \xi_3)^{k-1} = 0.$$

We see that $\xi_1^{k-1} = \xi_2^{k-1} = \xi_3^{k-1} = \xi_0^{k-1}$ where $\xi_0 = -\xi_1 - \xi_2 - \xi_3$. By taking $\xi_3 = 1$, say, we can assume that ξ_1, ξ_2, ξ_0 are $(k-1)$th roots of unity such that $\xi_0 + \xi_1 + \xi_2 + 1 = 0$. Note that four complex numbers of the same absolute value can only add up to zero if they form the sides of a parallelogram with equal sides. Hence one of the ξ_1, ξ_2, ξ_3 is -1 and the others are opposite. Suppose without loss of generality that $\xi_0 = -1$ and $\xi_1 = -\xi_2$. If k is even we see that $1 = \xi_3^{k-1} = -(-1) = -\xi_0^{k-1}$, contradicting $\xi_3^{k-1} = \xi_0^{k-1}$. Hence $\mathscr{C}$ is nonsingular if k is even. In particular $\mathscr{C}$ is irreducible in this case.

Now suppose that k is odd. Then we have $3k - 6$ singular points, namely $(\zeta, -\zeta, 1)$, $(\zeta, -1, 1)$, $(-1, \zeta, 1)$ where $\zeta^{k-1} = 1$. Note that we have a priori $3k - 3$ singular points, but some of them coincide. Consider such a singular point, say $(\zeta, -\zeta, 1)$. We study the singular point locally by introducing the coordinates $\xi_1 = \zeta + u, \xi_2 = -\zeta + v$. Up to third-order terms we find the local equation $(\zeta(u+v) - (u-v))(u+v) + \cdots$. Since the quadratic part factors in two distinct factors the singularity is simple, i.e. there are two distinct tangent lines through the point. Consider now the curves $(\xi_1 + \xi_2)(\xi_1 + \xi_3)(\xi_2 + \xi_3) = 0$ and $g_k^{(2)} = 0$. These curves intersect in $3(k-3)$ points. Moreover, the first curve has three singularities. This accounts for the $3k - 6$ singular points we found. Hence $g_k^{(2)} = 0$ cannot have any singular points and in particular it is irreducible. $\qquad\square$

2.9 The Filtered Lie Algebra Version of the Implicit Function Theorem

We give a filtered Lie algebra version of the implicit function theorem in Sect. 2.3. The proof is quite neat, but more abstract.

Consider a filtered Lie algebra $\mathscr{F} = \mathscr{F}^0 \supset \mathscr{F}^1 \supset \cdots \supset \mathscr{F}^n \supset \cdots$ and let $\mathscr{V}$ be a filtered $\mathscr{F}$-module $\mathscr{V} = \mathscr{V}^0 \supset \mathscr{V}^1 \supset \cdots \supset \mathscr{V}^n \supset \cdots$ (with $\bigcap_{i=0}^{\infty} \mathscr{V}^j = 0$), where the action of $\mathscr{F}$ on $\mathscr{V}$ is such that if $X^i \in \mathscr{F}^i$ and $v^j \in \mathscr{V}^j$, then $X^i \cdot v^j \in \mathscr{V}^{i+j}$.

Definition 28. We call $K \in \mathscr{F}$ **nonlinear injective** if for all $X^l \in \mathscr{V}^l, l > 0, K \cdot X^l \in \mathscr{V}^{l+1} \Rightarrow X^l \in \mathscr{V}^{l+1}$.

The nonlinear injectiveness of $K \in \mathscr{F}$ implies that $K \pmod{\mathscr{F}^1} \neq 0$.

Definition 29. We call $S \in \mathscr{F}$ **relatively l-prime** with respect to $K \in \mathscr{F}$ if $S \cdot X^j \in \operatorname{Im} K \pmod{\mathscr{V}^{j+1}} \Rightarrow X^j \in \operatorname{Im} K|_{\mathscr{V}^j} \pmod{\mathscr{V}^{j+1}}$ for all $j \geq l$ and $X^j \in \mathscr{V}^j$.

Theorem 30. *Let* $K, S \in \mathcal{F}$ *be linearly independent. Suppose there exists some* $\bar{Q} \in \mathcal{V}$ *such that*

- $[K, S] = 0$,
- K *is nonlinear injective,*
- S *is relatively l-prime with respect to* K

and there exists some $\bar{Q} \in \mathcal{V}$ *such that*

- $K \cdot \bar{Q} \in \mathcal{V}^l$ *and* $S \cdot \bar{Q} \in \mathcal{V}^1$.

Then there exists a unique $Q = \bar{Q} + Q^l, Q^l \in \mathcal{V}^l$ *such that*

$$K \cdot Q = S \cdot Q = 0.$$

Proof. We use the fact that we have an action of a Lie algebra on a module, i.e. $[K, S] \cdot = K \cdot S \cdot - S \cdot K \cdot$. It follows that

$$K \cdot S \cdot \bar{Q} = S \cdot K \cdot \bar{Q}$$

By the nonlinear injectiveness of K it follows that $S \cdot \bar{Q} \in \mathcal{V}^l$.

Now we prove by induction on p that there exists $\tilde{Q}$ satisfies that $K \cdot \tilde{Q} \in \mathcal{V}^p$ and $S \cdot \tilde{Q} \in \mathcal{V}^p$, $p \geq l$. For $p = l$ we can take $\tilde{Q} = \bar{Q}$. We have

$$K \cdot S \cdot \tilde{Q} = S \cdot K \cdot \tilde{Q}$$

and therefore $S \cdot K \cdot \tilde{Q} \in im\,K \pmod{\mathcal{V}^{p+1}}$. It follows from the relatively l-primeness that $K \cdot \tilde{Q} \in \mathrm{Im}\,K|_{\mathcal{V}^p} \pmod{\mathcal{V}^{p+1}}$. So we can define $Q^p \in \mathcal{V}^p$ by

$$K \cdot Q^p = -K \cdot \tilde{Q}.$$

By construction $\hat{Q} = \tilde{Q} + Q^p$ obeys $K \cdot \hat{Q} = 0 \pmod{\mathcal{V}^{p+1}}$. It then follows from the nonlinear injectiveness of K that $S \cdot \hat{Q} \in \mathcal{V}^{p+1}$. Therefore there exists a convergent (in the filtration topology) sequence with limit $Q = \tilde{Q} + \sum_{p=l+1}^{\infty} Q^p$ such that $K \cdot Q$ and $S \cdot Q$ vanish. Uniqueness follows from the assumption that $\bigcap_{p=0}^{\infty} \mathcal{V}^p = 0$. This proves the statement. □

Acknowledgments We would like to thank the Netherlands Organization for Scientific Research (NWO) for their financial support.

References

1. I.M. Bakirov, On the symmetries of some system of evolution equations, Technical report, Akad. Nauk SSSR Ural. Otdel. Bashkir. Nauchn. Tsentr, Ufa, 1991.
2. F. Beukers, On a sequence of polynomials, J. Pure Appl. Algebra, 117/118, 97–103, 1997. Algorithms for algebra (Eindhoven, 1996).
3. A.H. Bilge, A system with a recursion operator, but one higher local symmetry, Lie Groups Appl. 1(2), 32–139, 1994.

4. F. Beukers, J.A. Sanders, and J.P. Wang, On integrability of systems of evolution equations, J. Differ. Equations 172(2), 396–408, 2001.
5. F. Calogero, The evolution partial differential equation $u_t = u_{xxx} + 3(u_{xx}u^2 + 3u_x^2 u) + 3u_x u^4$, J. Math. Phys. 28(3), 538–555, 1987.
6. B. Fuchssteiner and A.R. Chowdhury, A new approach to the quantum KdV, Chaos Soliton Fract. 5(12), 2345–2355, 1995. Solitons in science and engineering: theory and applications.
7. A.S. Fokas, A symmetry approach to exactly solvable evolution equations, J. Math. Phys. 21(6), 1318–1325, 1980.
8. A.S. Fokas, Symmetries and integrability, Stud. Appl. Math. 77, 253–299, 1987.
9. I.M. Gel'fand and L.A. Dikiĭ, Asymptotic properties of the resolvent of Sturm-Liouville equations, and the algebra of Korteweg-de Vries equations, Uspehi Mat. Nauk 30(5(185)), 67–100, 1975. English translation: Russian Math. Surveys 30(5), 77–113, 1975.
10. V.P. Gerdt, N.V. Khutornoy, and A.Y. Zharkov, Computer algebra in physical research. In D.V. Shirkov, V.A. Rostovtsev, and V.P. Gerdt, editors, Solving algebraic systems which arise as necessary integrability conditions for polynomial–nonlinear evolution equations, World Scientific, Singapore, 321–328, 1991.
11. P.H. van der Kamp and J.A. Sanders, Almost integrable evolution equations, Selecta Math. (N.S.) 8(4), 705–719, 2002.
12. P.H. van der Kamp and J.A. Sanders, On testing integrability, J. Nonlinear Math. Phys. 8(4), 561–574, 2001.
13. P.H. van der Kamp, Symmetries of Evolution Equations: a Diophantine Approach, PhD thesis, Vrije Universiteit, Amsterdam, 2003.
14. C. Lech, A note on recurring sequences, Arkiv. Mat. 2, 417–421, 1953.
15. A.V. Mikhailov and V.S. Novikov, Perturbative symmetry approach, J. Phys. A 35(22), 4775–4790, 2002.
16. A.V. Mikhailov and V.S. Novikov, Classification of integrable Benjamin-Onotype equations, Moscow Math. J. 3(4), 1293–1305, 2003.
17. A.V. Mikhailov, V.S. Novikov, and J.P. Wang, Partially integrable nonlinear equations with one high symmetry, J. Phys. A 38, L337–L341, 2005.
18. A.V. Mikhailov, V.S. Novikov, and Jing Ping Wang, On classification of integrable non-evolutionary equations, Stud. Appl. Math. 118, 419–457, 2007.
19. A.V. Mikhailov, A.B. Shabat, and V.V. Sokolov, The symmetry approach to classification of integrable equations. In Zakharov [35], pages 115–184.
20. V.S. Novikov and J.P. Wang, Symmetry structure of integrable nonevolutionary equations, Stud. Appl. Math. 119(4), 393–428, 2007.
21. P.J. Olver, Applications of Lie groups to differential equations, volume 107 of Graduate Texts in Mathematics, Springer-Verlag, New York, second edition, 1993.
22. P.J. Olver, Classical Invariant Theory, volume 44 of London Mathematical Society Student Texts, Cambridge University Press, Cambridge, 1999.
23. P.J. Olver and V.V. Sokolov, Integrable evolution equations on associative algebras, Comm. Math. Phys. 193(2), 245–268, 1998.
24. P.J. Olver and J.P. Wang, Classification of integrable one-component systems on associative algebras, Proc. London Math. Soc. (3) 81(3), 566–586, 2000.
25. V.V. Sokolov and A.B. Shabat, Classification of integrable evolution equations. In Mathematical physics reviews, Vol. 4, volume 4 of Soviet Sci. Rev. Sect. C: Math. Phys. Rev., pages 221–280. Harwood Academic Publ., Chur, 1984.
26. J.A. Sanders and J.P. Wang, Combining Maple and Form to decide on integrability questions, Comput. Phys. Comm. 115(2–3), 447–459, 1998.
27. J.A. Sanders and J.P. Wang, On the integrability of homogeneous scalar evolution equations, J. Differ. Equations 147(2), 410–434, 1998.
28. J.A. Sanders and J.P. Wang, On the integrability of non–polynomial scalar evolution equations, J. Differ. Equations 166(1), 132–150, 2000.

29. J.A. Sanders and J.P. Wang, The symbolic method and co-symmetry integrability of evolution equations. In Equadiff'99, International Conference on Differential Equations, World Scientific, Singapore, 824–831, 2000.

30. J.A. Sanders and J.P. Wang, On a family of operators and their Lie algebras, J. Lie Theory 12(2), 503–514, 2002.

31. J.A. Sanders and J.P. Wang, On the integrability of systems of second order evolution equations with two components, J. Differ. Equations 203(1), 1–27, 2004.

32. G.Z. Tu and M.Z. Qin, The invariant groups and conservation laws of nonlinear evolution equations–an approach of symmetric function, Scientia Sinica 14(1), 13–26, 1981.

33. J.P. Wang, Symmetries and Conservation Laws of Evolution Equations. PhD thesis, Vrije Universiteit/Thomas Stieltjes Institute, Amsterdam, 1998.

34. J.P. Wang, On the structure of $(2 + 1)$–dimensional commutative and noncommutative integrable equations, J. Math. Phys. 47(11), 113508, 2006.

35. V.E. Zakharov (eds.), What is integrability? Springer-Verlag, Berlin, 1991.

Chapter 3
Four Lectures: Discretization and Integrability. Discrete Spectral Symmetries

S.P. Novikov

3.1 Introduction

In these lectures I am going to consider   the integrability phenomenon as a by-product of the hidden symmetry of the spectral theory of some famous linear operators. **Our objective is to apply it to the spectral theory** of these operators. This approach (not pretending to be universal) has indeed worked well since 1974 when the so-called finite-gap 1D periodic and quasi-periodic Schrodinger operators and corresponding solutions of KdV were discovered (see [1]). Recently we developed a theory based on the discrete symmetries of the continuous and discrete 1D and 2D Schrodinger operators (see [2, 3]). Some results for the 1D Schrodinger operators were obtained in the works [4–7, 14]. *New direction was developed by the present author in collaboration with I. Dynnikov in 2002–2008 years extending results of the last lecture dedicated to the Black and White Triangle Operators on triangulated manifolds. It involves completely new discretization of complex analysis on the equilateral triangle lattice of the euclidean plane R^2 and on the hyperbolic (Lobatchevski) plane H^2. We also developed completely new approach to discretization of the nonabelian differential-geometrical GL_n-connections – see the authors' homepage www.mi.ras.ru/snovikov (click publications, items 159,163).*

Going back to the famous discovery of the so-called inverse scattering transform for the KdV equation $u_t = 6uu_x - u_{xxx}$ in 1967 (see [8]), we know that it is based, in fact, on the interpretation of KdV as an isospectral deformation for the 1D Schrödinger operator $L_t = LA - AL, L = -\partial_x^2 + u(x,t)$ (see [9] where an infinite-dimensional commutative group of such deformations was found; people call it the KdV hierarchy).

We call this KdV hierarchy **a continuous spectral symmetry group** for the 1D Schrödinger operator L.

For the rapidly decreasing ("soliton-type") class of functions $u(x,t) \to 0$ when $x \to \pm\infty$, the inverse scattering problem was solved many years ago by Gel'fand,

S.P. Novikov (✉)
University of Maryland, College Park, Maryland 20742-2431, USA;
L.D.Landau Institute for Theoretical Physics, Kosygina 2, Moscow 117334, Russia,
novikov@ipst.umd.edu, novikov@itp.ac.ru

Novikov, S.P.: *Four Lectures: Discretization and Integrability. Discrete Spectral Symmetries.* Lect. Notes Phys. **767**, 119–138 (2009)
DOI 10.1007/978-3-540-88111-7_3   © Springer-Verlag Berlin Heidelberg 2009

Levitan, Marchenko and others. Therefore, the inverse scattering transform was considered as an application of this theory for solving the KdV equation.

However, for the x-periodic functions $u(x,t)$, no good solution of the inverse spectral problem was known before. The approach started in [1] was based on the connection of the 1D Schrödinger operator to KdV-type systems ("higher KdV"), generating a KdV hierarchy. It led to the effective solution of inverse spectral problems for the so-called "finite-gap" Schrödinger operators L and to the exact solutions of the nonlinear KdV equation. The spectral theory of finite-gap operators, its connection with Riemann surfaces and completely integrable Hamiltonian systems are (at least) as important as the solutions of KdV. So the continuous spectral symmetry group certainly played a fundamental role here.

During the last decade we started to study **discrete spectral symmetries**. In fact, some of these symmetries were known for many years. For example, the substitutions called today "Darboux transformations" for the 1D Schrödinger operator L were invented by Euler in 1742. There analogs for the 2D Schrödinger operators were found by Laplace. The association of the Darboux transformations with KdV was realized in the early 1970s under the name "Bäcklund transformations". The interesting conjecture concerning the connection of cyclic chains of such transformations with finite-gap periodic potentials was formulated in the work [4] in the 1980s.

However, the studies of the remarkable spectral properties of the low-dimensional Schrödinger operators based on the discrete spectral symmetries started only in 1990s. One can say that these investigations have roots also in the studies of the famous quantum physicists of 1930s and 1940s (Dirac and Schrödinger) who started to work with such transformations in the modern algebraic way and to use some examples of that kind for very important goals.

3.2 Continuous and Discrete Spectral Symmetries of 1D Systems and Spectral Theory of Operators. 1D Continuous Schrödinger Operator and Its Discrete Analogue

Let us consider a one-dimensional Schrödinger operator $L = -\partial_x^2 + u$. For the construction of the Darboux transformation B_c depending on the constant c, we factorize L in the form

$$L + c = QQ^+ = -(\partial_x + a)(\partial - a). \tag{3.1}$$

Such a factorization requires a solution for the Riccati equation

$$u + c = a_x + a^2. \tag{3.2}$$

For the real and bounded function $u(x)$ we can always find a constant c big enough such that this factorization is possible. We call it **strong factorization**. It depends on the parameter c and also on the solution of the Riccati equation.

Any strong factorization generates a **Darboux transformation** $\tilde{L} = B_c(L)$ of the operator L by the formula

$$\tilde{L} = Q^+ Q, L + c = QQ^+. \tag{3.3}$$

Lemma 1. *For any solution of the spectral equation* $L\psi = \lambda\psi$, *the new function* $Q^+\psi = \tilde{\psi}$ *is a solution of the new spectral equation* $\tilde{L}\tilde{\psi} = (\lambda + c)\tilde{\psi}$.

The proof of this lemma is trivial. Let us formulate some useful conclusions:

1. On the formal (local) level, the operator $\tilde{L}$ has "almost" the same eigenfunctions as L except maybe one function: the operator Q^+ has a kernel $Q^+\psi^0 = 0$ or $\psi_x^0 = a\psi^0$.
2. Let us assume that we are dealing with Hilbert space $L_2(R)$. The function ψ^0 belongs to this space (i.e., it is square integrable on the real line) if and only if the spectrum of the operator L starts from the point $-c$, i.e., $\lambda \geq -c$, and ψ^0 is a ground state. Therefore there is only one choice of the constant c if operator L is semibounded.

Example 2. Let $L = -\partial_x^2 + x^2$ is a quantum oscillator. We have a strong factorization here

$$L + 1 = -(\partial_x - x)(\partial_x + x); Q^+ = \partial_x + x; \psi_0 = \exp\{-x^2/2\} \in L_2(R).$$

In this case we have also the famous relations $QQ^+ - Q^+Q = -2$. All basic eigenfunctions ψ^n for this operator can be obtained by the iterations $\psi^n = Q^n\psi^0$ with eigenvalues $L\psi^n = (2n+1)\psi^n$.

As we can see for the opposite operator $L' = Q^+Q$ where $Q' = Q^+$, the equation $Q\psi = 0$ leads in this case to the function $\psi'_0 = \exp\{x^2/2\}$ which does not belong to the space $L_2(R)$. The operator L' is positive and strongly factorized but its spectrum does not start from 0 because the "instanton equation" $Q^+\psi = 0$ has no proper solutions.

As has been well known for many years in the Theory of Solitons, Darboux transformations generate multisoliton solutions and a more general class of "solitons on the given background". However, only recently their connection with periodic and quasi-periodic finite-gap solutions and finite-gap Schrödinger operators was revealed. Consider now a chain of Darboux transformations

$$\ldots, L_k, L_{k+1}, L_{k+2}, \ldots; L_{k+1} = B_{c_k}L_k = \tilde{L}_k. \tag{3.4}$$

We call chain periodic of the period N if $L_N + \sum c_k = L_0$. These chains were studied in the work [4] assuming all $c_k = 0$. In particular, an interesting conjecture was formulated that for the odd values of $N = 2M + 1$ the operators L_k in the periodic chain are the finite-gap ones. This conjecture was proved in the stronger form in [5]: Let $N = 2M + 1, \sum c_k = 0$. Then the operator L_k is **an algebraic operator**, i.e., there exists a differential operator A of the order $2M + 1$ such that $[L, A] = 0$. According to the result of [1], such operators are finite gap in the sense of the spectral theory if the coefficients are smooth periodic or quasi-periodic. For the case $N = 2M + 1, \sum c_k = c \neq 0$ it was proved in [5] that there is a differential operator A of the order $2M + 1$ such that

$$LA - AL = cA. \tag{3.5}$$

If all operators L_k in the cyclic chain are smooth, then the spectrum of all of them is equal to the union of N arithmetic progressions with the same difference. We can say that these operators are analogous to the quantum oscillator. For $N = 3$ we get new examples of operators with such remarkable properties of the spectrum. The equation for finding potential reduces in this case to the Painlevé' equation. Numerical calculations made by V. Adler in his PhD show that it really has such nonsingular solutions.

Our conclusion is that **even these simple discrete symmetries of the 1D Schrödinger operator on the line lead to new interesting results in the spectral theory.**

For the even values of $N = 2M$ we do not know of any classification of periodic Darboux chains. This problem is open.

The discrete analog of the "soliton-type" theory for the 1D Schrödinger operator appeared many years ago in the theory of the so-called Toda chain and discrete KdV systems (see [10–12]). The operator L here acts on the functions of the discrete variable $\psi_k, k \in Z$. It has a form (for the Toda chain) in terms of the unitary shift operators $T = \exp\{\partial_x\}, T : n \to n + 1, T^+ = T^{-1}$

$$L = c_n T + c_{n-1} T^{-1} + v_n; L\psi_n = c_n \psi_{n+1} + c_{n-1} \psi_{n-1} + v_n \psi_n = \lambda \psi_n \tag{3.6}$$

and reduction $v_n = 0$ for the discrete KdV [11, 12].

It is interesting to point out that the reduction to standard classical discretization $c_n = 1, n \in Z$, cannot be recognized in terms of the inverse spectral (scattering) data. It is noninvariant under the time dynamics of any nontrivial isospectral system. As we shall see, it is noninvariant also under the discrete Darboux transformations $B_c^{\pm}$. Therefore we come to the following important **conclusion: in order to construct a right ("good") discretization of the 1D Schrödinger operator $L = -\partial_x^2 + u(x)$, we need to replace derivative ∂_x by the "covariant shift" operator $c_k T = \exp\{\partial + s(x)\}$ instead of standard shift operator $T = \exp\{\partial\}$; otherwise the class of discretized operators will not have discrete (and continuous as well) spectral symmetry transformations.**

Let us construct them using the strong factorization of the first or of the second type.

The first-type discrete Darboux transformation B_c^+ has a form

$$L = QQ^+ + c; Q = a_n + b_n T; Q^+ = a_n + b_{n-1} T^{-1}; \tilde{L}_+ = Q^+ Q. \tag{3.7}$$

The second type B_c^- is defined in the same way but the role of T and T^{-1} are reversed:

$$L = RR^+ + c'; R = u_n + v_n T^{-1}, R^+ = u_n + v_{n+1} T; \tilde{L}_- = R^+ R. \tag{3.8}$$

The second-type transformations are inverse to the first ones.

These transformations were studied in the works [6] and [2], Appendix 2. In particular, we proved that **any cyclic sequence of the first-type transformations such that $\sum c_k = 0$ leads to the finite-gap (algebra-geometric) discrete operators with Riemann surfaces of the genus no more than half of the period of the chain.** However, if both types are involved, the classification of cyclic chains remains unclear. This problem is analogous to the classification of the periodic Darboux chains of the even length for the continuous Schrödinger operator.

Let us present here two interesting examples of the discrete 1D operators discussed from the algebraic point of view in the works [6, 7] and also in [2], Appendix 2 from the viewpoint of the spectral theory.

Example 3. Let $L = QQ^+ + c$ and $QQ^+ - Q^+Q = $ const. The operators Q, Q^+ can be easily found in the form

$$Q = 1 + \sqrt{a + bn}T; Q^+ = 1 + \sqrt{a + b(n-1)}T^{-1}.$$

However, these operators cannot be real and adjoint to each other on the whole lattice Z because linear function $a + bn$ cannot be positive for all $n \in Z$. We require the "quantization condition" $a/b = m \in Z$ and positivity $a > 0, b > 0$. Consider these operators acting on the subspace H_+ in the Hilbert space $L_2(Z)$ such that $\psi_n = 0$ for $n \leq m$. Let $n = m + k$ and $k > 0$. The operators Q, Q^+, L are well defined on the space H_+. The ground state $Q^+ \psi^0 = 0$ is such that

$$(\psi_k^0)^2 = \frac{b^{-k+1}}{(k-1)!}.$$

We can see that it is a Poisson distribution. The eigenfunctions $\psi^l = Q^l \psi^0$ are equal to the so-called Charlet polynomials in the discrete variable k multiplied by the ground state ψ_k^0. Our formula gives a good definition of these polynomials on the half-lattice Z_+ orthogonal corresponding to the Poisson weight. As far as I know, this discrete realization of the Dirac harmonic oscillator is not mentioned in the traditional literature in quantum mechanics. The eigenvalues, of course, are the same as in the standard realization of the commutation relations: $\lambda_l = lb, l \in Z$.

Example 4. Consider now a family of operators $L_c = Q_c Q_c^+ + $ const where $Q_c = 1 + ca^n T$, the constant $a \neq 0$ is fixed, $c \neq 0$. We have the following relations:

$$a^2 Q_c^+ Q_c = Q_{c'} Q_{c'}^+ + D, D = a^2 - 1, c' = ca^2, \tag{3.9}$$

where a is the same for all operators involved.

Theorem 5. *For $a > 1$ the operator $L = Q^+Q$ acting in the Hilbert space $L_2(Z)$ has a discrete spectrum $\lambda_n = 1 - a^{-2n}, n \geq 0$, for $\lambda < 1$.*
  For $a < 1$ the operator $L = Q^+Q$ has a discrete spectrum $\lambda_n = 1 - a^{2n}, n > 0$, for $\lambda < 1$.
  In both cases the spectrum is continuous for $\lambda \geq 1$
The investigation of the spectrum of this operator for $\lambda \geq 1$ is not done yet.

For the proof of this theorem, we solve equations $Q_c \psi^0 = 0$ and $Q_c^+ \psi^0 = 0$ for all $c \neq 0$. Selecting the cases when our solution belongs to the space $L_2(Z)$, we apply the "creation operators" $Q_{ca^{-2}}$ and get higher states for all values of c but the actual value of c is shifted every time when we apply the creation operator:

$$\psi_k^0 = (-1)^k c^{-k} a^{-(k-1)(k/2)}, k \in Z, \qquad (3.10)$$

$$\psi^l = Q_c Q_{ca^2} Q_{ca^4} \cdots Q_{ca^{2l-2}} \psi^0. \qquad (3.11)$$

3.3 2D Schrödinger Operator. Discrete Spectral Symmetries, Spectral Theory of the Selected Energy Level and Space/Lattice Discretization

Already in the eighteenth century Laplace invented the transformations which we are going to use later as **discrete spectral symmetries associated with one spectral level only.** Let us consider a **hyperbolic Laplace equation** on the plane x, y:

$$L\phi = \phi_{xy} + A\phi_x + B\phi_y + C\phi = 0, \qquad (3.12)$$

where A, B, C are some known functions. We can present it in the form (**a weak factorization of the first type**)

$$L\phi = (Q_1 Q_2 + 2W)\phi = \{(\partial_x + A)(\partial_y + B) + 2W\}\phi = 0, \qquad (3.13)$$

where $2W = C - AB - B_x$, or in the opposite form (**a weak factorization of the second type**)

$$L\phi = (Q_2 Q_1 + 2V)\phi = \{(\partial_y + B)(\partial_x + A) + 2V\}\phi = 0, \qquad (3.14)$$

where $2V - AB - A_y = C$. So we have $2V - 2W = A_y - B_x = 2H(x,y) = [Q_1, Q_2]$. We call the quantity H **a magnetic field or a curvature** for the operator L. There are natural **gauge transformations** for this operator

$$L \to e^f L e^{-f}, \phi \to e^f \phi \qquad (3.15)$$

for any function $f(x,y)$. The quantities W (or V) and H are only invariants of the gauge transformations.

By the **Laplace transformation** we call the following map

$$L \to \tilde{L} = W Q_2 W^{-1} Q_1 + 2W, \phi \to \tilde{\phi} = Q_2 \phi. \qquad (3.16)$$

By the **opposite Laplace transformation** we call the following map

$$L \to \tilde{L}' = V Q_1 V^{-1} Q_2 + 2V, \tilde{\phi}' = Q_1 \phi. \qquad (3.17)$$

Lemma 6. *For any solution $L\psi = 0$ we have $\tilde{L}\tilde{\psi} = 0$ and $\tilde{L}'\tilde{\psi}' = 0$. These transformations are inverse to each other modulo gauge transformation. For the case of the strong factorization $W = const$ or $V = const$ these transformations transform every eigenfunction $L\psi = \lambda\psi$ of the operator L into the eigenfunction $\tilde{\psi}$ or $\tilde{\psi}'$ for the operator $\tilde{L}$ or $\tilde{L}'$, correspondingly.*

The proof of this lemma is almost obvious: the equation $L\psi = 0$ implies $Q_1\tilde{\psi} = -2W\psi$ by definition. Therefore we have $W^{-1}Q_1\tilde{\psi} = -2\psi$. Applying Q_2 to both sides and multiplying by W after that, we get the desired result. Our lemma is proved for the first type. For the second type the proof is similar. Let us prove now that they are inverse to each other: Performing the second type after the first one, we come to the operator $\tilde{\tilde{L}}' = WLW^{-1}$. Taking $W = \exp\{f\}$ we get a gauge equivalence if $W \neq 0$. The lemma is proved.

Lemma 7. *The Laplace transformations are gauge invariant. In terms of the gauge-invariant quantities, they can be written in the form*

$$\tilde{W} = W + \tilde{H}; \tilde{H} = H + 1/2\partial_x\partial_y \log W. \tag{3.18}$$

Let us demonstrate this here by following a simple but important theorem (in fact, known already in the nineteenth century to Darboux).

Theorem 8. *Let an infinite Laplace chain be given:*

$$\ldots, L_k, L_{k+1}, \ldots : L_{k+1} = \tilde{L}_k. \tag{3.19}$$

Then this chain can be described by the 2D Toda lattice system and vice versa.

Proof. Let $W = e^f$. When we have

$$e^{f_{k+1}} = e^{f_k} + H_{k+1},$$
$$H_{k+1} = H_k + 1/2\partial_x\partial_y f_k$$

as a definition of the Laplace chain in terms of gauge-invariant quantities. So we exclude magnetic field using the first equation:

$$H_{k+1} = e^{f_{k+1}} - e^{f_k}.$$

After substitution of this expression into the second equation and making change of the dependent variables $f_k = g_{k+1} - g_k$, we come exactly to the famous 2D Toda lattice system

$$1/2\partial_x\partial_y g_k = e^{g_{k+1}-g_k} - e^{g_k-g_{k-1}}. \tag{3.20}$$

∎

People in soliton theory found the complete integrability of this system (see [13]) but did not know about its connection with the 2D Schrödinger equation (or Laplace equation in hyperbolic case).

Already in the nineteenth century, geometers like Darboux with his pupils and others started to use hyperbolic Laplace transformations for the needs of the theory of surfaces imbedded in the euclidean space R^3. They also considered Laplace chains and periodic chains in particular. Simple calculations show that for the period $N = 2$ where $L_2 = L_0$, we come to the equation

$$\partial_x \partial_y G(x,y) = -8 sinh\{G(x,y)\}. \tag{3.21}$$

For the period $N = 3$ assuming that the magnetic field is equal to zero $L_0 = L_3 = \partial_x \partial_y + 2W(x,y)$, we come to the equation

$$\partial_x \partial_y G = e^G - e^{-2G}. \tag{3.22}$$

Both of these systems are well known in the theory of completely integrable systems and were obtained in completely different way, with no relationship to the 2D (linear) Schrödinger operator.

We are going to apply these ideas to the spectral theory of the 2D Schrödinger operator. Let us consider now the **elliptic Schrödinger operator** written in the weakly factorized form through the complex derivatives $\partial = \partial_x - i\partial_y, \bar{\partial} = \partial_x + i\partial_y$:

$$L = (-\partial + A)(\bar{\partial} + B) + 2W. \tag{3.23}$$

We call operator L **physical** if magnetic field $H = 1/2(A_{\bar{z}} - B_z)$ and potential $W = exp\{f\}$ both are real. We call the operator **periodic** if both of them are smooth and double-periodic on the plane R^2. We call the periodic operator **topologically trivial** if the magnetic flux $[H]$ through the elementary cell K is equal to zero:

$$[H] = \bar{H}|K| = \int \int_K H(x,y)dxdy = 0. \tag{3.24}$$

We call the operator **quantized** if $[H] \in 2\pi Z$. From the formulas for the Laplace transformations written through the gauge-invariant quantities above, we deduce the following:

Lemma 9. *For the smooth physical double-periodic operators we have for the fluxes through the elementary cell*

$$[\tilde{H}] = [H]; [\tilde{W}] = [W] + [H]. \tag{3.25}$$

These changes lead only to the replacement of the operators ∂_x, ∂_y by the complex ones $\partial, \bar{\partial}$ in all formulas above for the Laplace transformations and Laplace chains. All formal calculations remain unchanged. However, the equations responsible for the periodicity property of chains became elliptic. In the global double-periodic problems on the plane R^2, this fact led to the important conclusions (see[2]):

Theorem 10. *Let a periodic elliptic Laplace chain be given such that all 2D Schrödinger operators L_k in this chain are smooth periodic in R^2 and physical. Then all these operators are topologically trivial. All of them have a family of Bloch–Floquet solutions $L\psi = 0$ parametrized by the points of some Riemann surface of finite genus with two marked points ("infinities"). These solutions can be found explicitly. This family contains a subfamily of the bounded functions on R^2 providing* *a basis for the spectral (i.e., energy) level $\lambda = 0$ in the Hilbert space $L_2(R^2)$. This class of 2D Schrödinger operators was invented in the work [15] in 1976.*

Nothing like that exists in the hyperbolic case. The reason for this is that in the smooth elliptic case any nonlinear system on compact manifold (2-torus here) may have only a finite-dimensional family of global solutions. For the 1+1-dimensional completely integrable systems describing the periodicity property of Laplace chains, this fact leads to the linear dependence of the higher flows in the corresponding hierarchy and finally to the Riemann surfaces of finite genus, exactly as it was found for KdV in 1974 (see in [16]).

Example 11. Let $N = 2$ be the period. We come to the equation

$$\Delta f_0 = -8 sinh\{f_0\}; W_0 = e^{f_0}; f_0 = -f_1; H_0 = 2 sinh\{f_0\}. \qquad (3.26)$$

Exactly this equation appeared in the theory of the toroidal surfaces in R^2 with constant mean curvature $k_1 + k_2 = $ const (see in [17] the details and the authors of this discovery). It was observed in this theory that all of them can be obtained from the Riemann surfaces of finite genus like in the periodic theory of solitons. Our theorem can be considered as a natural extension of that technical result with a completely different interpretation.

We also introduced a notion of **semi-cyclic chain** $L_0, \dots, L_N$ satisfying the identity:

$$L_0 = L_N + C. \qquad (3.27)$$

The most interesting new class of Laplace chains $L_0, \dots, L_N$ leading to the operators with very specific anomalous spectral properties is the class of **the quasi-cyclic chains** $L_0, \dots, L_N$, such that the boundary operators are strongly factorizable (all factorizations on the boundary are assumed to be of the first type, and the Laplace transformations are assumed to be of the second type):

$$L_0 = -(\partial + A)(\bar{\partial} + B), L_N = (\partial + A')(\bar{\partial} + B') + C_N, \qquad (3.28)$$

where $C = const$.

Lemma 12. *Both semi-cyclic and quasi-cyclic Laplace chains of the length equal to one $N = 1$ lead to the Landau operator QQ^+ with magnetic field equal to constant $H_0 = const$ and $W_0 = 0, V_0 = H_0$. Let $H_0 > 0$. Its spectrum consists of the infinite number of highly degenerate Landau levels $\Lambda_k, k \geq 0, \lambda_k = kH_0$, isomorphic to each other by the operator Q:*

$$Q = \partial + A(z, \bar{z}) : \Lambda_k \to \Lambda_{k+1}, \qquad (3.29)$$

where $Q^+ \Lambda_0 = 0$ and $Q^+ = \bar{\partial} + B(z, \bar{z})$.

Let a quasi-cyclic chain of the length N be given, $H_0 > 0$ and all operators $L_k = Q_k^+ Q_k + 2V_k$ in the chain are smooth physical and double-periodic on the plane R^2

where $Q_k = (\partial + A_k)$ and $Q^+ = (\bar{\partial} + B_k)$. It is convenient sometimes in the physical case to choose gauge conditions such that Q and $-Q^+$ are adjoint to each other, i.e., $\bar{A} = -B$. We have always

$$V_0 - 1/2\Delta \log V_0 + H_0 = V_1; H_1 = H_0 - 1/2\Delta \log V_0$$

for the second-type Laplace transformation.

Theorem 13. *The operator* $L_N - C_N = Q_N Q_N^+$ *has a highly degenerate space of ground states*

$$Q_N^+ \psi^0 = 0; \lambda_0 = 0$$

isomorphic to the Landau level Λ_0. *It has also a second highly degenerate level* $\Lambda_N; \lambda_N = C_N = N[H_0]$, *isomorphic to the Landau level. The second level can be obtained from the solutions* $Q_0^+ \phi^0 = 0$ *belonging to the space* $L_2(R^2)$, *by the formula*

$$\psi = (\partial + A_{N-1}) \ldots (\partial + A_0)\phi^0.$$

The exact elliptic formulas for the functions ϕ^0 and ψ^0 can be extracted from the work [2] for the case where the magnetic flux is quantized. This formula is based on the result of [18] for the strongly factorized Schrödinger operators where these eigenfunctions of the ground level were calculated. The result itself remains true in the case of the irrational fluxes as well, because we may use a completely localized basis in the space of groundstates instead of the magnetic Bloch functions used in these works.

Example 14. Consider the case $N = 2$. The condition of the strong factorization of the boundary operators leads to the equation

$$\Delta g(z, \bar{z}) = 4e^g - 2C_2; V_0 = H_0 = \exp\{g\}. \tag{3.30}$$

We have also $C_2 = W_2 = V_1$ and $H_2 = H_1 = H_0 - 1/2\Delta g = C_2 - H_0$. We can see that this equation has a lot of periodic smooth solutions depending on one variable. It is not hard to prove that it has a lot of smooth double-periodic solutions essentially dependent on both variables x, y.

3.4 Discretization of the 2D Schrödinger Operators and Laplace Transformations on the Square and Equilateral Lattices

In the continuous case all formal calculations for the hyperbolic and elliptic cases were identical. The difference between them originated in the global properties only. For the difference operators these cases look completely different even on the formal level.

I. Let us start with the **hyperbolic case**. The discrete Schrödinger (or Laplace) equation is defined for the function ψ_n where $n = (n_1, n_2)$ on the square lattice $n \in Z^2$

on the plane by the formula

$$0 = L\psi_n = a_n\psi_n + b_n\psi_{n+T_1} + c_n\psi_{n+T_2} + d_n\psi_{n+T_1+T_2}, \qquad (3.31)$$

where $n + T_1 = (n_1 + 1, n_2); n + T_2 = (n_1, n_2 + 1)$. The operator L is well-defined modulo **gauge transformations**

$$L \to f_n L g_n; \psi_n \to g_n^{-1}\psi_n, \qquad (3.32)$$

where f_n, g_n are nonzero functions.

There exists a unique **weak factorization** of this operator written in the form

$$L = f_n[(1 + u_n T_1)(1 + v_n T_2) + w_n] = f_n[Q_1 Q_2 + w_n] \qquad (3.33)$$

(this is a first-type factorization). It generates a (first-type) Laplace transformation

$$L \to \tilde{L} = w_n Q_2 w_n^{-1} Q_1 + w_n; \tilde{\psi} = Q_2 \psi \qquad (3.34)$$

up to gauge transformation. As in the continuous case, the coefficients u_n, v_n, w_n can be easily found by elementary algebraic formulas. It was observed, in fact, in 1985 (see [19]) that the equation $L\psi = 0$ on the square lattice (above) has a nice family of algebra-geometric exactly solvable cases. Such solvable cases and discrete spectral symmetries normally appear exactly for the same classes of operators.

There are many orthonormal bases T_1', T_2' equivalent to each other in the square lattice. We can take any one of them: $(T_1', T_2') = (T_i^{\pm 1}, T_j^{\pm 1})$ where $i = \neq j$ and $i, j = 1, 2$. Any choice of basis defines a Laplace transformation

$$L \to \tilde{L}'; \tilde{\psi}' = Q_2'\psi$$

through the weak first-type factorization of the form

$$L = f_n'[Q_1'Q_2' + w_n']; Q_1' = 1 + u_n'T_1'; Q_2' = 1 + v_n'T_2'. \qquad (3.35)$$

We have a total number of eight for the Laplace transformations defined in this way.

Lemma 15. *The Laplace transformations defined above generate a group with four generators* $B_{\pm,\pm}$ *corresponding to the bases* $T_1' = T_1^{\pm 1}, T_2' = T_2^{\pm 1}$. *The Laplace transformations correspondent to the basis* T_1', T_2' *are inverse to the Laplace transformation correspondent to the basis* T_2', T_1' *modulo gauge transformations.*

This statement can be checked by elementary calculation.

As in the continuous case, we have gauge-invariant quantities.

Lemma 16. *A pair of gauge-invariant quantities (the "discrete curvatures") is defined as*

$$K_{1n} = \frac{b_n c_{n+T_1}}{d_n a_{n+T_1}},$$

$$K_{2n} = \frac{c_n b_{n+T_2}}{d_n a_{n+T_2}}.$$

All other gauge invariants, including the potential w_n, can be expressed through them. In particular, the potential w_n has a form $K_{1n} = (1 + w_n)^{-1}$. A "magnetic field"

$$H_n = \frac{v_n u_{n+T_2}}{u_n v_{n+T_1}}$$

can be expressed through the quantities K_1, K_2. They can also be expressed in terms of w_n, H_n.

As in the continuous case, it is convenient to write Laplace transformation in terms of w_n, H_n:

Lemma 17. *The Laplace transformation can be written in the form*

$$1 + \tilde{w}_{n+T_1} = (1 + w_{n+T_2}) \frac{w_n w_{n+T_1+T_2}}{w_{n+T_1} w_{n+T_2}} H_n^{-1},$$

$$\tilde{H}_n = \frac{1 + w_{n+T_2}}{1 + \tilde{w}_{n+T_2}}.$$

For the infinite Laplace Chain

$$\tilde{H}^k = H^{k+1}, \tilde{w}^k = w^{k+1},$$

we can express H^k through w^k, w^{k+1} as in the continuous case. It leads to the completely discrete 2D Toda lattice (it is a discrete 3D system found by Hirota many years ago from completely different ideas)

$$\frac{\left(1 + w_{n+T_1}^{k+2}\right)\left(1 + w_{n+T_2}^{k+1}\right)}{\left(1 + w_{n+T_1}^{k+1}\right)\left(1 + w_{n+T_2}^{k}\right)} = \frac{w_{n+T_1}^{k} w_{n+T_2}^{k}}{w_n^k w_{n+T_1+T_2}^{k}}.$$

Its reduction for the periodic Laplace chains of the length $N = 2$ leads to the nice analog of the sinh-Gordon equation (see [3]). In the discrete case we have a big group of Laplace transformations generated by the four generators (above). This group has not been studied yet.

II. **The elliptic case** is especially interesting. It turns out that in this case the right discretization of the second-order elliptic real self-adjoint operators (i.e., operators of the form $L = -\Delta + U(x,y)$) admitting Laplace transformations should be constructed on the **equilateral triangle lattice**. So in this case even the form of discretized elliptic operators has nothing to do with the hyperbolic case described above.

For the equilateral triangle lattice we have a basis T_1, T_2 such that the shift operator $T_1 T_2^{-1}$ has the same length. Therefore any vertex $n = (n_1, n_2)$ in the lattice has exactly six closest neighbors $n + T'$ where $T' = T_1^{\pm 1}$ or $T' = T_2^{\pm 1}$ or $T' = (T_1 T_2^{-1})^{\pm 1}$. We write a real self-adjoint operator in the form

$$L = a_n + b_n T_1 + c_n T_2 + d_{n-T_2} T_1 T_2^{-1} + ad\,joint.$$

We consider the **zero-level gauge transformations** preserving a form of the operator and a zero spectral level $L\psi = 0$:

$$L \to f_n L f_n, \psi_n \to f_n^{-1} \psi_n.$$

Lemma 18. *Any real self-adjoint operator of this form with nonzero coefficients* b_n, c_n, d_n *can be presented in the weakly factorized form of the first type*

$$L = QQ^+ + w_n; Q = x_n + y_n T_1 + z_n T_2,$$

where $T_i^+ = T_i^{-1}, i = 1, 2; (AB)^+ = B^+ A^+$. *This form is unique if the coefficients* c, b, d, x, y, z *are positive.*

Any equivalent basis T_1', T_2' with angle equal to $2\pi/3$ defines the analogous Laplace transformation. There is no difference between the pairs T_1', T_2' and T_2', T_1' in this factorization. So we have six different pairs:

$$(T_1, T_2), (T_2, T_2 T_1^{-1}), (T_2 T_1^{-1}, T_1^{-1}), (T_1^{-1}, T_2^{-1}), (T_2^{-1}, T_1 T_2^{-1}), (T_1 T_2^{-1}, T_2^{-1}).$$

Lemma 19. *For the nonzero potential* w_n *a Laplace transformation is defined as*

$$\tilde{L} = w_n^{1/2} Q_1 w_n^{-1} Q_2 w^{1/2} + w_n; \tilde{\psi} = w^{-1/2} Q_1 \psi$$

and the operator $\tilde{L}$ *is real self-adjoint. The Laplace transformations correspondent to the inverse bases* T_1, T_2 *and* T_1^{-1}, T_2^{-1} *are inverse to each other. Therefore the group of Laplace transformations is generated by three generators.*

In the work [3] we calculated how these three generators can be expressed through the first one and rotations of the lattice. Therefore there is essentially one Laplace transformation only in this case.

Let us consider now a special class of the purely factorizable operators in the strong sense:

$$L = QQ^+ + const$$

("white factorization") or

$$L = Q^+ Q + const,$$

("black factorization") where $Q = x_n + y_n T_1 + z_n T_2$. Especially interesting here is the case when **the white triangle equation**

$$Q^+ \psi = 0 \tag{3.36}$$

for the first case, or **the black triangle equation**

$$Q\psi = 0 \tag{3.37}$$

for the second case, has nontrivial solutions belonging to the space $L_2(Z^2)$.

Example 20. Let $Q_{c,d} = 1 + ce^{l_1(n)}T_1 + de^{l_2(n)}T_2$ where l_1, l_2 are the linear forms in the variables n_1, n_2 with real coefficients

$$l_i = \sum_j l_{ij}n_j; i,j = 1,2; n_j \in Z. \tag{3.38}$$

Theorem 21. *The black triangle equation $Q\psi = 0$ has an infinite-dimensional subspace of solutions belonging to the space $L_2(Z^2)$, if one of the following conditions is satisfied*

$$(a)\ l_{ii} > 0, i = 1,2; l_{11}l_{22} - l_{12}^2 > 0,$$
$$(b)\ l_{ii} > 0, i = 1,2; l_{11}l_{22} - l_{21}^2 > 0,$$
$$(c')\ l_{11} > 0; l_{11}l_{22} - l_{12}^2 > l_{11}(l_{21} - l_{12}),$$
$$(c'')\ l_{22} > 0; l_{11}l_{22} - l_{21}^2 > l_{22}(l_{21} - l_{12}).$$

The operator $L = Q^+Q$ has a zero point $\lambda = 0$ as a point of discrete spectrum in these cases, such that its multiplicity is infinite.
There is also a similar statement for the white triangle equation.

    For the proof of the theorem, we make a substitution

$$\psi_n = e^{-K_2(n)}\eta_n,$$

where $K_2(n)$ is a quadratic form in the variables n_1, n_2. After that we assume that coefficients of the equation for the quantity η_n either depend on the variable n_1 only (this is the case (a) above) or depend on the variable n_2 only (this is the case (b) above) or depend on the variable $n_1 + n_2$ only (this assumption leads to the cases (c') or (c") above).

    In case (a) we are looking for the solutions of the form

$$\eta_n = w^{n_2}\phi_{n_1}.$$

Let $l_{21} > l_{12}$. We choose the value of $w = w_q$ such that $\phi_{n_1} = 0$ for $n_1 > q; q \in Z$. This assumption leads to the solutions belonging to the space $L_2(Z^2)$. Other cases can be considered in a similar way – see details in [3].

    Consider now a special subcase of this example where

$$2l_{11} = 2l_{22} = l_{12} + l_{21}. \tag{3.39}$$

Lemma 22. *The operators Q, Q^+ satisfy the following relations*

$$Q_{c,d}Q_{c,d}^+ - 1 = u^{-2}(Q_{c',d'}^+Q_{c',d'} - 1), \tag{3.40}$$

where $u = e^{l_{11}}, v = e^{l_{12}}, c' = u^2c, d' = u^2d$.
Using these relations and the groundstates found before, we come to the following:

Theorem 23. *The spectrum of operators $L = QQ^+$ and $\tilde{L} = Q^+Q$ under the conditions above is discrete for $\lambda < 1$ and can lie in the following points only:*

$$(a)\ \lambda_j = 1 - u^{2j}, j \geq 0, u < 1,$$
$$(b)\ \lambda_j = 1 - u^{-2j}, u > 1.$$

In the following cases the spectrum of operator L occupies all these points, and the spectrum of operator $\tilde{L}$ occupies all these points except $\lambda_0 = 0$:

$$u^{-3} > v^{-1} > u^{-1} > 1,$$
$$u^{-1} > max(v, v^{-1}) \geq 1.$$

The replacement $u \to u^{-1}$ in these conditions leads to the interchange between L and $\tilde{L}$ in the theorem. All these levels are infinitely degenerate ("The discrete analogs of Landau levels").

Nothing is known about the spectrum for $\lambda \geq 1$. It is certainly continuous. The interesting multi-dimensional analogs of the operators satisfying the relation above were found in the work [3] for the multi-dimensional analogs of the equilateral lattice, but their spectrum is not found yet.

In the special case $u = 1$ of the example above, we have

$$Q_{c,d} Q_{c,d}^+ = Q_{c,d}^+ Q_{c,d}.$$

Here we should consider both (white and black triangle equations) simultaneously:

$$Q\psi = 0; Q^+\psi = 0.$$

This situation can be naturally extended to the more general pair of equations (black and white):

$$Q_1\psi = 0; Q_2\psi = 0.$$

This pair leads to the "discrete curvature" making an obstacle for the local existence of solutions around every vertex. These ideas were developed in [3] in a much more general situation.

3.5 2D Manifolds with the Colored Black–White Triangulation. Integrable Systems on a Trivalent Tree

In the work [3] a theory of Laplace transformations was developed on the **2D manifolds with the colored "black–white triangulation"**. We assume that a color (black or white) is assigned to every triangle in the triangulation such that any triangles with common edge should have opposite colors. The black triangle operator Q can be defined by the field associating number $b_{P:T}$ to the pair P, T where T is a black triangle and P is its vertex $P \in T$. We define operator Q by the formula

$$\tilde{\psi}_T = Q\psi_T = \sum_P b_{P:T} \psi_{P:T}.$$

It maps the space of functions on the set of vertices into the space of functions on the set of black triangles. The factorized operators have a form $L = Q^+ Q$; their zero modes satisfy the black triangle equation

$$Q\psi_T = 0.$$

This structure permits to define **combinatorial geodesics** consisting of edges and passing every vertex "as a straight line" (i.e., the numbers of triangles from both its sides should be equal to each other). The right (left) horocycles are such lines that there is exactly three triangles from the right (left) side of it in every vertex. The right (left) curvature of the combinatorial line is measured by the number of triangles from the left (right) side of it in the vertices. This structure imitates somehow conformal geometry. In particular, the black (or white) triangle equation can be considered as reasonable discrete analogs of the complex (covariant) $\partial + A$ and $\bar\partial + B$ operators: they factorize the second-order elliptic operators (it does not matter that complex numbers are not involved in their definition); they are "more elliptic" than any other first-order discrete operators known until now.

Example 24. Let $b_{P;T} = 1$ for the operator Q. The operator $L = Q^+ Q$ can be compared with Laplace–Beltrami operator $L_0 = dd^*$ where d is a standard boundary operator and d^* is a coboundary operator. If R_P is the number of triangles entering the vertex P, we have

$$L_0 \psi_P = -\sum_{P'} \psi_{PP'} + R_P \psi_P,$$
$$L \psi_P = \sum_{P'} \psi_{PP'} + R_P \psi_P.$$

Therefore we conclude that there is an equality

$$L = -L_0 + 2R_P.$$

The case $R_P = 6$ corresponds to euclidean geometry. In principle, a quantity like R_P corresponds to something like the **scalar curvature**.

Boundary problems of the Dirichlet type for the triangle equations can be posed for the bounded simply connected domains on the plane with the black–white triangulation. However, careful analysis of the admissible boundary functions is required.

Example 25. Let me remind here that in the euclidean plane with equilateral lattice Z^2 the black triangles have a form $n, n + T_1, n + T_2$ for all $n \in Z^2$. Consider any lattice straight line Z' dividing Z^2 into the parts

$$Z^2 = R_+ \bigcup R_-; R_+ \bigcap R_- = Z',$$

where R_+ touches its boundary Z' by the black triangles. Starting with arbitrary data $\phi_n, n \in Z'$, we can always find unique solution to the black triangle equation

$Q\psi_n = 0$ in the domain $n \in R_+$ such that $\psi = \phi$ on the boundary. This initial value problem is hyperbolic. However, the initial value problem in the other direction R_- is parabolic: for finding a solution in the domain R_- to the black triangle equation $Q\psi = 0$ such that $\psi_n = \phi_n$ for $n \in Z'$ we should require some decay for the Cauchy data ϕ_n on the line Z'. The operator expressing the solution ψ on any line parallel to R through the initial value ϕ became nonlocal in this domain: you have to integrate along the whole line Z'.

Now let us consider **a plane R^2 with a colored black–white triangulation**. Studying the Dirichlet-type boundary problems, we start with some simply connected bounded triangulated sub-domain D in it with the **thin** boundary polygon $\Gamma = \partial D$. It means that there are no triangles in D whose vertices all belong to the polygon Γ. We call a boundary edge **white** if its white side lies inside of the domain D, otherwise we call a boundary edge **black**. We have

$$|\Gamma| = \Gamma_b + \Gamma_w,$$

where Γ_b and Γ_w are exactly the number of black (white) boundary edges in Γ.

The elliptic-type Dirichlet boundary problem is to find a solution to the black triangle equation $Q\psi = 0$ in the domain D such that $\psi_P = \phi_P$ on the boundary $P \in \partial D = \Gamma$. It turns out that for the correct solution of this problem we should start with the boundary function ϕ given in some part of the boundary only:

1. The total number of known values $\phi_P, P \in \Gamma$, should be equal to the number $V - T_b$ where V is the number of vertices in D, T_b (T_w) is the number of black (white) triangles and $T_b + T_w = T$, $T_w = T_b + \Delta$ by definition.

Lemma 26.

$$\Delta = -(\Gamma_b - \Gamma_w)/3,$$
$$V - T_b = 1 + (|\Gamma| + \Delta)/2.$$

The proof of this statement follows easily from the topology of the plane. Let us denote by the letters $V, E, T = 2T_b + \Delta$ the numbers of vertices, edges, triangles and black triangle T_b correspondingly in the domain D. From the Euler identity and elementary combinatorics we have

$$V - E + T = 1; E = 3/2T + |\Gamma|/2; \Delta = -(\Gamma_b - \Gamma_w)/3.$$

The total number of unknown quantities is equal to V. The number of equations is equal to T_b where $T = T_b + T_w$, $T_w = T_b + \Delta$. So the number Q of independent data should be equal to

$$Q = V - T_b = 1 + (|\Gamma| + \Delta)/2.$$

The lemma is proved.

2. For the "elliptic-type" boundary problems the set of known values should never contain both boundary vertices $P_1 \cup P_2 = \partial l$ of any black edge l on the curve $\Gamma = \partial D$. We are going to develop this subject in the next work.

III. Let us consider now **a trivalent tree** following the work [20]. Many people studied the second-order (Laplace–Beltrami) difference operators on the trees, but nothing like hidden integrability of the soliton type was found for them. We are going to consider graphs (one-dimensional simplicial complexes) with the natural geodesic metric such that the length of every edge $d(PP')$ is equal to one, and every edge has exactly two vertices PP'. There are no cycles in the trees by definition.

The operator L acting on the functions of vertices

$$L\psi_P = \sum_Q b_{PQ}\psi_Q$$

is **real** if all coefficients are real. It has **an order** k equal to the maximal diameter of the interaction domain in the vertices P, i.e., $k = \max_P d(Q_1 Q_2)$ such that $b_{PQ_1} \neq 0, b_{PQ_2} \neq 0$. The real operator is symmetric or **self-adjoint** if $b_{PQ} = b_{QP}$. A self-adjoint operator should have an even order $k = 2l, l = 0, 1, 2, \ldots$. For the second- and fourth-order cases we frequently numerate the highest order coefficients by the pair of adjusting edges $b_{PP''} = b_{RR'}$ and the second-order terms by one edge $b_{PP'} = b_R$. Consider now the set of all fourth-order real self-adjoint operators L on the trivalent tree such that the highest order coefficients are always positive:

$$b_{PP''} > 0; d(P, P'') = 2,$$
$$L\psi_P = \sum b_{PP''}\psi_{P''} + b_{PP'}\psi_{P'} + w_P\psi_P,$$

where $d(PP'') = 2, d(PP') = 1$. Let me remind that in 1976 the so-called **L–A–B-triples** were invented and studied in the works [15, 21] as completely integrable soliton systems associated with the zero level of the 2D Schrödinger operator on the Euclidean plane R^2. Their discretization on the regular lattices Z^2 was discussed above.

Trivalent tree Γ has a geodesic structure analogous to the 2D hyperbolic (noneuclidean) plane. As we shall see, nontrivial $L - A - B$ triples appear here for the fourth-order self-adjoint operators. Nothing like that exists here for the second-order difference operators.

Theorem 27. *There exists a nontrivial time dynamics of the form*

$$L_t = LA - BL,$$

where the difference operators A, B have second order and $B = A^t$

$$A\psi_P = \sum c_{PP'}\psi_{P'}.$$

The coefficients $c_{PP'}$ for the edges $R = PP'$ can be calculated by the following formula. Fix some "initial" point $P_0 \in \Gamma$; for every point $P \in \Gamma$ there is a unique simple path $\gamma = [P_0, \ldots, P]$ consisting of the edges $R_0, \ldots, R_k$ and joining the initial point *with point P. We introduce a multiplicative one-cocycle $\Psi(R)$ whose value for the oriented edge $R = Q_1 Q_2$ can be described in the following way. Let the edges R'_1, R'_2 enter the first vertex Q_1, and the edges R''_1, R''_2 come out of the second vertex Q_2, not one of these edges coincides with R. We define this cocycle and the coefficients c*

$$\Psi(R) = -\frac{b_{RR''_1} b_{RR''_2}}{b_{R'_1 R} b_{R'_2 R}},$$

$$c_R = -\frac{1}{b_{R'_1 R'_2}} \left(\prod_{R_i \in \gamma} \Psi(R_i) \right),$$

where $R = PP'$.

There is nothing surprising here that this expression is nonlocal: let me remind that for the best-known hierarchy (the so-called "Novikov–Veselov" hierarchy [22, 23]) associated with the 2D Schrödinger operator L, such nonlocality is also presented. It is presented also in the famous KP hierarchy, so it always appears for the $2 + 1$-systems.

Theorem 28. *The generic real fourth-order operator L on the trivalent tree Γ admits a one-parametric family of factorizations through the second-order operators*

$$L = Q^t Q + u_P,$$

where $Q\psi_P = \sum_Q d_{PQ} \psi_Q + v_P \psi_P$ and $d_{PQ} > 0$,

$$b_{PP''} = d_{P'P} d_{P'P''}; b_{PP'} = d_{P'P} v_{P'} + d_{PP'} v_P,$$
$$w_P = v_P^2 + \sum_{P'} d_{P'P}^2 + u_P.$$

Therefore the Laplace transformation are defined for this class of operators.

Recently in the work [24] these results were extended to all trees: the last theorem is not true anymore for the generic operators, but for the subclass of factorizable real self-adjoint fourth-order operators L the analog of the first theorem remains true.

References

1. S.P. Novikov, Funct. Anal. Appl. 8(3), 54–66, 1974.
2. S.P. Novikov and A.P. Veselov, AMS Translations, Series 2, v 179; Solitons, Geometry and Topology: On the Crossroads, 109–133, 1997.
3. S.P. Novikov, I.A. Dynnikov, Russian Math Surveys., 52(5), 175–234, 1997.
4. J. Weiss, J. Math. Phys. 27, 2647–2656, 1986.
5. A.B. Shabat and A.P. Veselov, Funct. Anal. Appl. 2, 1993.
6. V. Spiridonov, L. Vinet, and A. Zedanov, Lett. Math. Phys. 29, 63–73, 1993.
7. N.M. Atakishev and S.K. Suslov, Theor. Math. Phys. 85(1), 64–73, 1990.

8. C.S. Gardner, J. Green, M. Kruskal, and R. Miura, Phys. Rev. Lett. 19, 1095–1098, 1967.
9. P. Lax, Comm. Pure. Appl. Math. 21(5), 467–490, 1968.
10. H. Flashka, Progr. Theor. Phys. 51, 703–716, 1974.
11. S.V. Manakov, JETP, 67, 543–555, 1974.
12. J. Moser, Adv. Math. 16, 354, 1975.
13. A.V. Mikhailov, JETP 30, 443–448, 1979.
14. B.A. Dubrovin, I.M. Krichever, and S.P. Novikov, Dynamical Systems 4–Completely Integrable Systems, Encyclopedia Math Sciences, Arnold V. and Novikov S. (eds.), second edition, Springer Verlag, vol. 4, 177–133, 2000.
15. B.A. Dubrovin, I.M. Krichever, and S.P. Novikov, Soviet Math Doklady 229, 15–18, 1976.
16. B.A. Dubrovin, V.B. Matveev, and S.P. Novikov, Russian Math Surveys 31(2), 55–136, 1976.
17. A.I. Bobenko, Russian Math Surveys 46(4), 3–42, 1991.
18. B.A. Dubrovin and S.P. Novikov, JETP 3, 1006–1016, 1980.
19. I.M. Krichever, Soviet Math. Doklady 285(1), 31–36, 1985.
20. I.M. Krichever and S.P. Novikov,Russian Math. Surveys 54(6), 149–150, 1999.
21. S.V. Manakov, Russian Math. Surveys 31(5), 245–246, 1976.
22. S.P. Novikov and A.P. Veselov, Soviet Math. Doklady 279(5), 1984.
23. S.P. Novikov, A.P. Veselov, and D. Fisica 18, 267–273, 1986.
24. L.O., Chekhov and N.P. Pusyrnikova, Russian Math. Surveys 55, 2000.

Chapter 4
Symmetries of Spectral Problems

A. Shabat

Abstract Deriving abelian KdV and NLS hierarchies, we describe non-abelian symmetries and "pre-Lax" elementary approach to Lax pairs. Discrete symmetries of spectral problems are considered in Sect. 4.2. Here we prove Darboux classical theorem and discuss a modern theory of dressing chains.

4.1 Lie-Type Symmetries

4.1.1 Cross-Differentiation

We discuss below the consistency of the spectral problem

$$\psi_{xx} = U(x,\lambda)\psi, \quad U(x,\lambda) = \lambda^m + u_1(x)\lambda^{m-1} + \cdots + u_m(x), \quad (4.1)$$

with the evolutionary linear equation of the general form

$$D_t(\psi) \overset{\text{def}}{=} \psi_t + \lambda_t\psi_\lambda = A(x,\lambda)\psi_x + B(x,\lambda)\psi. \quad (4.2)$$

The spectral parameter λ does not depend on x and therefore we assume that

$$\lambda_t = k(\lambda).$$

Proper forms of $k(\lambda)$ will be defined using necessary conditions of compatibility of Eqs. (4.1) and (4.2) for a single function $\psi = \psi(x,t,\lambda)$.

The action of D_t on the potential U is defined by the equations

$$D_t(U) = U_t + \lambda_t U_\lambda = B_{xx} + 2A_x U + AU_x, \quad 2B_x = -A_{xx}. \quad (4.3)$$

In order to derive these we have to differentiate (4.1) with respect to time t. If there is a common solution $\psi(x,t,\lambda)$ of (4.1), (4.2) then $(\psi_x)_t = (\psi_t)_x$ and, therefore,

A. Shabat (✉)
L.D. Landau Institute for Theoretical Physics, Kosygina 2, Moscow 117334, Russia,
shabat@itp.ac.ru

Shabat, A.: *Symmetries of Spectral Problems*. Lect. Notes Phys. **767**, 139–173 (2009)
DOI 10.1007/978-3-540-88111-7_4

$$D_t(\psi_x) = A_x \psi_x + A \psi_{xx} + B_x \psi + B \psi_x = A_1 \psi_x + B_1 \psi,$$

where $A_1 = A_x + B$ and $B_1 = B_x + AU$. Equating now the terms with ψ_x and ψ, respectively, in the equation

$$D_t(\psi_{xx}) = A_2 \psi_x + B_2 \psi = D_t(U)\psi + UD_t(\psi)$$

one obtains (4.3). Hereafter, we exclude B and rewrite (4.2) and (4.3) as follows:

$$D_t(\psi) = \psi_t + \lambda_t \psi_\lambda = A \psi_x - \frac{1}{2} A_x \psi, \qquad (4.4)$$

$$2(U_t + k(\lambda)U_\lambda) = 4UA_x + 2U_x A - A_{xxx}. \qquad (4.5)$$

We shall consider the polynomial (in λ) case

$$A = A(x,t;\lambda) = a_0(x,t)\lambda^n + a_1(x,t)\lambda^{n-1} + \cdots + a_n(x,t)$$

and try to define coefficients of A and the function $k(\lambda) = \lambda_t$ by equating different powers in λ in Eq. (4.5).

The consistency condition (4.5) has numerous unexpected applications. To highlight the general procedure we begin with a description of obvious symmetries of the classical linear Schrödinger equation

$$\psi_{xx} = (u - \lambda)\psi. \qquad (4.6)$$

in terms of (4.4).

Example 1. Let $U = u - \lambda, A \equiv a$ in (4.5). Then

$$2u_t - 2k(\lambda) = 4(u - \lambda)a_x + 2u_x a - a_{xxx} \Rightarrow \lambda_t = \varepsilon_1 \lambda + \varepsilon_0, \quad a_{xx} = 0.$$

In the generic case we find

$$a = x, \quad \lambda_t = 2\lambda \Rightarrow \psi_t + 2\lambda \psi_\lambda = x\psi_x, \quad u_t = 2u + xu_x.$$

In order to integrate these first-order PDEs one can use the corresponding system of characteristic differential equations

$$\frac{dt}{1} = \frac{d\lambda}{2\lambda} = -\frac{dx}{x} = \frac{du}{2u}. \qquad (4.7)$$

Thus one obtains the underlying scaling transformation

$$\psi(x,\lambda) \mapsto \psi\left(\hat{x}, \hat{\lambda}\right), \quad \hat{x} = x/q, \quad \hat{u} = q^2 u, \quad \hat{\lambda} = q^2 \lambda.$$

Other obvious transformations

$$\psi(x,\lambda) \mapsto \psi\left(\hat{x}, \hat{\lambda}\right), \quad \hat{x} = x + \tau, \quad \hat{u} = u + \varepsilon, \quad \hat{\lambda} = \lambda + \varepsilon$$

which leave the Schrödinger spectral problem invariant correspond to the case $a_x = 0$. In particular,[1]

$$a = 1, \quad \lambda_t = 0 \Rightarrow \psi_t = \psi_x, \quad u_t = u_x.$$

In this case the characteristic differential equations (cf. (4.7)) define the vector field

$$D = \partial_x + u_x \partial_u + u_{xx} \partial_{u_x} + \dots,$$

which we call the operator of total differentiation with respect to x. Analogous vector field in the case (4.7)

$$\hat{D} = 2\lambda \partial_\lambda - x\partial_x + 2u\partial_u + 3u_x \partial_{u_x} + \dots$$

satisfies the commutation relation as follows: $[\hat{D}, D] = D$.

Much more intriguing applications are related to the case

$$U = \lambda^2 + u_1(x)\lambda + u_2(x), \quad A(x,\lambda) = a_0(x)\lambda + a_1(x). \tag{4.8}$$

Equation (4.5) yields in this case

$$\lambda_t = k(\lambda) = \varepsilon_0 \lambda^2 + \varepsilon_1 \lambda + \varepsilon_2, \quad a_{0,x} = \varepsilon_0, \quad (2a_1 + a_0 u_1)_x = 2\varepsilon_1, \tag{4.9}$$

$$u_{1,t} + \varepsilon_1 u_1 + 2\varepsilon_2 = a_1 u_{1,x} + 2u_1 a_{1,x} + a_0 u_{2,x} + 2u_2 a_{0,x}, \tag{4.10}$$

$$u_{2,t} + \varepsilon_2 u_1 = a_1 u_{2,x} + 2u_2 a_{1,x} - \frac{1}{2} a_{1,xxx}. \tag{4.11}$$

Now, let us consider the potentials $u_{1,t} = u_{2,t} = 0$ which are invariant under the t-evolution. In this case the first two of the above equations allow to find u_1, u_2 in terms of a_1 and the last one can be reduced to the second-order ODE for a_1. Choosing in (4.9)

$$k(\lambda) = \quad 1, \quad \lambda, \quad \lambda^2, \quad \lambda(\lambda - \lambda_0),$$

we get, respectively, the list of Painlevé equations as follows:

$$y_{xx} = 2y^3 + xy + \alpha, \tag{4.12}$$

$$y_{xx} = \frac{y_x^2}{2y} + \frac{3y^3}{2} + 4xy^2 + 2(x^2 - \alpha)y + \frac{\beta}{y}, \tag{4.13}$$

$$y_{xx} = \frac{y_x^2}{y} - \frac{y_x}{x} + \frac{\alpha y^2 + \beta}{x} + \gamma y^3 + \frac{\delta}{y}, \tag{4.14}$$

$$y_{xx} = \left(\frac{1}{2y} + \frac{1}{y-1} \right) y_x^2 - \frac{y_x}{x} + \frac{2}{x^2}(y-1)^2 \left(\frac{\alpha y + \beta}{y} \right) + \frac{\gamma y}{x} + \frac{\delta y(y+1)}{y-1}. \tag{4.15}$$

For instance, in the case $k(\lambda) = 1$ we find by $a_0 = 1$

---

[1] Any spectral problem (4.1) possesses this symmetry.

$$u_1 = -2a_1 + c_1, \quad u_2 = 2x + 3a_1^2 + 2c_1a_1 + c_2,$$

where c_1, c_2 are constants of integration. The Painlevé equation $P2$ in the canonical form (4.12) is now obtained from (4.11) by appropriate rescaling.

One can, thus, reformulate results about zero-curvature representations for Painlevé equations $P1$–$P5$ known in literature as the Garnier theorem. Namely, the list above corresponds to the Lax pairs

$$k(\lambda)\psi_\lambda = A(x,\lambda)\psi_x - \frac{1}{2}A_x(x,\lambda)\psi, \quad \psi_{xx} = U(x,\lambda)\psi,$$

where the general form of A, U is shown in (4.8). The Painlevé equation $P1$:

$$y_{xx} = 6y^2 + x, \tag{4.16}$$

which is absent in the above list, corresponds to the Lax pair as follows:

$$\psi_\lambda = (4\lambda + 2u)\psi_x - u_x\psi, \quad \psi_{xx} = (u(x) - \lambda)\psi$$

and describes stationary solutions of the evolutionary equation

$$u_t + u_{xxx} = 6uu_x + 1.$$

In general, the compatibility condition (4.5) for Schrödinger spectral problem (4.6) with $A = a_0\lambda + a_1$ describes degenerated cases of Painlevé equations with the vanishing of some parameters in the above list (see [5]). In particular, in the case of Example 1 potentials of spectral problem (4.6) which are invariant under scaling are as follows:

$$xu_x + 2u = 0 \Leftrightarrow u(x) = \alpha x^{-2}.$$

It should be noticed that when $\lambda_t = 0$, $A = \lambda + a(x)$ stationary equations (4.10), (4.11) with $u_{1,t} = u_{2,t} = 0$ are reduced to the first-order ODE as follows:

$$a_x^2 = \varepsilon_4 a^4 + \varepsilon_3 a^3 + \varepsilon_2 a^2 + \varepsilon_1 a + \varepsilon_0.$$

Thus, invariant potentials in this case are just elliptic functions.

Closing this terse introduction of the theory of symmetries we have to define the Lie bracket for t-derivations defined by (4.4). Namely, for any two symmetries of the spectral problem (4.1)

$$D_l(\psi) = A_l(x,\lambda)\psi_x - \frac{1}{2}A_{l,x}(x,\lambda)\psi, \quad D_l(\lambda) = k_l(\lambda), \quad l = i, j,$$

we introduce $D_{ij} = [D_i, D_j]$ as follows

$$D_{ij}(\psi) = A(x,\lambda)\psi_x - \frac{1}{2}A_x(x,\lambda)\psi = (D_iD_j - D_jD_i)\psi,$$

where

$$A = D_i(A_j) - D_j(A_i) + A_j A_{i,x} - A_i A_{j,x}, \quad D_{ij}(\lambda) = k_{ij}(\lambda) = k_{j,\lambda} k_i - k_{i,\lambda} k_j. \quad (4.17)$$

It is straightforward yet not easy to check out that the compatibility condition (4.5) for D_{ij} follows from ones for D_i and D_j.

4.1.2 Isospectral Symmetries

The basic role further on will play the bilinear form of the spectral problem (4.1). Namely, if ψ_1, ψ_2 are linear independent solutions (4.1) then

$$\beta(\lambda) = \langle \psi_1, \psi_2 \rangle = \psi_1 \psi_{2,x} - \psi_{1,x} \psi_2 = \text{const}$$

and

$$A = \psi_1 \psi_2 \implies \frac{\beta}{A} = f_2 - f_1, \quad \frac{A_x}{A} = f_2 + f_1, \quad f_j = \frac{\psi_{j,x}}{\psi_j}.$$

In virtue of the Riccati equation

$$f_x + f^2 = U, \quad f = \psi^{-1} \psi_x$$

and the above formulae for f_j in terms of A we find that the function $A(x, \lambda) = \psi_1 \psi_2$ satisfies the equation as follows:

$$-2AA_{xx} + A_x^2 + 4UA^2 = \alpha(\lambda) \quad (4.18)$$

with $\alpha = \beta^2$. Thus for a given solution A of (4.18) using the formulae for f_1 and f_2 we can find ψ_1 and ψ_2 by quadratures. One has to notice that we can normalize the right-hand side of Eq. (4.18) by multiplication of solutions upon appropriate function on λ.

Definition 2. We shall call the nth degree polynomial A in λ an isospectral symmetry of order n of the spectral problem (4.1) iff $deg(4UA_x + 2U_xA - A_{xxx}) < deg(U)$.

The aim of this definition is to eliminate ambiguity in the process of defining the polynomial $A(x, \lambda)$ by Eq. (4.5) using the isospectrality constraint $D_t(\lambda) = 0$. Thus the evolutionary differentiation D_t acts now as follows:

$$2D_t(U) = 4UA_x + 2U_xA - A_{xxx}, \quad D_t(\psi) = A\psi_x - \frac{1}{2}A_x\psi. \quad (4.19)$$

We shall show that the isospectrality constraint cancels the explicit x-dependence of the coefficients of the polynomial A (cf. Sect. 1.1).

Definition 3. We shall call the generating function (of isospectral symmetries) of spectral problem (4.1) the formal power series in $1/\lambda$:

$$Y(x, \lambda) = 1 + \sum_{k=1}^{\infty} (\lambda)^{-k} y_k(x), \quad (4.20)$$

which is defined by the equation

$$-2Y_{xx}Y + Y_x^2 + 4UY^2 = 4\lambda^m. \tag{4.21}$$

Equation (4.21) yields $2y_1 + u_1 = 0$ by equating coefficients by λ^{N-1}. One can similarly extract from Eq. (4.21) the exact recurrency formula for coefficients y_j in terms of y_k, $k < j$. For example, in the quadratic case $U = \lambda^2 + u_1\lambda + u_2$ we find

$$2y_1 + u_1 = 0, \quad 2y_2 + u_2 = 3y_1^2, \quad 4(y_3 - 3y_1y_2 + 2y_1^3) = y_{1,xx}, \ldots. \tag{4.22}$$

For any spectral problem (4.1), we can assume that coefficients y_k in (4.20) are defined using (4.21) in terms of the potential U and, thus, are given as differential polynomials in $u_1, \ldots, u_N$. Obviously,

$$-Y_{xxx} + 4UY_x + 2U_xY = 0 \tag{4.23}$$

and after multiplication by λ^n we obtain readily

Lemma 4. *For any* $n = 0, 1, 2, \ldots$ *the formula*

$$A_n = \lambda^n + y_1\lambda^{n-1} + \cdots + y_n \equiv (\lambda^n Y)_+ \tag{4.24}$$

defines the nth order isospectral symmetry of corresponding spectral problem.

It follows from Definition 2 that the sum of two isospectral symmetries is symmetry as well and it is easy to see that leading coefficient a_0 of any nth order symmetry $A = a_0\lambda^n + \cdots + a_n$ has to be constant. Invoking Lemma 4 and induction in n we see that polynomials (4.24) form a basis in the linear space of isospectral symmetries. These polynomials A_n in λ are homogeneous in the sense that

$$\hat{D}(A_n) = nA_n, \tag{4.25}$$

where $\hat{D}$ is a vector-field related to the scaling symmetry of spectral problem (4.1) (cf. Sect. 1.1). For instance, in the quadratic case $U = \lambda^2 + u_1\lambda + u_2$

$$\hat{D} = \lambda\partial_\lambda + u_1\partial_{u_1} + 2u_2\partial_{u_2} + 2u_{1,x}\partial_{u_{1,x}} + 3u_{2,x}\partial_{u_{2,x}} + \cdots.$$

Using (4.25), one can prove (see also [3])

Theorem 5. *The Lie algebra of isospectral symmetries of the spectral problem* (4.1) *with the bracket* (4.17) *is abelian and symmetries* (4.24) *define the basis of this algebra.*

Proof. Let A_i and A_j be two isospectral symmetries (4.24). Then the formula (4.17) yields an isospectral symmetry A and we have

$$\hat{D}(A_i) = iA_i, \quad \hat{D}(A_j) = jA_j \implies \hat{D}(A) = (i+j)A.$$

In the case $A \neq 0$ it is $(i+j)$th order isospectral symmetry and hence $A = a_0\lambda^{i+j} + \cdots + a_{i+j}$, where $a_0 \neq 0$. Yet the formula (4.17) yields $a_0 = 0$. Therefore $A = 0$. ∎

In the next subsection we will discuss examples of isospectral symmetries of basic spectral problems. But let us consider now an interesting generalization of the theory which gives rise to a universal model irrelevant to specific properties of the spectral problem (4.1). Roughly speaking it corresponds to evolutionary equations (4.19) rewritten in terms of coefficients y_k of the generating function (4.20). The governing equations of this model are as follows ($n = 0, 1, 2, \ldots$):

$$D_n(Y) = < \lambda^n + y_1 \lambda^{n-1} + \cdots + y_n, Y >, \quad < a, b > \overset{\text{def}}{=} ab_x - ba_x. \tag{4.26}$$

It is easy to see that

$$Y = \psi_1 \psi_2, \quad \psi_{j,t} = A \psi_{j,x} - \frac{1}{2} A_x \psi_j \Rightarrow Y_t = < A, Y >.$$

In that sense equations $D_n(Y) = < A_n, Y >$, $A_n = (\lambda^n Y)_+$ can be considered as corollary of Lemma 4.

Particularly, we have $D_0 y_k = y_{k,x}$ and denoting $D_1 = \partial_t$ we get in virtue of (4.26) the infinite system of equations

$$y_{1,t} = y_{2,x}, \quad y_{2,t} = y_{3,x} + < y_1, y_2 >, \quad y_{3,t} = y_{4,x} + < y_1, y_3 >, \ldots . \tag{4.27}$$

The consistency with similar equations for D_n, $n > 1$, can be stated now as follows:

Theorem 6. *Defined by (4.26) evolutionary derivations D_n, $n = 1, 2, \ldots$, in the set of functions of variables $y_1, y_2, \ldots$ mutually commutate.*

Proof. It is sufficient to prove (as in Theorem 5) that for any i, j

$$D_i(A_j) - D_j(A_i) = < A_i, A_j > \tag{4.28}$$

and that follows now straightforwardly from the formulae

$$D_n(y_m) = \sum_{k=1}^{k=m} < y_{m-k}, y_{n+k} >, \quad n \geq 1, \quad y_0 \overset{\text{def}}{=} 1,$$

which are equivalent with (4.26). ∎

The analogue of non-isospectral symmetries (4.8), (4.9) considered in Sect. 1.1 can be introduced by the evolutionary derivation D_τ, $[D_\tau, D_0] = 0$ such that

$$D_\tau(Y) = (xD_1 - \lambda^2 \partial_\lambda - y_1)Y.$$

It yields

$$D_\tau(y_j) = xD_1(y_j) + (j+1)y_{j+1} - y_1 y_j, \quad j = 1, 2, \ldots, \tag{4.29}$$

and one can prove that $[D_\tau, D_1] = 2D_2$. In other words the vector-field (4.29) is *master-symmetry* (see for example [5]) for the system (4.27) of the first-order PDEs. At least in principle, it allows to reconstruct by the recurrence $[D_\tau, D_n] = 2D_{n+1}$ the whole hierarchy of Eqs. (4.26).

The hierarchy of derivations D_n, $n = 0, 1, \ldots$, exhibits an interesting type of symmetry transformations [9]. First, we introduce new derivations

$$\hat{D}_i \overset{\text{def}}{=} D_i - y_i D_0, \quad i \geq 1, \tag{4.30}$$

which commute since (4.28) implies

$$D_i(y_j) - D_j(y_i) = \langle y_i, y_j \rangle. \tag{4.31}$$

Second, we define the new wronskian $\langle a, b \rangle_1 \overset{\text{def}}{=} a\hat{D}_1(b) - b\hat{D}_1(a)$ and find that

$$< y_i, y_j >_1 = y_i y_{j+1,x} - y_j y_{i+1,x}.$$

Now, it is easy to verify that Eq. (4.26) can be rewritten as follows:

$$\hat{D}_n Y = \langle (\lambda^{n-1} Y)_+, Y \rangle_1, \quad i \geq 1, \tag{4.32}$$

and, thus, kept invariant.

If $Y = Y(\lambda; x, \mathbf{t})$, $\mathbf{t} = (t_1, t_2, \ldots)$ is a solution of (4.26), then we can define solution $Y' = Y'(\lambda; x', \mathbf{t}')$, $x' = t_1$, $\mathbf{t}' = (t_2, t_3, \ldots)$ as follows:

$$Y'\left(\lambda; x', \mathbf{t}'\right) = Y\left(\lambda; X(x', \mathbf{t}'), x', \mathbf{t}'\right), \tag{4.33}$$

where $X = X(x', \mathbf{t}') = X(t_1, t_2, \ldots)$ is a solution of the associated with (4.26) system of differential equations

$$D_i(X) + y_i(X, t_1, t_2, \ldots) = 0, \quad i \geq 1. \tag{4.34}$$

Equations (4.31) provide compatibility conditions for it.

Proposition 7. *Let* $Y = Y(\lambda; x, \mathbf{t})$ *satisfy*

$$2Y_{xx} Y - Y_x^2 - 4UY^2 + 4\lambda^m = 0, \quad U(\lambda, x) := \lambda^m + \sum_{i=0}^{m-1} \lambda^i u_i(x). \tag{4.35}$$

Then, the transformed function $Y' = Y'(\lambda; x', \mathbf{t}')$ *is a solution of*

$$2Y'_{x'x'} Y' - (Y'_{x'})^2 - 4U'Y'^2 + 4\lambda^{m+2} = 0, \tag{4.36}$$

where

$$U' = \lambda^2 U - \frac{1}{2}(\lambda y_{1,xx} + y_{2,xx} - y_1 y_{1,xx}) + \frac{1}{4} y_{1,x}^2.$$

Proof. One finds

$$Y_x = \frac{1}{\lambda}(Y'_{x'} + y_{1,x}Y'),$$

$$Y_{xx} = \frac{1}{\lambda^2}\left(Y'_{x'x'} + y_{1,x}Y'_{x'} + (\lambda y_{1,xx} + y_{2,xx} - y_1 y_{1,xx})Y'\right).$$

Substituting these formulae in (4.35) we get (4.36). ∎

The proposition means that the transformation $Y \to Y'$ establishes some relationships between different energy-dependent Schrödinger problems (see also [8]). In particular, this result proves that the whole family of hierarchies of integrable models associated with spectral problems (4.1) can be generated from its two first members, namely the KdV hierarchy ($m = 1$) and NLS hierarchy ($m = 2$).

We shall consider besides (4.20) the *modified generating functions* which are formal series $\tilde{Y} = \alpha(\lambda)Y$ where

$$\alpha(\lambda) = 1 + \frac{\alpha_1}{\lambda} + \frac{\alpha_2}{\lambda^2} + \cdots$$

are formal series with arbitrary constant coefficients. Obviously, Eqs. (4.26) imply that

$$D_n(\tilde{Y}) = <A_n, \tilde{Y}>, \quad \tilde{Y} = 1 + \frac{\tilde{y}_1}{\lambda} + \frac{\tilde{y}_2}{\lambda^2} + \cdots, \quad A_n = (\lambda^n Y)_+ \qquad (4.37)$$

and we shall define the *N-polynomial reduction* of (4.26) by the condition that $\lambda^N \tilde{Y}$ has to be polynomial in λ for some $N > 0$. Since $\tilde{y}_k = 0, k > N$ in the case of polynomial reduction derivations $D_n, n \geq N$, can be expressed as linear combination of $D_n, n < N$, and thus (4.37) is reduced to a finite system of first-order PDEs (cf. [6]). For instance, in the case $N = 2$ Eqs. (4.37) are reduced to the system as follows:

$$\tilde{y}_{1,t} = \tilde{y}_{2,x}, \quad \tilde{y}_{2,t} = \tilde{y}_1\tilde{y}_{2,x} - \tilde{y}_2\tilde{y}_{1,x}, \quad D_t \overset{\text{def}}{=} D_1 + \alpha_1 D_x. \qquad (4.38)$$

It can be shown that in the case of N-polynomial reduction the corresponding system of equations for $\mathbf{u} = (u_1, \ldots, u_N), u_i = \tilde{y}_i$ can be integrated in quadratures. Namely, introducing the vector-field $\mathbf{u}_x = X_0(\mathbf{u})$ one gets readily N vector-fields $X_0, \ldots, X_{N-1}$ using (4.37) and we rewrite it as follows:

$$D_0(\mathbf{u}) = X_0(\mathbf{u}), \quad D_1(\mathbf{u}) = X_1(\mathbf{u}), \ldots, D_{N-1}(\mathbf{u}) = X_{N-1}(\mathbf{u}). \qquad (4.39)$$

The problem of commutativity[2]

$$[X_i, X_j] = 0, \quad \forall i, j,$$

---

[2] It is equivalent to equating mixed derivatives.

can be effectively resolved in this case (4.26) (see next section and [7]). Commutativity conditions guarantee, as is known, the existence of the function $\mathbf{u} = \mathbf{u}(x, t_1, \ldots, t_{N-1})$ of N independent variables satisfying (4.39). That allows to rewrite Eqs. (4.39) in a "differential form":

$$d\mathbf{u} = ds P(\mathbf{u}), \quad ds \overset{\text{def}}{=} (dx, dt_1, \ldots, dt_{N-1}),$$

where components of vector-fields X_k constitute $N \times N$ matrix $P = P(\mathbf{u})$. Thus, we get the famous formula from Liouville's theorem:

$$ds = d\mathbf{u} Q(\mathbf{u}), \quad Q = (q_{ij}) = P^{-1}. \tag{4.40}$$

The main point is

$$[X_i, X_j] = 0 \;\Rightarrow\; \partial_{u_j} q_{ki} = \partial_{u_i} q_{kj}, \; \forall\, i, j, k,$$

and thus all N differential forms $dx, dt_1, \ldots, dt_{N-1}$ defined by the right-hand side of (4.40) are closed ones. These differential forms generate N time-dependent Liouville's first integrals of (4.39). For instance,

$$dx = q_{11} du_1 + q_{21} du_2 + \cdots + q_{N1} du_N \;\Leftrightarrow\; x - q(\mathbf{u}) = \text{const},$$

where the first derivatives of the function q are just $q_{11}, \ldots, q_{N1}$. It is easy to see also that for this function $D_0(q) = 1, D_i(q) = 0, i > 0$.

4.1.3 Differential Constraints

Examples. Reformulating results about N-phase solutions of KDV by the classical paper [6] (see also [7, 10]) we can see that these solutions correspond to N-polynomial reduction of (4.27). In accordance with cited papers (see also [12]) we shall use besides coefficients $\tilde{y}_i, i = 1, \ldots, y_N$, zeros $\gamma = -(\gamma_1, \ldots, \gamma_N)$ of modified generating function $\tilde{Y}$:

$$\tilde{Y}(\lambda) = \alpha(\lambda) Y(\lambda) = \left(1 + \frac{\gamma_1}{\lambda}\right) \cdots \left(1 + \frac{\gamma_N}{\lambda}\right). \tag{4.41}$$

Putting $\lambda = -\gamma_j$ in (4.37) it is easy to rewrite these in the form of dynamical systems for γ. In particular, in the case (4.38) we readily get

$$\gamma_{1,t} = \gamma_2 \gamma_{1,x}, \quad \gamma_{2,t} = \gamma_1 \gamma_{2,x}, \tag{4.42}$$

where $\tilde{y}_1 = \gamma_1 + \gamma_2, \tilde{y}_2 = \gamma_1 \gamma_2$.

Remark 8. In general case (4.37), (4.41) it can be proved (cf. [7, 10]) that γ are *riemannian invariants* of systems of first-order PDEs under consideration. In turn that allows to realize the integrating scheme (4.40) more constructively.

In order to highlight interesting possibilities provided by Eqs. (4.37), (4.41) let us find differential constraints

$$y_{1,x} = h(y_1, y_2), \tag{4.43}$$

which are compatible with the system of first-order PDEs

$$y_{1,t} = y_{2,x}, \quad y_{2,t} = y_1 y_{2,x} - y_2 y_{1,x},$$

and Eqs. (4.42) related to these by the change of variables $y_1 = \gamma_1 + \gamma_2, y_2 = \gamma_1 \gamma_2$. Following the scheme of separation of variables outlined in the end of Sect. 1.2 we have to find for (4.42) two functions $f_i(\gamma_1, \gamma_2), i = 1, 2$, such that the pair of dynamical systems

$$\gamma_{1,x} = f_1, \ \gamma_{2,x} = f_2; \qquad \gamma_{1,t} = \gamma_2 f_1, \ \gamma_{2,t} = \gamma_1 f_2$$

satisfies the condition $[D_x, D_t] = 0$. The latter yields

$$(\gamma_1 - \gamma_2)\partial_{\gamma_2} f_1 = f_1, \quad (\gamma_2 - \gamma_1)\partial_{\gamma_1} f_2 = f_2$$

and therefore $f_1 = a_1/\gamma_{12}, f = a_2/\gamma_{21}$ where functions $a_1 = a_1(\gamma_1)$ and $a_2 = a_2(\gamma_2)$ are arbitrary and $\gamma_{ij} = \gamma_i - \gamma_j$. Thus

$$y_{1,x} = \gamma_{1,x} + \gamma_{2,x} = \frac{a_1(\gamma_1) - a_2(\gamma_2)}{\gamma_1 - \gamma_2}$$

and if $a_2(\gamma) = a_1(\gamma)$ it is symmetrical in γ_1, γ_2 and, hence, can be rewritten in the form (4.43). In that case Eqs. (4.42) are reduced to scalar Burgers-type equations

$$u_t = F(u, u_x, u_{xx}), \quad u \equiv y_1 = \gamma_1 + \gamma_2,$$

which possess solutions described by formulae (4.40):

$$dt = \frac{d\gamma_1}{a(\gamma_1)} + \frac{d\gamma_2}{a(\gamma_2)}, \quad -dx = \frac{\gamma_1 d\gamma_1}{a(\gamma_1)} + \frac{\gamma_2 d\gamma_2}{a(\gamma_2)}. \tag{4.44}$$

The simplest choice $a(\gamma) = \gamma^3$ leads to Burgers equation $u_t = u_{xx} + 2uu_x$. The next choice $a(\gamma) = \gamma^4$ gives rise to the following equation:

$$u_t = \left(\frac{u_x + \beta}{u} + \frac{1}{2}u^3 \right)_x,$$

with a rich family of exact solutions defined by (4.44). Although essentially the same formulae (4.44) give two-phase solutions of solitonic equations, almost all built up Burgers-type equations do not correspond to any spectral problem (4.1) and are non-integrable.

The above enlargement of the class of nonlinear evolution equations with standard form of two-phase solutions appears very intriguing. The next example

$$u_t = \left(\frac{u_{xx}}{u} + \frac{2}{3}u^2\right)_x, \quad u \equiv y_1 = \gamma_1 + \gamma_2,$$

corresponds to the constraint of the form $y_{1,xx} = \alpha y_1 y_2 + \beta y_1^3$ compatible with (4.42). It suggests future developments and generalizations of the above scheme of integration of "non-integrable" equations. First of all, it concerns three-phase solutions corresponding to differential constraints compatible with the equations

$$\gamma_{1,t} = (\gamma_2 + \gamma_3)\gamma_{1,x}, \quad \gamma_{2,t} = (\gamma_3 + \gamma_1)\gamma_{2,x}, \quad \gamma_{3,t} = (\gamma_1 + \gamma_2)\gamma_{3,x}.$$

Most important problems, it appears, are related with generalizations of the basic system (4.26) into multi-dimensional case.

In conclusion, we describe tersely KDV and NLS hierarchies corresponding to the basic spectral problems (4.1) with $m = 1, 2$. In the case of linear Schrödinger equation (4.6) we find, using (4.5), that KdV equation

$$u_t = 6uu_x - u_{xxx} \tag{4.45}$$

corresponds to the Lax pair

$$\psi_t = (4\lambda + 2u)\psi_x - u_x\psi, \quad \psi_{xx} = (u - \lambda)\psi.$$

On the other hand, a straightforward substitution of u from the second equation into the first one yields

$$\phi_t + \phi_{xxx} - 6\lambda\phi_x = 2\phi_x^3, \quad \phi = \log\psi. \tag{4.46}$$

Solutions of this modified KDV equation generate solutions of (4.45) by the Miura transformation:

$$u = \phi_{xx} + \phi_x^2 + \lambda.$$

I think there is interesting possibility to define λ properties of ψ directly from (4.46) using general solvability theory for the specific boundary value problem under consideration.

Coming back to our universal model (4.26) we find using (4.21) the differential constraint (cf. (4.43)) related to KDV hierarchy

$$y_{1,xx} = 6y_1^2 - 4y_2, \quad u = 2y_1. \tag{4.47}$$

Obviously, the equation $y_{1,t} = y_{2,x}$ coincides now with (4.45) up to scaling. Compatibility of this constraint with N-polynomial reductions of (4.27) follows by Lemma 4 and the general formula (4.40) can be used for defining N-phase solutions of KDV (cf. [6]).

In case (4.6) the third-order equation (4.23) for the generation function implies that

$$4y_{j+1,x} = (-D^3 + 4uD + 2u_x)y_j, \quad j = 1, 2, \dots.$$

These additional constraints, it appears, have to be corollaries of the first one (4.47) and Eqs. (4.27). On the other hand, in the case $m = 2$ in virtue of (4.22), the constraint looks as follows:

$$y_{1,xx} = 4\left(y_3 - 3y_1y_2 + 2y_1^3\right). \tag{4.48}$$

Thus, both basic hierarchies ($m = 1$, $m = 2$) correspond to particular choices of differential constraints of the form

$$y_{1,xx} = h(y_1, y_2, y_3)$$

and that gives rise to the question: do there exist, different from (4.43), (4.48), constraints of such kind? In any case the problem of constraints compatible with N-polynomial reductions of (4.27) seems to be a well-posed mathematical problem.

Few interesting solitonic models are closely related with the constraint (4.48). We start with variational problem

$$\delta \int dt\, dx\, \Phi(q_t, q_{xx}, q_x, q) = 0 \tag{4.49}$$

for the potential $q_x = 2y_1$, $q_t = 2y_2$. It is easy to see that this variational problem with

$$\Phi = q_t^2 - q_t q_x^2 + \frac{1}{4}\left(q_{xx}^2 + q_x^4\right)$$

represents the first two equations (4.27) closed by constraint (4.48). Rewriting (4.49) in the Hamiltonian form

$$q_t = \frac{\delta H[p, q]}{\delta p}, \quad p_t = -\frac{\delta H[p, q]}{\delta q} \tag{4.50}$$

we can use common formulae

$$H = q_t \Phi_{q_t} - \Phi, \quad \Phi = pH_p - H, \quad (p = \Phi_{q_t}),$$

which link Hamiltonian density H with density Φ of the Lagrangian. Noticing that the rescaling of p and substitution

$$p \rightarrow p + \alpha q_{xx}$$

keep the original q-equation invariant we arrive at

$$H = p_x q_x + p^2 + p q_x^2.$$

At last, differentiating with respect to x the first of corresponding equations (4.50) we can rewrite it in a more symmetrical form

$$z_t = \left(2p + z^2 - z_x\right)_x, \quad p_t = (2zp + p_x)_x, \quad z \equiv q_x = 2y_1. \tag{4.51}$$

The above-described change of variables $(y_1, y_2) \to (z, p)$ corresponds to transformation of the spectral problem (4.1) with quadratic in λ potential into the following one:

$$\psi_{xx} + (z(x) - 2\lambda)\psi_x + p(x)\psi = 0. \tag{4.52}$$

Indeed, the cross-differentiation with

$$\psi_t = (z(x) + 2\lambda)\psi_x - p(x)\psi \tag{4.53}$$

yields (4.51) and the substitution

$$\psi \to e^{\lambda x}\sqrt{v}\,\psi, \quad (\log v)_x + z = 0$$

transforms (4.52) into (4.1) with

$$U(x, \lambda) = \lambda^2 - \lambda z(x) + p(x), \quad \rho = \frac{1}{4}z^2 + \frac{1}{2}z_x - p.$$

In a certain sense (4.51) plays the role of Korteveg–de Vries equation (4.45) and in order to get the modified version of (4.51) similar to (4.46) one can introduce "projective" coordinates as follows:

$$a = i\frac{\varphi}{\psi}, \quad b = i\frac{\varphi_x}{\psi_x}.$$

Here φ and ψ are two solutions of (4.52), (4.53) with non-vanishing wronskian $\langle \varphi, \psi \rangle = \varphi\psi_x - \psi\varphi_x$. This yields

$$a_t - 4\lambda a_x = a_{xx} - 2\frac{a_x^2}{a - b}, \quad b_t - 4\lambda b_x = -b_{xx} - 2\frac{b_x^2}{a - b}. \tag{4.54}$$

Using the stereographic projection

$$\mathbf{S} = \mathbf{S}(a, b) = \frac{1}{a - b}(1 - ab, i + iab, a + b), \tag{4.55}$$

one can now prove that

$$\mathbf{S}_t + 4\lambda \mathbf{S}_x = i[\mathbf{S}, \mathbf{S}_{xx}], \quad S_1^2 + S_2^2 + S_3^2 = 1. \tag{4.56}$$

This is a complexified form of the well-known isotropic Landau–Lifshitz model.

Last, not the least, the nonlinear Schrödinger equation (NLS)

$$iu_t - u_{xx} = u^2 v, \quad iv_t + v_{xx} = v^2 u \tag{4.57}$$

is related to the Hamiltonian dynamical system (4.50), (4.51) by a triangular point transformation $v = Q(q)$, $u = P(q, p)$ and the substitution $\partial_t \to i\partial_t$. This corresponds to a transition from (4.52) to the Zakharov–Shabat spectral problem

$$\psi_x^1 - \lambda\psi^1 = u\psi^2, \quad \psi_x^2 + \lambda\psi^2 = v\psi^1 \tag{4.58}$$

with

$$z = q_x = -(\log v)_x, \qquad p = -uv.$$

Namely, rewriting Eqs. (4.58) in the following terms

$$\varphi = e^{-\lambda x}\psi^1, \quad \psi = e^{\lambda x}\psi^2$$

one gets

$$\varphi_x = u e^{-2\lambda x}\psi, \quad \psi_x = v e^{2\lambda x}\varphi$$

and thus we find (cf. (4.52))

$$\varphi_{xx} = \left(\frac{u_x}{u} - 2\lambda\right)\varphi_x + uv\varphi, \qquad \psi_{xx} = \left(\frac{v_x}{v} + 2\lambda\right)\psi_x + uv\psi. \tag{4.59}$$

Summing up, we can see that, formulated at the very beginning, ansatz (4.8) describes important solitonic models (4.56), (4.57) as well as the list of Painlevé ODEs (4.12), (4.13), (4.14), (4.15). In order to highlight the more deep interconnection we can rewrite equations for y_1, y_2 in (4.29) in terms of z, p variables used in (4.51). That yields

$$z_\tau + (xz)_{xx} = \left[x\left(2p + z^2\right)\right]_x + 2p, \quad p_\tau = (xp)_{xx} + [2xzp]_x + p_x + 2zp.$$

On the one hand it represents the master-symmetry (4.29) in terms of coefficients of spectral problem (4.52) and on the other hand it corresponds to non-isospectral symmetries with $\lambda_\tau = 4\lambda^2$ in Sect. 1.1. Furthermore, the cross-differentiation with (4.51) and the formula $[D_\tau, D_t] \overset{\text{def}}{=} D_{t_2}$ (cf. Sect. 1.2) results in

$$D_{t_2}(z) = \left(6pz + z^3 + z_{xx} - 3zz_x\right)_x, \quad D_{t_2}(p) = \left(3zp_x + 3z^2 p + p_{xx} + 3p^2\right)_x. \tag{4.60}$$

In terms of z, p it represents the next (after (4.51)) member of NLS hierarchy (Theorem 5) and corresponds to D_2 in the hierarchy (4.26) (Theorem 6). Finally, we notice that reductions

$$p = 0, \qquad z = 0$$

transform the system of equations (4.60) into Burgers and KdV equations, respectively. Obviously, the spectral problem (4.1) with $U = \lambda^2 + u_2$ (or (4.52) with $z = 0$) is equivalent to (4.6) and, thus, KDV-hierarchy could be considered as the reduction of NLS-hierarchy.

4.2 Discrete Symmetries

4.2.1 Matrix Representations

Considering discrete symmetries of spectral problems (4.1) we choose as starting point the formula analogous to (4.2):

$$\hat{\psi} = A(x,\lambda)\psi_x + B(x,\lambda)\psi, \tag{4.61}$$

where

$$\psi_{xx} = U(x,\lambda)\psi, \quad \hat{\psi}_{xx} = \hat{U}(x,\lambda)\hat{\psi}.$$

Similar to Sect. 1.1 we get

$$\hat{\psi}_x = (A_x + B)\psi_x + (B_x + UA)\psi$$

and further differentiation yields

$$\hat{U} = U + \frac{A_{xx}}{A} + 2\frac{B_x}{A}, \quad B_{xx} + 2A_xU + AU_x = B(\hat{U} - U). \tag{4.62}$$

Definition 9. We shall call by Darboux transformation (or shortly DT) the linear mapping $\psi \mapsto \hat{\psi}$ defined by (4.61), (4.62) and call the potential U DT-invariant if there exists DT (4.61) with A, B polynomial in λ such that $\hat{U} = U$.

Theorem 10. DT-invariant potentials are stationary points for corresponding iso-spectral Lie symmetries.

Proof. Let U be DT-invariant with respect to (4.61) and operator D_t defined by (4.19) *with same coefficient A*. Since $\hat{U} = U$ in (4.62) we have

$$2B_x = -A_{xx}, \quad B_{xx} + 2A_xU + AU_x = 0 \Rightarrow -A_{xxx} + 4UA_x + 2U_xA = 0.$$

Therefore D_t is the isospectral symmetry and $D_t(U) = 0$. ∎

Let us now consider Eqs. (4.62) with $\hat{U} \neq U$. In the case

$$U = u - \lambda, \quad \hat{U} = \hat{u} - \lambda, \quad \hat{\psi} = a(x)\psi_x + b(x)\psi,$$

these equations imply that $a_x = 0$ and putting $a = 1$ one obtains

$$b_x - b^2 + u = \mu, \quad \hat{u} - u = 2b_x, \tag{4.63}$$

where μ is the constant of integration. Thus, in order to define DT one has to solve the Riccati equation for Schrödinger spectral problem (4.6) for some value μ of the spectral parameter λ. That is "bad" news in comparison with the analogous example of Lie-type symmetry considered in Sect. 1.1. The "good" news is that starting with trivial potential $u = 0$ we can build up non-trivial ones solving this Riccati equation (4.63).

The definition implies that DT superposition is Darboux transformation as well. In order to define this operation explicitly we can rewrite the formula (4.61) in matrix form as follows:

$$\begin{bmatrix} \hat{\psi} \\ \hat{\psi}_x \end{bmatrix} = \begin{bmatrix} B, & A \\ B_x + UA, & A_x + B \end{bmatrix} \begin{bmatrix} \psi \\ \psi_x \end{bmatrix}. \tag{4.64}$$

Obviously, the DT superposition is realized now as matrices multiplication. Moreover, since wronskians of solutions of spectral problem (4.1) are x-independent the same is true for the determinant of the matrix in (4.64):

$$\chi(\lambda) \equiv \chi(\lambda; A, B) \stackrel{\text{def}}{=} A^2 U + AB_x - BA_x - B^2. \tag{4.65}$$

The Darboux transformation corresponding to the matrix

$$\begin{bmatrix} A_x + B, & -A \\ -B_x - UA, & B \end{bmatrix}.$$

can be seen as inverse transformation $\hat{U} \to U$ (c.f. (4.61)).

For example, in the case (4.63) we find $\chi(\lambda) = \mu - \lambda$ and

$$\hat{\psi} = \psi_x + b\psi \iff (\lambda - \mu)\psi = b\hat{\psi} - \hat{\psi}_x.$$

The theory of Darboux transformations will be based upon an idea of factorization of the matrix in (4.64) in the product of simplest ones corresponding to DT like (4.63). We shall now find these elementary DT for our basic spectral problem (4.1) with $m = 2$. The "gauged" form of it indicated in Sect. 1.3 by (4.52) appears most convenient and, rectifying notations, we rewrite it as follows:

$$\psi_{xx} + (z(x) - \lambda)\psi_x + p(x)\psi = 0. \tag{4.66}$$

Lemma 11. *For Eq. (4.66) there exist two elementary Darboux transformations*

$$\psi \mapsto f\psi_x + g\psi$$

such that f and g do not depend on λ:

$$T: \quad \psi \mapsto \tilde{\psi} = \psi_x/p, \quad \tilde{z} - z = (\log p)_x, \quad \tilde{p} - p = \tilde{z}_x, \tag{4.67}$$

$$\hat{T}: \quad \psi \mapsto \hat{\psi} = \psi + \psi_x/a, \quad \hat{z} - z = (\log a)_x, \quad \hat{p} - p = b_x, \tag{4.68}$$

where $b = p/a$ and a is a solution of the Riccati equation

$$a_x = a^2 + (\mu - z)a + p, \quad a + (\log \varphi(\mu)))_x = 0. \tag{4.69}$$

Here $\varphi(\mu)$ denotes a fixed solution of (4.52) at $\lambda = \mu$.

Proof. We have

$$\hat{\psi} = f\psi_x + g\psi, \quad \hat{\psi}_x = F\psi_x + G\psi, \tag{4.70}$$

where

$$F = f_x + g + f(\lambda - z), \quad G = g_x - pf.$$

Let us consider the wronskian of two arbitrary linear independent solutions φ and ψ of (4.52)

$$w = \psi_x \varphi - \psi\varphi_x, \quad w_x = (\lambda - z)w, \quad \hat{w} = hw,$$

where $\hat{w}$ denotes the wronskian of $\hat{\psi}$, $\hat{\varphi}$. Thus

$$h = Fg - Gf = A\lambda + B, \quad A = fg, \quad B = f_x g - g_x f + g^2 + pf^2 - zfg$$

and

$$\hat{z} - z + (\log h)_x = 0 \Rightarrow fg = 0, \quad \text{or} \quad B = \mu A, \quad \mu = \text{const}.$$

In the case $f = 0$ it is easy to see that $g = \text{const}$ and the case $g = 0$ yields (4.67) with $\tilde{w} = w/p$.

In the case $fg \neq 0$ the condition $B = \mu A$ yields the Riccati equation (4.69) for $a = g/f$ and one finds further on that $(af)_x = 0$ by collecting the terms with $\lambda \psi$ in the transformed equation for $\hat{\psi}$. ∎

The given solution a of Riccati equation (4.69) allows to build up in addition to $\hat{T}$ another Darboux transformation

$$S: \quad \psi \mapsto \psi^\dagger = (\hat{\lambda} - a)\psi - \psi_x, \quad \hat{\lambda} = \lambda - \mu. \tag{4.71}$$

In this case

$$p^\dagger = p - a_x, \quad z^\dagger = z - [\log(a - z - \mu)]_x.$$

Lemma 12. *The DT relationship $\hat{T} = TS = ST$ is equivalent to the following conditions:*

$$z = a + T^{-1}b + \mu, \quad \hat{T}z = a + b + \mu; \quad p = ab, \quad \hat{T}p = bTa. \tag{4.72}$$

Proof. Let Ψ denote the vector with components ψ, ψ_x and $\hat{T}\Psi = A\Psi$, $S\Psi = B\Psi$, $T\Psi = C\Psi$ where A, B and C are corresponding matrices:

$$A = \frac{1}{a}\begin{pmatrix} a, 1 \\ -ab, \hat{\lambda} - b \end{pmatrix}, \quad B = \begin{pmatrix} \hat{\lambda} - a, -1 \\ aT^{-1}(b), T^{-1}(b) \end{pmatrix}, \quad C = \frac{1}{p}\begin{pmatrix} 0, 1 \\ -p, \lambda - T(z) \end{pmatrix},$$

where $\hat{\lambda} = \lambda - \mu$. Since $ST\Psi = T(B)C\Psi$ and $TS\Psi = S(C)B\Psi$ we get (4.72) as a result of corresponding matrices multiplication. ∎

We now consider a few primary applications of elementary DTs concentrating on the simplest one denoted by T. Notice that the picture here is richer in comparison with the case of Schrödinger spectral problem and this transformation T is well defined in terms of the coefficients z, p of (4.52) (i.e. does not invoke the solution of Riccati equation (4.63)). Considering iterations T^j, $j \in \mathbb{Z}$, we discuss the case $T^N = E$ related with Theorem 10.

First, we notice that in order to define T^{-1} it is sufficient, like in the case (4.63), to differentiate (4.67). Thus, we obtain[3]

$$\tilde{\psi}_x + (\tilde{z} - \lambda)\tilde{\psi} + \psi = 0, \quad \psi_x = p\tilde{\psi}.$$

---

[3] This coupled system of equations is gauge equivalent to the spectral problem (4.58).

It is easy to verify now that

$$T^{-1}T = TT^{-1} = E \iff (pT + T^{-1} + z - \lambda)\psi = 0$$

and hence we can rewrite the above equations for T^{-1} and T as follows:

$$pT\psi + (z - \lambda)\psi + T^{-1}\psi = 0, \quad \psi_x = pT(\psi). \tag{4.73}$$

Second, revising our stand we will consider the pair of equations above as the basis of theory. From this point of view, the first of equations (4.73) is now the basic spectral problem and the second one is the Lie-type symmetry of it. Compatibility condition (cf. Sect. 1.1) of this Lax pair yields the equations from Lemma 4:

$$z_x = p - T^{-1}p, \quad (\log p)_x = Tz - z \tag{4.74}$$

and original spectral problem (4.66) is the corollary of (4.73) as well (cf. Footnote 3). In order to support this new point of view let us find the modified version of Eqs. (4.74) similar to (4.46). Denoting $\phi = -\log \psi$ we find by (4.73)

$$p = -\phi_x e^{T\phi - \phi}, \quad z - \lambda = \phi_x - e^{\phi - T^{-1}\phi}. \tag{4.75}$$

Direct substitution of these formulae into (4.66) results in the modified version of (4.74):

$$\phi_{xx} + \phi_x(e^{T\phi - \phi} - e^{\phi - T^{-1}\phi}) = 0. \tag{4.76}$$

Considering the periodic closure $T^N = E$ of the system of equations (4.74) we shall use lower indices $a_j \overset{\text{def}}{=} T^j(a)$, $j \in \mathbb{Z}$, for the discrete variable related to powers of T. Thus, we have $2N$-dimensional dynamical system

$$z_{j,x} = p_j - p_{j-1}, \quad p_{j,x} = p_j(z_{j+1} - z_j), \quad j \in \mathbb{Z}_N, \tag{4.77}$$

which describes DT-invariant potentials under consideration. This system possesses two obvious first integrals

$$z_1 + z_2 + \cdots + z_N = \text{const}, \quad p_1 p_2 \cdots p_N = \text{const}$$

and in order to get more we can use the Lax pair (4.73). Rewriting it in the matrix form

$$p_j \psi_{j+1} = B_j \psi_j, \quad \psi_{j,x} = U_j \psi_j, \tag{4.78}$$

where

$$\psi_j = \begin{pmatrix} \psi_j \\ \psi_{j-1} \end{pmatrix}, \quad B_j = \begin{pmatrix} \lambda - z_j, & -1 \\ p_j, & 0 \end{pmatrix}, \quad U_j = \begin{pmatrix} \lambda - z_j, & -1 \\ p_{j-1}, & 0 \end{pmatrix},$$

we get

Proposition 13. *The generating function* $P(\lambda) = P_N(\lambda)$ *for first integrals of (4.77) is an Nth degree polynomial in* λ *as follows:*

$$P_N(\lambda) = \text{trace}\, B_N B_{N-1} \cdots B_1 = \prod_{i=1}^{N} (I - p_i \partial_i \partial_{i+1}) \left(\prod_{k=1}^{N} (\lambda - z_k) \right), \qquad (4.79)$$

where $\partial_i = \partial / \partial z_i$.

Sketch of Proof. The Lax pair (4.78) implies

$$B_{j,x} = V_{j+1} B_j - B_j V_j, \quad V_j \overset{\text{def}}{=} U_j + z_j I,$$

where I denotes identity matrix. Thus

$$(B_2 B_1)_x = V_3 B_2 B_1 - B_2 B_1 V_1 \quad \cdots \quad G_{N,x} = V_{N+1} G_N - G_N V_1,$$

where $G_N = B_N \cdots B_1$. Since $V_{N+1} = V_1$ we have

$$\frac{d}{dx} P_N(\lambda) = \text{trace}(V_1 G_N - G_N V_1) = 0$$

and thus it is proved that the coefficients of $P_N(\lambda)$ are first integrals of (4.77) indeed.

In order to obtain the exact expression of the generation function in terms of dynamical variables one can use induction in N. More exactly, one has to prove by induction in N the more general formula for all four components of the matrix $G_N = B_N \cdots B_1$:

$$G_N = \begin{pmatrix} 1 + p_N \partial_N \partial_1, & \partial_1 \\ -p_N \partial_N, & -p_N \partial_N \partial_1 \end{pmatrix} (P_N).$$

Here it is assumed that P_N is defined by the right-hand side of (4.79). ∎

Generally, Eqs. (4.77) correspond by Theorem 10 to polynomial in λ solutions (4.18), and integrating scheme (4.40) involves hyperelliptic integrals. But now the dynamical system (4.77) allows to formulate explicitly the constraint $p(kN) = 0$, $k \in \mathbb{Z}$, which yields solutions in terms of rational functions in exponents. We shall call this type of reductions *solitonic* ones.

Example 14. In the case $N = 2$ the formula (4.79) yields

$$P(\lambda) = (\lambda - z_1)(\lambda - z_2) - p_1 - p_2 = (\lambda - \beta)^2 - \gamma^2.$$

The solitonic reduction $p(2) = 0$ gives $y_x + y^2 = \gamma^2$ where $z_{1,2} = \beta \pm y$, $p_1 = y_x$ and

$$y(x) = \gamma \tanh \gamma(x - x_0).$$

On the other hand the quasi-periodic closure of (4.74) with

$$T^2(z) = z - \varepsilon, \quad T^2(p) = p$$

leads to dynamical system for z_1, p_1, z_2, p_2 like (4.77). It can be reduced to ODE for $y = p_1$ as follows:

$$yy_{xx} - y_x^2 + 2y^3 = 2yR(x), \quad R' + \varepsilon R = 0, \quad p_1 p_2 = R(x).$$

This equation is equivalent to degenerated Painlevé equation (4.14) from Sect. 1.1 with

$$\alpha = -2\varepsilon^{-2}, \quad \beta = 2\varepsilon^{-2}, \quad \gamma = \delta = 0.$$

In the case of Schrödinger spectral problem and elementary Darboux transformation T defined by (4.63) one obtains similar to (4.73)

$$T^2 \psi + (fT + Tf)\psi = (\mu - \lambda)\psi, \quad \psi_x = (T + f)\psi, \qquad (4.80)$$

where $f = -b$, $f_x + f^2 + \mu = u$. Although in this case DT is implicit (one cannot resolve the Riccati equation for f) it gives rise to the discrete analog of Schrödinger spectral problem (4.6), and (4.80) yields *dressing chain*

$$f_{j,x} + f_{j+1,x} = f_j^2 - f_{j+1}^2 + \alpha_j, \quad j \in \mathbb{Z}. \qquad (4.81)$$

Here

$$\alpha_j \stackrel{\text{def}}{=} \mu_{j+1} - \mu_j, \qquad f_{j,x} + f_j^2 + \mu_j = u_j.$$

This chain contains in itself all needed information (cf. Sect. 2.2) about Darboux transformations of Schrödinger spectral problem (4.6). Since $f_j(x) = (\log \psi_j)_x$ where $\psi_{j,xx} = (u_j - \mu_j)\psi_j$ the dressing chain can be considered as modified version, like (4.76), of the equations as follows:

$$(q_{j+1} + q_j)_x + (q_{j+1} - q_j)^2 = \mu_j, \quad u_j = q_{j,x}, \quad f_j = q_{j+1} - q_j. \qquad (4.82)$$

In the case of N-periodic closure related with DT-invariant potentials of Schrödinger operator the dressing chain (4.81) provides the system of differential equations for $f_1, \ldots, f_N$ with the left-hand matrix $E + T$ where T

$$T = \begin{pmatrix} 0 & 1 & & & \\ & \ddots & \ddots & & \\ & & & \ddots & 1 \\ 1 & & & & 0 \end{pmatrix}.$$

For odd N this system is well defined since the matrix $E + T$ is non-singular:

$$T^N = E \implies (E + T)(E - T + T^2 - \cdots - T^{N-1}) = 2E.$$

In order to build up first integrals of dressing chain we have to rewrite Lax pair (4.80) in the matrix form similar to (4.78):

$$\psi_{j,x} + U_j \psi_j = 0, \quad \psi_{j+1} = B_j \psi_j, \quad B_j = \begin{pmatrix} g_j & \lambda_j \\ 1 & 0 \end{pmatrix}, \quad U_j = \begin{pmatrix} f_j, & \lambda_j \\ 1, & -f_j \end{pmatrix}. \quad (4.83)$$

Here

$$\psi_j = (\psi_{j+1}, \psi_j)^\tau, \quad g_j = f_j + f_{j+1}, \quad \lambda_j \overset{\text{def}}{=} \mu_j - \lambda.$$

Slight difference in comparison with (4.78) is not important for proving this and like in the above-considered case (4.77) we have

Proposition 15. *The formula*

$$P_N(\lambda) = \text{trace } B_N B_{N-1} \cdots B_1 = \prod_{i=1}^{N}(I + \lambda_i \partial_{i-1} \partial_i)\left(\prod_{k=1}^{N} g_k\right), \tag{4.84}$$

where $\partial_i = \partial/\partial g_i$ yields the generating function $P_N(\lambda)$ for first integrals of $\mathbb{Z}_N$ closure of (4.81) with $f_{i+N} = f_i$ if

$$\gamma = \alpha_1 + \cdots + \alpha_N = 0. \tag{4.85}$$

The general case with $\gamma \neq 0$ is considered in [5]. It has been proved that the corresponding spectra of Schrödinger operators are constituted by collection of arithmetic progressions.

Example 16. The periodic $\mathbb{Z}_3$ closure equations (4.81) reduces these to dynamical system for three functions $g_j = f_j + f_{j+1}$, $j \in \mathbb{Z}_3$:

$$g_{1,x} = g_1(g_3 - g_2) + \alpha_1, \quad g_{2,x} = g_2(g_1 - g_3) + \alpha_2, \quad g_{3,x} = g_3(g_2 - g_1) + \alpha_3. \tag{4.86}$$

The obvious first integral

$$g_1 + g_2 + g_3 = a(x), \quad a' = \gamma = \alpha_1 + \alpha_2 + \alpha_3$$

reduces the order of the system and any of functions g_j, $j = 1,2,3$, satisfies the second-order equation as follows:

$$2g_j g_{j,xx} - g_{j,x}^2 = g_j^2 \left(3g_j^2 - 4ag_j + a^2 + 2\alpha_{j+2} + 2\alpha_{j+1}\right) - \alpha_j^2. \tag{4.87}$$

In the general case, that corresponds to Painlevé ODE (4.13) with full set of parameters.

By $g_3 = 0$ Eqs. (4.86) are reduced to Riccati equation

$$g_x + g^2 = \gamma x g + \alpha,$$

where $g \equiv g_2$, $\alpha \equiv \alpha_2$ and $g_1 = \gamma x - g$. By $g_2 = g_3 = 0$ we have $g_{1,x} = \gamma$ what corresponds to quantum harmonic oscillator. At last, in the case $\gamma = 0$ this system of equations (4.86) has additional first integral (4.84)

$$g_1 g_2 g_3 + g_1 \mu_3 + g_2 \mu_1 + g_3 \mu_2$$

and the solution can be expressed in terms of elliptic functions.

4.2.2 Wronskian Determinants and Dressing Chains

In the case of Darboux transformations as well as in the case of Lie-type symmetries (Sect. 1.2) there is a universal model. It is based upon identity

$$\langle \varphi_1, \dots, \varphi_k \rangle = \varphi_1 \langle \hat{\varphi}_2, \dots, \hat{\varphi}_m \rangle, \quad \hat{\varphi}_j = (D - f_1)\varphi_j, \quad f_1 = D\log\varphi_1, \qquad (4.88)$$

where φ_j, $j = 1, \dots, k$ are arbitrary smooth functions and

$$\langle \varphi_1, \dots, \varphi_l \rangle \overset{\text{def}}{=} \det\left(\partial_x^{k-1}(\varphi_j) \right), \quad j, k = 1, \dots, l.$$

We are going to prove that any DT is a superposition of transformations $\varphi \mapsto \hat{\varphi}$ defined by this identity (4.88).

Instead of spectral problems (4.1) with λ-dependent potentials we consider now spectral problems $L\psi = \lambda\psi$ where L is higher order scalar differential operator

$$L = D^m + u_1 D^{m-1} + u_2 D^{m-2} + \dots + u_{m-1}D + u_m, \quad D \equiv d/dx. \qquad (4.89)$$

Generalizing (4.63) we prove the next lemma.

Lemma 17. *For differential operators (4.89) with $u_1 = 0$ the elementary DT $\psi \mapsto \hat{\psi}$ is defined as follows:*

$$\hat{\psi} = a(x)\psi_x + b(x)\psi = \psi_x - f\psi = (D - f)\psi,$$

where $f = D\log\varphi$, $L\varphi = \mu\varphi$.

Proof. Let $M = aD + b$ then, since $\hat{L}\hat{\psi} = \lambda\hat{\psi}$, we have

$$(\hat{L}M - ML)\psi = 0$$

for all ψ such that $L\psi = \lambda\psi$. Since the differential operator $\hat{L}M - ML$ has finite-dimensional null space it is possible only if

$$ML = \hat{L}M. \qquad (4.90)$$

This operator equation implies that

$$M\psi = 0, \quad \tilde{\psi} = L\psi \Rightarrow M\tilde{\psi} = 0 \Rightarrow \tilde{\psi} = \mu\psi$$

since a null space $\mathrm{Ker}\,M$ is one-dimensional. Thus, $\mathrm{Ker}\,M$ is generated by an eigenfunction of L which is denoted by φ and, thus,

$$M = a(D - f), \quad f = D\log\varphi, \quad L\varphi = \mu\varphi.$$

On the other hand, equating leading coefficients in (4.90) we find

$$a(\hat{u}_1 - u_1) + ma_x = 0,$$

where m is the order of operators L, $\hat{L}$ and $\hat{u}_1$ denotes the coefficient by D^{m-1} in $\hat{L}$. By conditions of the lemma $u_1 = \hat{u}_1 = 0$ and, therefore, $a_x = 0$. ∎

The "universal" model for Darboux transformations is based upon the well-known formula

$$\frac{\langle \varphi_1, \dots, \varphi_n, \psi \rangle}{\langle \varphi_1, \dots, \varphi_n \rangle} = M\psi, \tag{4.91}$$

which is valid for any differential operator M of order n with unitary leading coefficient. The functions $\varphi_1, \dots, \varphi_n$ here build up the basis in $\mathrm{Ker}\, M$. The identity (4.88) and formula (4.91) imply the factorization formula as follows:

Proposition 18. *Let M be a nth order operator with unitary leading coefficient and functions $\varphi_1, \dots, \varphi_n$ constitute basis in $\mathrm{Ker}\, M$. Then*

$$M = (D - f_n) \cdots (D - f_1), \quad f_j = D \log \frac{\langle \varphi_1, \dots, \varphi_j \rangle}{\langle \varphi_1, \dots, \varphi_{j-1} \rangle}. \tag{4.92}$$

Generally speaking, an application of Proposition 18 to the operator L and the transpositions of factors in the corresponding L formula (4.92) describe Darboux transformations of this operator. Namely, by conditions of Lemma 17 the direct application of formulae (4.91), (4.88) to the operator $L - \mu$ yields $L - \mu = \tilde{L}(D - f)$ and thus we find

$$\hat{L} - \mu = (D - f)\tilde{L} = (D - f)(L - \mu)(D - f)^{-1}, \quad f = D \log \varphi, \quad L\varphi = \mu\varphi. \tag{4.93}$$

Considering iteration of this transformation $L \mapsto \hat{L}$ and introducing lower indices for numbering of iterations we can rewrite the above formula as follows:

$$(D - f_j)L_j = L_{j+1}(D - f_j). \tag{4.94}$$

In essence, these two formulae are equivalent and the latter one expresses in a very terse form the "polynomial" approach to Darboux transformations. In particular, in the case of Schrödinger operator $L = u - D^2$ the formula (4.93) yields

$$\hat{L} - \mu = (f - D)(f + D), \quad L - \mu = (f + D)(f - D)$$

and one can verify that Eqs. (4.94) are equivalent to dressing chain (4.81) considered in Sect. 2.1.

We have to now notice that in the case of Schrödinger spectral problem (4.6)

$$\hat{\psi} = A\psi_x + B\psi \iff \hat{\psi} = M(\psi),$$

where the differential operator M is uniquely defined by the polynomials A, B in λ and the formulae

$$\lambda^k \psi = (u - D^2)^k \psi, \quad \lambda^k \psi_x = D(u - D^2)^k \psi.$$

That connects the next definition with Definition 9 of Darboux transformation.

Definition 19. The differential operator M is called Darboux transformation operator (or transformation operator $L \mapsto \hat{L}$) if $ML = \hat{L}M$.

It follows from the proof of Lemma 17 that transformation operators $M : L \mapsto \hat{L}$ of the first order are defined by formula (4.93). The condition of Lemma 17 is that the coefficients by D^{m-1} in $L, \hat{L}^4$ should vanish. This condition is not restrictive since a simple gauge transformation $L \mapsto a^{-1}La$ allows to kill the coefficient u_1 in (4.89).

Theorem 20. *Transformation operator $M : L \mapsto \hat{L}$ of order $n > 1$ can be represented as superposition of n elementary one*

$$M = (D - f_n) \cdots (D - f_2)(D - f_1).$$

It is assumed that the coefficients by D^{m-1} in $L, \hat{L}$ (see Footnote 4) vanish.

Proof. Since $ML = \hat{L}M$ the operator L maps $V = \mathrm{Ker}\,M$ into itself and defines n-dimensional operator $\tilde{L} = L|_V$. There is, due to a linear algebra, eigenfunction φ_1 such that $\tilde{L}\varphi_1 = \mu_1\varphi_1$. Let $f_1 = D\log\varphi_1$. We have

$$M\varphi_1 = 0 \;\Rightarrow\; M = M_1(D - f_1),$$

where M_1 is a differential operator of order $n - 1$. That yields (cf. (4.93))

$$(D - f_1)L = L_1(D - f_1), \quad M_1L_1 = M_1\hat{L}.$$

Therefore, $M_1 : L_1 \mapsto \hat{L}$ is the Darboux transformation operator of the order $n - 1$ and one can apply an induction procedure. ∎

In order to build up a transformation operator M of nth order for given operator L it is sufficient (see the proof of Theorem 20) to choose the basis $\varphi_1, \ldots, \varphi_n$ in n-dimensional vector space V invariant under action of L and apply the formula (4.91). In a generic case[5] it is sufficient to fix n linearly independent solutions $\varphi_j, j = 1, \ldots, n$, of the spectral problem under consideration:

$$L\varphi_j = \mu_j\varphi_j, \quad j = 1, \ldots n.$$

Example 21. In the case $L = -D^2, n = 2$ denoting

$$a(y) = e^y + e^{-y}, \quad b(y) = e^y - e^{-y}, \quad y = k(x - \tau)$$

we obtain

$$\langle \varphi_1, \varphi_2 \rangle = (k_2 - k_1)\,a(y_1 + y_2) + (k_1 + k_2)\,b(y_1 - y_2), \tag{4.95}$$

where $0 < k_1 < k_2$,

---

[4] m denotes the order of operators $L, \hat{L}$.

[5] One can use "eigenfunctions" in the basis of the invariant space V.

$$\varphi_1 = a(y_1), \quad \varphi_2 = b(y_2); \quad y_j = k_j(x - \tau_j), \quad \varphi_{j,xx} + \mu_j \varphi_j = 0, \quad \mu_j = -k_j^2.$$

Thus, using the formulae for f_j from (4.92), we obtain for the potential $\hat{u}$ of transformed operator $\hat{L} = -D^2 + \hat{u}$ the explicit formula as follows:

$$\hat{u} = -2(f_1 + f_2)_x = -2D^2 \log(\langle \varphi_1, \varphi_2 \rangle).$$

This formula describes, as one can verify, smooth reflectionless potentials $\hat{u}(x) \to 0, x \to \pm\infty$ with two eigenvalues $\lambda = \mu_j = -k_j^2, j = 1, 2$. Dynamics in t of this potential, corresponding to KdV equation (4.45), are defined by KdV Lax pair $\varphi_{j,t} = 4\mu_j \varphi_{j,x}$ (see Sect. 1.3) which yields

$$y_j = k_j(x - k_j^2 t - x_j^0). \tag{4.96}$$

A direct generalization of (4.95) looks as follows:

$$< \varphi_1, \ldots, \varphi_n >= \langle a(y_1), b(y_2), a(y_3), \ldots \rangle = \sum_\varepsilon |\alpha_\varepsilon| a(\epsilon_n y_n + \ldots + \epsilon_1 y_1), \tag{4.97}$$

where y_j are defined by (4.96) and $a(y) = e^y + e^{-y}$. The summation goes over the set of vectors $\{\epsilon = (\epsilon_1, \ldots, \epsilon_n), \epsilon_i = \pm 1\}$. Vectors ϵ and $\tilde{\epsilon} = -\epsilon$ are considered as identical, so we have 2^{n-1} different terms in this sum. The formula (4.97) represents sum of the elementary wronskians

$$\langle \exp(\epsilon_1 y_1), \exp(\epsilon_2 y_2), \ldots \rangle = \alpha_\epsilon \exp(\epsilon_1 y_1 + \ldots + \epsilon_n y_n),$$

where $\epsilon_i = \pm 1$ and

$$\alpha_\epsilon = \prod_{i>j} (\epsilon_i k_i - \epsilon_j k_j).$$

We are now going to discuss dressing chains for spectral problems $L\psi = \lambda\psi$ for operators L of the order $m > 2$. The point is that classical formulae (4.91), (4.92) leave open the problem of choice of "eigenfunctions" $\varphi_1, \ldots, \varphi_n$ of the initial spectral problem $L\varphi_j = \mu_j \varphi_j$. On the other hand in the case $m = 2$ we have solved this problem (Sect. 2.1) using the notion of DT invariance and rewriting formulae (4.94) in terms of the dressing chain equations (4.81) for f_j. That allowed us to consider (see Example 14) finite-gap and Painlevé-type potentials as well. Generally speaking, the change of variables from φ_j to f_j which is linked by (4.92) reminds, in certain sense, the transition to Euler-type equations in fluid mechanics instead of Lagrange ones referred to initial state of the flow.

We reformulate the DT-invariance property (see Definition 9) in terms of Definition 19 as follows

$$ML = (L + \gamma)M \iff [M, L] = \gamma M. \tag{4.98}$$

One can verify that in Example 14 the proper DT-invariance corresponds to $[M, L] = 0$ and the case $\gamma \neq 0$ gives rise to Painlevé-type potentials. Our discussion below will be concentrated upon operators L of order $m = 3$.

In the case $m = 2$ the dressing chain (4.81), as it was mentioned before, is equivalent to the basic DT relation (4.94). Similar to the case $m = 3$ and $L_j = D^3 + u_j D + v_j$ it yields

$$u_{j+1} - u_j = 3f_{j,x}, \quad v_{j+1} - v_j = \left(3f_{j,x} + \frac{3}{2}f_j^2 + u_j\right)_x,$$

$$f_{j,xx} + 3f_j f_{j,x} + f_j^3 + u_j f_j + v_j = \mu_j. \tag{4.99}$$

The exclusion of potentials u and v leads to the dressing chain

$$\left(q_{j-1} + q_j + q_{j+1}\right)_{xx} = 3\left(q_j - q_{j-1}\right)q_{j-1,x} - 3\left(q_{j+1} - q_j\right)q_{j+1,x}$$
$$+ \left(q_j - q_{j-1}\right)^3 + \left(q_j - q_{j+1}\right)^3 + \alpha_j, \tag{4.100}$$

where

$$\alpha_j = \mu_j - \mu_{j-1}, \quad u_j = 3q_{j,x}, \quad f_j = q_{j+1} - q_j.$$

Namely, the last two formulae resolved the first of Eq. (4.99) and allowed to express v in terms of q by the third one which plays the role of Riccati equation in the case $L = D^3 + uD + v$.

In order to derive Lax pair for this dressing chain (4.100) one can start with the general form of difference operator (cf. (4.80))

$$\mathscr{L}\psi \stackrel{\text{def}}{=} \left(T^3 + aT^2 + bT + c\right)\psi = \mu(\lambda)\psi, \quad \psi_x = (T + f)\psi.$$

We are going to prove that $c_x = 0$. By x-differentiation of the difference spectral problem $\mathscr{L}\psi = \mu\psi$ and equating the terms with $T^2\psi$, $T\psi$, ψ we find, respectively,

$$a_x = a(A + f - f_2) + b_1 - b, \quad b_1 = T(b), \quad f_2 = T^2(f),$$
$$b_x = b(A + f - f_1) + c_1 - c + \mu - \mu_1, \quad \mu_1 = T(\mu),$$
$$c_x = (\mu - c)A, \quad A = a - a_1 + f_3 - f.$$

Since μ is λ-dependent, $A = 0$ and hence $c_x = 0$. Redefining μ in the right-hand side of the spectral problem we put $c = 0$ and obtain

$$\mu = \mu(\lambda, n) = \varepsilon(\lambda) - \beta(n), \quad a = f + f_1 + f_2.$$

At last the consistence of two spectral problems $\mathscr{L}\psi = \mu\psi$ and $L\psi = \lambda\psi$ yields $\varepsilon(\lambda) = \lambda$.

Rewriting the above equations in the final form we get

$$\mathscr{L}\psi \stackrel{\text{def}}{=} \left(T^3 + aT^2 + bT\right)\psi = (\lambda - \mu)\psi, \quad \psi_x = (T + f)\psi, \tag{4.101}$$

$$\begin{cases} a_j &= f_j + f_{j+1} + f_{j+2}, \quad f_j \stackrel{\text{def}}{=} T^j(f), \\ a_{j,x} &= a_j(f_j - f_{j+2}) + b_{j+1} - b_j, \\ b_{j,x} &= b_j(f_j - f_{j+1}). \end{cases} \tag{4.102}$$

The equation for a in terms of f yields

$$\left(1+T+T^2\right)(f_{j,x}) = (1-T)\left(f_j^2 + f_j f_{j+1} + f_{j+1}^2 - b_j\right)$$

and we find that

$$f_j = q_{j+1} - q_j \;\Rightarrow\; b_j = \left(1+T+T^2\right)(q_{j,x}) + f_j^2 + f_j f_{j+1} + f_{j+1}^2.$$

It is easy to now see that these formulae for a and b and (4.101) provide the Lax pair for the dressing chain (4.100) in the case $m=3$. It has been shown by [4] that Lax pair for dressing chains for operators of order $m > 3$ looks like (4.101) and, for example, in the case $m=4$ we have

$$\mathscr{L}\psi \overset{\text{def}}{=} (T^4 + aT^3 + bT^2 + cT)\psi = (\lambda - \mu)\psi, \qquad \psi_x = (T+f)\psi.$$

The corresponding dressing chain in terms of coefficients of $\mathscr{L}$ looks quite similar to (4.102):

$$\begin{cases} a_{j,x} = a_j(f_j - f_{j+3}) + b_{j+1} - b_j, \\ b_{j,x} = b_j(f_j - f_{j+2}) + c_{j+1} - c_j, \\ c_{j,x} = c_j(f_j - f_{j+1}), \end{cases}$$

where $a_j = f_j + f_{j+1} + f_{j+2} + f_{j+3}$. The missing coefficient of $\mathscr{L}$ at T^0 corresponds to vanishing $u_1 = 0$ coefficient of (4.89) which are both responsible for wronskian conservation laws for $\mathscr{L}\psi = (\lambda - \mu)\psi$ and $L\psi = \lambda\psi$, respectively.

We now discuss $\mathbb{Z}_2$-closure of (4.100):

$$(2q_1 + q_2)_{xx} = 6q_{1,x}(q_2 - q_1) + 2(q_2 - q_1)^3 + \alpha_1,$$

$$(2q_2 + q_1)_{xx} = 6q_{2,x}(q_1 - q_2) + 2(q_1 - q_2)^3 + \alpha_2.$$

The first integral

$$(q_1 + q_2)_x + (q_1 - q_2)^2 = a(x), \quad 3a' = \gamma = \alpha_1 + \alpha_2$$

allows to reduce this system to Painlevé ODE (4.12) for $y = q_2 - q_1$:

$$y_{xx} = 2y^3 - 6a(x)y + \alpha_2 - \alpha_1, \quad f_1 = y.$$

In order to apply the general formulae (4.98) let us consider a chain of operators $L_j = D^3 + u_j D + v_j$, $j = 1, 2, 3$, linked by Eqs. (4.93), (4.99):

$$(D - f_1)L_1 = L_2(D - f_2), \quad (D - f_2)L_2 = L_3(D - f_2) \;\Rightarrow\; ML_1 = L_3 M,$$

where $M = (D - f_2)(D - f_1)$ is the second-order transformation operator. In virtue of the periodicity $q_3 = q_1$ and thus $f_2 = q_3 - q_2 = -f_1 = -y$. Coefficients of the operators L_1, L_2 and L_3 can be defined directly in terms of q_1, q_2 and using (4.99) one finds that

$$M = (D+y)(D-y), \quad L_3 = L_1 + \gamma, \quad L_j = D^3 + 3q_{j,x}D + \frac{3}{2}q_{j,xx}, \ j = 1, 2.$$

Summing up, we see that $\mathbb{Z}_k$-closures of dressing chains ($k = 3$ for $m = 2$ and $k = 2$ for $m = 3$) describe solutions of equations as follows:

$$AB - BA = \alpha A + \beta B,$$

where the order of differential operators A and B is two and three, respectively. The case $\beta \neq 0$, related with ODE (4.12), has been discussed above; the case $\beta = 0$ is related with ODE (4.13) and with periodic closure of dressing chain for Schrödinger operator. A systematic study of spectral problems (4.52) DT-invariant up to the shift of spectral parameter λ has been fulfilled by [5]. DT-invariance has appeared more efficient in comparison with Lie-type symmetries from Sect. 1.1. In particular, it was shown that the missed general Painlevé sixth equation in Sect. 1.1 can be characterized by the condition $S^2 \hat{T}^2 = E$ where $\hat{T}$ and S are Darboux transformations (4.68) and (4.71), respectively.

4.2.3 Symmetry Approach

In conclusion we are going to derive, in certain sense, a complete list of chain equations

$$q_{n,xx} = A(q_{n+1,x}, q_{n,x}, q_{n-1,x}, q_{n+1}, q_n, q_{n-1}) \tag{4.103}$$

related with elementary Darboux transformations for the spectral problem (4.52) introduced in Sect. 2.1. These chain equations play the role of models in mathematical theory of solitons and have few important applications. The classical example (going up to Darboux) of so-called Toda chain can be written as follows:

$$q_{n,xx} = f(q_{n+1} - q_n) - f(q_n - q_{n-1}), \quad f' = \alpha f. \tag{4.104}$$

It corresponds to (4.67) and is just another form of Eqs. (4.74) considered in Sect. 2.1.

The starting point here is **not the spectral problems** but a rather general variational problem

$$\delta \int dt \, dx \, \Phi(q_t, q_{xx}, q_x, q) = 0 \tag{4.105}$$

and its dual Hamiltonian coupled dynamical system of equations

$$q_t = \frac{\delta H[p, q]}{\delta p}, \quad p_t = -\frac{\delta H[p, q]}{\delta q}, \tag{4.106}$$

where

$$H = q_t \Phi_{q_t} - \Phi, \quad \Phi = pH_p - H, \quad p = \Phi_{q_t}.$$

In most of the examples below the Lagrangian density Φ is quadratic in q_t and, hence the transformation $q_t \leftrightarrow p$ is well defined. Proper definition (see below) of the notion of discrete symmetries directly in terms of variational problems (4.105) will allow us to reveal the universal form of chain equations which goes out of domain of applications of the spectral problem (4.52).

Primarily, we will address ourselves to

$$H = p_x q_x + \varepsilon_0 p^2 q_x^2 + \varepsilon_1 p q_x^2 + \varepsilon_2 p^2 q_x + \varepsilon_3 p^2 + \varepsilon_4 p q_x + \varepsilon_5 q_x^2 + \varepsilon_6 p, \qquad (4.107)$$

where ε_i are arbitrary constants. We remind that preliminaries of the variational approach and the coupled system (4.51) of equations

$$z_t + z_{xx} = (2p + z^2)_x, \quad p_t - p_{xx} = (2zp)_x, \qquad z \equiv q_x, \qquad (4.108)$$

which represent (4.106) in the case

$$H[p,q] = p^2 + p q_x^2 + p_x q_x \equiv p^2 + p(q_x^2 - q_{xx}),$$

have been discussed in Sect. 1.3. Two more examples of evolutionary equations connected by hamiltonian density (4.107) are given below.

Example 22. By "elementary" transformations like $q_x \to q_x + \text{const}$ the hamiltonian density (4.107) can be reduced to three canonical forms. The first one gives rise to (4.108) and the other two are as follows:

$$H = p_x q_x + p q_x^2 + q_x p^2 \Rightarrow \qquad (4.109)$$
$$z_t + z_{xx} = \left(2pz + z^2\right)_x, \quad p_t - p_{xx} = \left(2zp + p^2\right)_x;$$

$$H = p_x q_x + p^2 q_x^2 + \varepsilon_3 p^2 + \varepsilon_5 q_x^2 \Rightarrow \qquad (4.110)$$
$$z_t + z_{xx} = 2\left(pz^2 + \varepsilon_3 p\right)_x, \quad p_t - p_{xx} = 2\left(zp^2 + \varepsilon_5 z\right)_x.$$

All these coupled system of equations for z, p possess three local conservation laws of zero order

$$z_t, \; p_t, \; (zp)_t \in \Im D_x.$$

This characteristic feature allows to relate (4.109) with the other two by Miura-type transformations (see references in review [5]).

The interconnection of generalizations of the Toda chain (4.104) with Hamiltonians (4.107) is formulated in the next definition of *canonical Bäcklund transformations*.

Definition 23. [4] The function $F[q,\hat{q}] = F(q, q_x, \hat{q}, \hat{q}_x)$ is called the canonical Bäcklund transformations generation function (or shortly BCT-function) for Hamiltonian system (4.106) if there exists $\sigma[q,\hat{q}] = \sigma(q, q_x, \hat{q}, \hat{q}_x)$ such that

$$p = \frac{\delta F[q,\hat{q}]}{\delta q}, \quad \hat{p} = -\frac{\delta F[q,\hat{q}]}{\delta \hat{q}} \;\Rightarrow\; \hat{H}[\hat{p},\hat{q}] - H[p,q] = \frac{d}{dx}\sigma[q,\hat{q}].$$

In this case the Lagrangian system of equations with one discrete and one continuous variable

$$\delta \mathscr{F} = \delta \int dx \sum F_n = 0, \quad F_n \stackrel{\text{def}}{=} F[q_n, q_{n+1}], \tag{4.111}$$

is called the associated system.

Rewriting (4.111) in more explicit form we find

$$\frac{\delta F[q_n, q_{n+1}]}{\delta q_n} + \frac{\delta F[q_{n-1}, q_n]}{\delta q_n} = 0 \quad \Rightarrow \tag{4.112}$$

$$p_n = \frac{\delta F[q_n, q_{n+1}]}{\delta q_n} = -\frac{\delta F[q_{n-1}, q_n]}{\delta q_n} = \hat{p}_{n-1}.$$

One can now readily verify that Definition 23 implies that the chain of transformations

$$\dots (p_{n-1}, q_{n-1}) \rightarrow (p_n, q_n) \rightarrow (p_{n+1}, q_{n+1}) \dots$$

maps solutions of Hamiltonian equations (4.106) in solutions again.

Theorem 24. *The BCT-function* $F[q, \hat{q}]$ *for Hamiltonian density* (4.107) *is as follows:*

$$F[q, \hat{q}] = F(z, y) = W(z) + zV(y) + U(y), \quad y \stackrel{\text{def}}{=} \hat{q} - q, \quad z \equiv q_x, \tag{4.113}$$

where $W''(z) = -(\varepsilon_0 z^2 + \varepsilon_2 z + \varepsilon_3)^{-1}$ *and the functions* U, V *satisfy the system of ODE*

$$a' + \varepsilon_2 a^2 - 2\varepsilon_0 ab + \varepsilon_1 a = 0, \quad b' + \varepsilon_3 a^2 - \varepsilon_0 b^2 + \varepsilon_1 b = \varepsilon_5, \quad V' = a, \quad U' = b. \tag{4.114}$$

Sketch of Proof. In the case (4.113) we find

$$p = cq_{xx} - a\hat{q}_x - b, \quad \hat{p} = -aq_x - b, \quad c = \varepsilon_0 z^2 + \varepsilon_2 z + \varepsilon_3.$$

The substitution of the above expressions in (4.107) and integration by parts in the difference $\hat{H} - H$ yields the equations (4.114) which are equivalent to the condition $\hat{H} - H \in \mathfrak{I}D_x$. ∎

Theorem 24 describes discrete symmetries of dynamical systems (4.106), (4.107). On the other hand, Eqs. (4.114) and formulae (4.113) fully define a family of Lagrangian chains (4.111) depending on five parameters $\varepsilon_0, \varepsilon_1, \varepsilon_2, \varepsilon_3, \varepsilon_5$. which we call *generalized Toda chains*. It can be proved that the exclusion of p by (4.112) in the first part of dynamical equations (4.106) results in variational symmetry (see [11]) of the Lagrangian (4.111). Thus, the next theorem is just a reformulation of Theorem 24 which states that evolution equations (4.106) are consistent with (4.111).

Theorem 25. *Let* $W''(z) = -(\varepsilon_0 z^2 + \varepsilon_2 z + \varepsilon_3)^{-1}$ *and functions* U, V *are defined by* (4.114). *Then the Lagrangian* (4.111), (4.113) *and corresponding chain equations with* $y_n = q_{n+1} - q_n$

$$W''(z_n)z_{n,x} + z_{n+1}V'(y_n) - z_{n-1}V'(y_{n-1}) + U'(y_n) - U'(y_{n-1}) = 0, \quad z_n = q_{n,x}$$
$$(4.115)$$

possess variational symmetry as follows:

$$W''(z_n)(q_{n,t} + \varepsilon_1 z_n^2) + z_{n+1}V'(y_n) + z_{n-1}V'(y_{n-1}) + U'(y_n) + U'(y_{n-1}) = 0. \quad (4.116)$$

For each generalized Toda chain (4.115) there is, due to E. Noether theorem [11], besides common conservation laws of momentum and energy

$$\frac{d}{dx}\sum\{W'(z_n) + V'(y_n)\} = 0, \quad \frac{d}{dx}\sum\{z_nW'(z_n) - W(z_n) - U(y_n)\} = 0,$$

higher conservation law related with the variational symmetry (4.116). The difficult problem of classification "arbitrary" chains with higher order conservation laws is considered by Ravil Yamilov (see references in review [5]).

The next two examples should illustrate an application of the above theorems to a given chain of the form (4.103). In the case of Theorem 25 one has to first derive Lagrangian form of (4.103) and find the "integrating multiplier" denoted by $W''(z)$. Second, one should check out Eqs. (4.114) and find the exact form of Hamiltonian density H from Theorem 24.

Example 26. For Toda chain (4.104) $W'' = -1$ and Lagrangian density is

$$F_n = -\frac{1}{2}z_n^2 + \beta f(y_n), \quad f'(y) = \alpha f(y), \quad \beta\alpha = 1.$$

The comparison with (4.114) yields $\varepsilon_0 = \varepsilon_2 = \varepsilon_5 = 0$, $\varepsilon_1 = \alpha$, $\varepsilon_3 = 1$. Thus

$$H = p_xq_x + \alpha p^2 q_x^2 + p^2 + \varepsilon_4 pq_x + \varepsilon_6 p$$

and Theorem 24 implies that Toda chain (4.104) defines Bäcklund transformations for the Hamiltonian system (4.106) corresponding to (4.108) when $\alpha = 1$.

For Volterra chain we have

$$z_{n,x} = z_n(z_{n+1} - z_{n-1}) \Rightarrow W''(z) = z^{-1} \quad \varepsilon_0 = \varepsilon_3 = \varepsilon_5 = 0, \quad \varepsilon_1 = \varepsilon_2 = 1.$$

In this case Lagrangian equations (4.111) define canonical Bäcklund transformations for Hamiltonian system (4.106), (4.109). In virtue of Theorem 25 the variational symmetry is

$$q_{n,t} = z_n(z_{n+1} + z_{n-1}) + z_n^2$$

and it is a good exercise to verify that independently. Notice that the same Hamiltonian density

$$H = p_xq_x + pq_x^2 + q_xp^2$$

corresponds as well to modified Toda chain (4.76) from Sect. 2.2.

There is the analogy of Theorem 25 with the list of Painlevé second-order differential equations in Sect. 1.1. First, as well as differential equations (4.12), (4.13),

(4.14), (4.15), the chain equations from Theorem 25 are related to the same spectral problem (4.52). Second, there is, missed in Theorem 25, the so-called "elliptic Toda chain"

$$q_{n,xx} = (r(q_n) - z_n^2) \left(\frac{1}{q_{n+1} - q_n} - \frac{1}{q_n - q_{n-1}} \right) + \frac{1}{2} r'(q_n), \qquad (4.117)$$

which is the analog of the famous Painlevé sixth equation discussed at the end of Sect. 2.2.

The chain (4.117) is Lagrangian (4.111) with

$$F[q, \hat{q}] = W(z, q) + U(q, \hat{q}), \quad z \equiv q_x, \quad W_{zz} = \frac{1}{r(q) - z^2} \qquad (4.118)$$

for any function $r(q)$ and possesses variational symmetry

$$q_{n,t} = (r(q_n) - z_n^2) \left(\frac{1}{q_{n+1} - q_n} + \frac{1}{q_n - q_{n-1}} \right),$$

if and only if $r(q) = \alpha q^4 + \beta q^3 + \gamma q^2 + \delta q + \varepsilon$ is an arbitrary polynomial of the fourth degree [2]. In terms of Definition 23 it means that (4.118) defines the generation function of BCT for Hamiltonian density (cf. (4.110)) as follows:

$$H[p, q] = p_x q_x + q_x^2 p^2 - p^2 r(q) + \frac{1}{2} p r'(q) - \frac{1}{12} r''(q). \qquad (4.119)$$

The point is that Hamiltonian system (4.106) coincides with complexified form of anisotropic Landau–Lifshitz model

$$\mathbf{S}_t = i[\mathbf{S}, \mathbf{S}_{xx} + J\mathbf{S}], \quad S_1^2 + S_2^2 + S_3^2 = 1 \qquad (4.120)$$

up to change of variables (cf. (4.55)):

$$S_3 = 1 + 2pq, \quad iS_2 + S_1 = 2q(1 + pq), \quad iS^2 - S^1 = 2p.$$

In the case of diagonal matrix $J = \text{diag}(J_1, J_2, J_3)$ in (4.120) one finds that

$$4r(q) = (J_2 - J_1)q^4 + 2(J_1 + J_2 - 2J_3)q^2 + J_2 - J_1.$$

It is important that applying Definition 23 one can find, as in Theorem 24, the general form of BCT-function

$$F[q, \hat{q}] = W(z, q) + zV(q, \hat{q}) + U(q, \hat{q}), \quad z \equiv q_x$$

for the Hamiltonian (4.119). The exact connection (4.119) with Sklyanin chain [14] has been uncovered in this way [1, 5].

The general problem of classification of Lagrangians (4.105) consistent with canonical Bäcklund transformations (Definition 23) is discussed in the paper by

[4]. In particular, it has been demonstrated that the Lagrangian densities

$$\Phi = q_t q_x + q_{xx}^2 + 4q_x^2; \quad \Phi = \frac{q_t}{q_x} + \frac{r(q) - q_{xx}^2}{q_x^2}, \quad r^{(V)}(q) = 0$$

give rise, respectively, to the KdV dressing chain (4.82) and its analog for Krichever–Novikov equation

$$q_t = q_{xxx} - \frac{3}{2q_x}(q_{xx}^2 - r(q)).$$

This evolutionary equation plays the role of universal model (4.120) in the case of KdV hierarchy.

The dressing chain (4.100) (see Sect. 2.2) for the third-order differential operator $L = D^3 + uD + v$ is Lagrangian (4.111) with the density

$$F_n = 2q_{n,x}^2 + 4q_{n,x}q_{n+1,x} + 3(q_{n+1} + q_n)_x f_n^2 + f_n^4 - 4\mu_n f_n, \quad f_n = q_{n+1} - q_n. \quad (4.121)$$

It corresponds to variational problem (4.105) with

$$\Phi = (q_t + q_{xx})^2 + q_x^2 \Rightarrow H[p,q] = p^2 + 4pq_{xx} + 16q_{xx}^2 - 48q_x^3.$$

Acknowledgments My work was supported by the Russian Foundation for Basic Research (Grant Nos. 96-15-96093 and 98-01-01161), INTAS (Grant No. 99-1782) and a Rotschild professorship at the Isaac Newton Institute, Cambridge.

References

1. V.E. Adler, Theor. Math. Phys. 124, 897–908, 2000.
2. V.E. Adler and A.B.Shabat, Generalized Legendre transformations, Theor. Math. Phys. 112(5), 935–948, 1997.
3. V.E. Adler and A.B. Shabat, Model equation of the theory of solitons, Theor. Math. Phys. 153(1), 1373–1383, 2007.
4. V.E. Adler, V.G. Marikhin, A.B. Shabat, Canonical Bäcklund transformations and Lagrangian chains, Theor. Math. Phys. 129(2), 163–183, 2001.
5. V.E. Adler, A.B. Shabat, and R.I. Yamilov, Symmetry approach to the integrability problem, Theor. Math. Phys. 125(3), 1603–1661, 2000.
6. B.A. Dubrovin, V.B. Matveev, and S.P. Novikov, Nonlinear equations of KdV type, finite-zone linear operators and abelian varieties. Russ. Math. Surv. 31(1), 59–146, 1976.
7. E. Ferapontov Phys. Lett. A 158, 112, 1991.
8. A.N.W. Hone, Phys. Lett. A 249, 46, 1998.
9. L. Martínez Alonso and A.B. Shabat, Energy dependent potentials revisited. A universal hierarchy of hydrodynamic type, Phys. Lett. A, to appear.
10. V.G. Mikhalev, Hamiltonian formalism of Korteveg-de Vries hierarchies, Funct. Anal. Appl. 26(2), 140, 1992.
11. P.J. Olver Applications of Lie groups to Differential Equations, 2nd Ed., Graduate Texts in Mathematics Vol. 107 Springer-Verlag, New York, 1993.
12. A.B. Shabat, Transparent potentials, (in russian) Dinamika sploshnoi sredy, Institute of Hydrodynamics, Novosibirsk, No.5, 130–145, 1970.

13. A.B. Shabat, Third version of the dressing method, Theor. Math. Phys. 121(1), 1397–408, 1999.
14. E.K. Sklyanin, Funkts. Anal. Prilozen. 16, 263–270, 1983.
15. A.P. Veselov and A.B. Shabat Dressing chain and spectral theory of Schrödinger operator, Funkts. Anal. Prilozen. 27(2), 1–21, 1993.

Chapter 5
Normal Form and Solitons

Y. Hiraoka and Y. Kodama

Abstract We present a review of the normal form theory for weakly dispersive nonlinear wave equations where the leading order phenomena can be described by the KdV equation. This is an infinite-dimensional extension of the well-known Poincaré–Dulac normal form theory for ordinary differential equations. In particular, the normal form theory shows that the perturbed equations given by the KdV equation with higher order corrections are asymptotically integrable up to the first-order correction, and the first-order corrections can be transformed into a symmetry of the KdV equation called the fifth-order KdV equation. We then give the explicit conditions for the asymptotic integrability up to the third-order corrections. As an important example, we consider the Gardner–Miura transformation for the modified KdV equation and show that the inverse of the transformation is a normal form transformation. We also provide a detailed analysis of the interaction problem of solitary waves as an important application of the normal form theory. Several explicit examples are discussed based on the normal form theory, and the results are compared with their numerical simulations. Those examples include the ion acoustic wave equation, the shallow water wave equation and the Hirota bilinear equation having a seventh-order linear dispersion.

5.1 Introduction

In this chapter, we review the normal form theory developed in [18, 19] for weakly nonlinear and weakly dispersive wave equations where the leading order equation is given by the KdV equation in an asymptotic perturbation sense. The chapter is based on the report [20] and the master thesis of the first author at Osaka University (February 2002, [12]).

The lectures started with a brief summary of the Poincaré–Dulac normal form theory for a system of ordinary differential equations [3]. The main point in the lectures is to present the normal form theory for *near*-integrable system where the

Y. Kodama (✉)

Department of Mathematics, The Ohio State University, Columbus OH 43210,
kodama@math.ohio-state.edu

Hiraoka, Y., Kodama, Y.: *Normal Form and Solitons*. Lect. Notes Phys. **767**, 175–214 (2009)
DOI 10.1007/978-3-540-88111-7_5

leading order system is given by a nonlinear wave equation. This may be considered as an infinite-dimensional extension of the Poincaré–Dulac normal form theory. The basic technique of the normal form theory based on the Lie transformation can be formally extended to the infinite-dimensional case. Unfortunately the convergence theorem of the normal form series has no extension to the present theory. However, one should emphasize that the leading order in the present theory is given by a nonlinear equation, and it is not clear how one defines a resonant surface for the leading order equation. This may provide a good problem for a future project.

Let us briefly summarize a background of the normal form theory for near-integrable systems of nonlinear dispersive equations. It is well known that for a wide class of nonlinear dispersive wave equations, the leading order nonlinear equation in an asymptotic expansion turns out to be given by an integrable system, such as the Korteweg–de Vries (KdV) equation in weak dispersion limit and the nonlinear Schrödinger equation in strong dispersion limit (see for example [38]). This implies that most of the nonlinear dispersive wave equations are integrable at the nontrivial leading orders in an asymptotic sense. Then a natural question is to ask how the higher order corrections affect the integrability of the leading order equations. In [24], the effect of the higher order corrections on one-soliton solution of the KdV equation was studied, and it was shown that the velocity of soliton is shifted by the secular terms in the higher order terms. Those secular terms or resonant terms are given by the symmetries of the KdV equation. The nonsecular terms then contribute to modify the shape of the soliton. However, multisoliton interactions were not studied in that paper. In [18], the normal form for weakly dispersive equations was first introduced up to the second-order corrections, and it was found that the integrable approximation can be extended beyond the KdV approximation but not to the second order. This shows that asymptotic equations for the weakly dispersive and weakly nonlinear wave systems are integrable not only at the leading order KdV approximation but also up to the next order corrections. Those corrections are then shown to be transformed into a symmetry of the KdV equation by normal form transformation. In this sense, those perturbed KdV equations with higher order corrections are asymptotically equivalent up to the first-order corrections (see for example [8, 9, 25, 28, 29] for a further development of the asymptotic equivalence). The obstacle to the *asymptotic* integrability appearing at the higher order corrections plays no rule for one-soliton solution, but provides a crucial effect for two-soliton interactions. This was found in [19, 20]. The obstacles are defined as the nonexistence of the integrals of perturbed equation in the form of the power series in a small parameter. This was also recognized as the nonexistence of approximate symmetries of the perturbed equation [31]. (The normal form for strongly dispersive wave equations has also been studied in [15, 22].) Then in [19], the effect of the obstacle on the interaction of two solitary waves was studied for the regularized long wave equation (although the method can apply to other equations of weakly dispersive system). An inelasticity due to the obstacle was found, and it leads to the shifts of the soliton parameters and the generation of a new soliton as well as radiation through the interaction.

In this chapter, we present a comprehensive study of the normal form theory for weakly dispersive wave equations: We start in Sect. 5.2 to define the perturbed KdV equation as an asymptotic expansion of a weakly dispersive wave equation whose leading order equation is given by the KdV equation. We give a recursion formula to generate the higher order corrections which may be obtained by an asymptotic perturbation method (see for example [38]). The set of those higher order terms forms an extended space of differential polynomials which includes some nonlocal terms. The space is denoted by $\widehat{\mathscr{P}}_{\text{odd}}$, where "odd" implies the odd weight of those integro-differential polynomials.

In Sect. 5.3, the conserved quantities or integrals of the KdV equation are reviewed, and we discuss *approximate* integrals of the perturbed KdV equation. We then obtain the conditions for the existence of approximate integrals in each order (Proposition 6). We also discuss a connection of the conserved quantities and the N-soliton solutions of the KdV equation.

In Sect. 5.4, we review the symmetries of the KdV equation and discuss the *approximate* symmetries for the perturbed KdV equation. Here we also define the space of $\widehat{\mathscr{P}}_{\text{even}}$, which together with $\widehat{\mathscr{P}}_{\text{odd}}$ provides the appropriate spaces for the normal form transform defined in the next section.

In Sect. 5.5, we describe the normal form theory. The normal form transformation is then obtained by a linear equation of an adjoint map defined as

$$\text{ad}_{K^{(0)}} : \widehat{\mathscr{P}}_{\text{even}} \longrightarrow \widehat{\mathscr{P}}_{\text{odd}},$$

where $K^{(0)}$ is the KdV vector field (Theorem 10). The explicit form of the normal form is given for the perturbed KdV equation which contains the first three lowest weight approximate conserved quantities (Theorem 12). The normal form then admits one-soliton solution of the KdV equation, which confirms the result in [24]. We also discuss the Gardner–Miura transformation, which is an invertible version of the Miura transformation, and show that the inverse Gardner–Miura transformation is nothing but the normal form transformation after removing the symmetries of the KdV equation (Theorem 14).

In Sect. 5.6, we consider the interaction problem of two solitary waves under the influence of the obstacles of asymptotic integrability and provide the explicit formulae for the shifts of the soliton parameters (Theorem 15). We also give the formulae of the radiation energy and additional phase shifts which are used to compare with the numerical simulations for some examples in the next section.

Finally in Sect. 5.7, we present explicit examples including ion acoustic wave equation, the Boussinesq equation as a model of shallow water waves and the regularized long wave equation. We show the good agreements with the results obtained by the normal form theory. We also consider a seventh-order Hirota bilinear equation which admits an exact two-soliton solution but is known to be nonintegrable. We look for an obstacle to the asymptotic integrability and find that the obstacles appear at the fourth order. This implies that the fourth-order obstacles play no rule for two-soliton solution, just like all obstacles have no rule for one-soliton solution for the system with the first three approximate integrals.

5.2 Perturbed KdV Equation

Under the assumption of weak nonlinearity and weak dispersion, the wave propagation in a one-dimensional medium can be described by the KdV equation in
the leading order of an asymptotic expansion. Using an appropriate asymptotic perturbation method (see, e.g., [38]) one can show that the higher order correction to
the KdV equation has the following expansion form with a small parameter ϵ with
$0 < \epsilon \ll 1$,

$$u_t + K(u;\epsilon) = O\left(\epsilon^{N+1}\right),$$
$$\text{with } K(u;\epsilon) = K^{(0)}(u) + \epsilon K^{(1)}(u) + \epsilon^2 K^{(2)}(u) + \cdots + \epsilon^N K^{(N)}(u), \quad (5.1)$$

where $K^{(0)}(u)$ gives the KdV flow and the higher order corrections $K^{(n)}(u)$ are generated by a recursion formula starting from $n = -1$,

$$K^{(n)}(u) = \sum_{i=1}^{M(n)} a_i^{(n)} X_i^{(n)}(u)$$
$$= a_1^{(n)} u_{(2n+3)x} + \sum_{\substack{n_1+n_2=n-2 \\ 1 \le i \le M(n_1) \\ 1 \le j \le M(n_2)}} c_{ij}^{(n)} \left(X_i^{(n_1)} D^{-1} X_j^{(n_2)}\right)(u). \quad (5.2)$$

Here $a_i^{(n)}$, $c_{ij}^{(n)}$ are the real constants determined by the original physical problem,
$u_{nx} = \partial^n u/\partial x^n$, and D^{-1} indicates an integral over x, $D^{-1}(\cdot) := \int_{-\infty}^x dx'(\cdot)$. Each
$X_i^{(n)}(u)$ is a monomial in the polynomial $K^{(n)}(u)$, and $M(n)$ is the total number
of independent monomials of the order n. The first few terms of $K^{(n)}(u)$ are then
given by

$$K^{(-1)} = a_1^{(-1)} u_x, \qquad\qquad M(-1) = 1,$$
$$K^{(0)} = a_1^{(0)} u_{3x} + a_2^{(0)} u u_x, \qquad\qquad M(0) = 2,$$
$$K^{(1)} = a_1^{(1)} u_{5x} + a_2^{(1)} u_{3x} u + a_3^{(1)} u_{2x} u_x + a_4^{(1)} u_x u^2, \qquad M(1) = 4,$$
$$K^{(2)} = a_1^{(2)} u_{7x} + a_2^{(2)} u_{5x} u + a_3^{(2)} u_{4x} u_x + a_4^{(2)} u_{3x} u^2 + a_5^{(2)} u_{3x} u_{2x}$$
$$\quad + a_6^{(2)} u_{2x} u_x u + a_7^{(2)} u_x u^3 + a_8^{(2)} u_x^3, \qquad M(2) = 8,$$
$$K^{(3)} = a_1^{(3)} u_{9x} + a_2^{(3)} u_{7x} u + a_3^{(3)} u_{6x} u_x + a_4^{(3)} u_{5x} u_{2x}$$
$$\quad + a_5^{(3)} u_{5x} u^2 + a_6^{(3)} u_{4x} u_{3x} + a_7^{(3)} u_{4x} u_x u + a_8^{(3)} u_{3x} u_{2x} u$$
$$\quad + a_9^{(3)} u_{3x} u_x^2 + a_{10}^{(3)} u_{3x} u^3 + a_{11}^{(3)} u_{2x}^2 u_x + a_{12}^{(3)} u_{2x} u_x u^2$$
$$\quad + a_{13}^{(3)} u_x^3 u + a_{14}^{(3)} u_x u^4 + a_{15}^{(3)} u_x D^{-1}(u_x^3), \qquad M(3) = 15.$$

We normalize the KdV flow so that $K^{(0)}(u)$ is given by the standard form which we
denote by $K_0^{(0)}(u)$, i.e.,

$$K_0^{(0)} = u_{3x} + 6uu_x.$$

Each polynomial $K^{(n)}$ has the scaling property: Assign the weight 2 to $u(x,\cdot)$, and 1 to $\partial/\partial x$. Then if $u(x,\cdot) = \delta^2 v(\delta x,\cdot) = \delta^2 v(\xi,\cdot)$, we have

$$K^{(n)}(u(x,\cdot)) = \delta^{2n+5} K^{(n)}(v(\xi,\cdot)) = \delta^{2n+5}\left(a_1^{(n)} v_{(2n+5)\xi} + \cdots\right).$$

Thus each polynomial $K^{(n)}(u)$ has the homogeneous weight "$2n+5$". We denote by $\widehat{\mathscr{P}}_{\mathrm{odd}}[u]$ the set of all the odd weight polynomials generated by the formula (5.2). Then we have

$$\widehat{\mathscr{P}}_{\mathrm{odd}}[u] = \bigoplus_{n=-1}^{\infty} \widehat{\mathscr{P}}_{\mathrm{odd}}^{(n)}[u],$$

where $\widehat{\mathscr{P}}_{\mathrm{odd}}^{(n)}[u]$ is the finite-dimensional subspace of the polynomials with the homogeneous weight $2n+5$, and the dimension of the space is given by $\dim \widehat{\mathscr{P}}_{\mathrm{odd}}^{(n)}[u] = M(n)$,

$$\widehat{\mathscr{P}}_{\mathrm{odd}}^{(n)}[u] = \mathrm{Span}_{\mathbb{R}}\left\{ X_i^{(n)}(u) : 1 \le i \le M(n) \right\}.$$

As shown in the examples, the subspaces $\widehat{\mathscr{P}}_{\mathrm{odd}}^{(n)}[u]$ up to $n=2$ are given by the differential polynomial of u and its derivatives, and we denote

$$\widehat{\mathscr{P}}_{\mathrm{odd}}^{(n)}[u] = \mathscr{P}_{2n+5}[u] \oplus_{\mathbb{R}} \mathscr{Q}_{2n+5}[u],$$

where $\mathscr{P}_k[u]$ is the space of homogeneous differential polynomials of weight k,

$$\mathscr{P}_k[u] = \mathrm{Span}_{\mathbb{R}}\left\{ u^{l_0} u_x^{l_1} \cdots u_{nx}^{l_n} : \sum_{j=0}^{n}(j+2)l_j = k,\ l_j \in \mathbb{Z}_{\ge 0} \right\},$$

and $\mathscr{Q}_k[u]$ consists of the polynomials of weight k containing the integral operator(s) D^{-1},

$$\mathscr{Q}_{2n+5}[u] \subset \mathrm{Span}_{\mathbb{R}}\left\{ X_{i_0}^{(n_0)} D^{-1} X_{i_1}^{(n_1)} \cdots D^{-1} X_{i_l}^{(n_l)} : \begin{array}{l} \sum_{j=0}^{l} n_j + 2l = n \\ X_{i_j}^{(n_j)} \in \mathscr{P}_{n_j}[u] \end{array} \right\}.$$

The space $\mathscr{P}_k[u]$ can also be extended for $k =$ even, and we will later define $\mathscr{Q}_k[u]$ for $k =$ even.

Remark 1. In order to verify the expansion (5.1), one may need to impose the following conditions for the initial data on $u(x,t=0)$:

a) $|u(x,0)| \le C\exp(-\epsilon^{1/2}|x|),$ as $|x| \to \infty,$

b) $\|u(x,0)\|_{H^\infty(\mathbb{R})}^2 = \sum_{n=0}^{\infty} \int_{\mathbb{R}} |u_{nx}(x,0)|^2 dx < \infty.$

The rigorous justification of the KdV equation from a physical model such as the shallow water waves has been discussed in [7, 14].

5.3 Conserved Quantities and *N*-Soliton Solutions

The integrability of the KdV equation implies the existence of an infinite number of conserved quantities or integrals. Here we briefly summarize those quantities and discuss *approximate* integrals of the perturbed equation (5.1). The approximate integrals will play a fundamental role for the normal form theory discussed in Sect. 5.5 and provide an analytical tool to study the nonintegrable effect on the solution of the perturbed KdV equation.

5.3.1 Conserved Quantities

Let us first recall the definition of a conserved quantity for the evolution equation in the form

$$u_t = f(u), \quad \text{with} \quad f(u) \in \widehat{\mathscr{P}}_{\text{odd}}[u]. \tag{5.3}$$

Definition 2. An integral of a differential polynomial $\rho(u) \in \mathscr{P}[u] = \oplus_k \mathscr{P}_k[u]$,

$$I[u] = \int_{\mathbb{R}} \rho(u) \, dx, \quad \text{so that} \quad \rho(u) \in \mathscr{P}[u]/\text{Im}(D),$$

is a conserved quantity of (5.3) if

$$\rho_t \in \text{Im}(D), \quad \text{with} \quad D = \partial/\partial x.$$

The polynomial $\rho(u)$ is called a conserved density for (5.3).

We also define a vector field generated by $f(u)$ and its action on the space $\mathscr{P}[u]$.

Definition 3. A vector field generated by $f(u)$ is defined by

$$V_f := \sum_{i=0}^{\infty} D^i(f) \frac{\partial}{\partial u_{ix}},$$

which acts on the space $\mathscr{P}[u]$ as a differential operator,

$$V_f : \mathscr{P}[u] \longrightarrow \mathscr{P}[u],$$
$$g \longmapsto V_f \cdot g.$$

With this definition, the condition for $\rho(u)$ being a conserved quantity can be expressed by

$$\rho_t = V_f \cdot \rho \in \text{Im}(D).$$

The KdV equation with $f(u) = K^{(0)}(u) \in \mathscr{P}_5[u]$ has an infinite sequence of conserved quantities,

$$I_k^{(0)}[u] = \int_{\mathbb{R}} \rho_k^{(0)}(u)\, dx, \quad \text{with} \quad \rho_k^{(0)} \in \mathscr{P}_{2k+2}[u] \quad \text{for } k = 0, 1, 2, \cdots,$$

which are generated by the bi-Hamiltonian relation,

$$D\nabla I_{l+1}^{(0)}(u) = \Theta \nabla I_l^{(0)}(u), \quad \text{with} \quad \Theta := D^3 + 2(Du + uD). \tag{5.4}$$

The gradient $(\nabla I)(u)$ is defined by

$$\int_{\mathbb{R}} v(\nabla I)(u)\, dx = \lim_{\delta \to 0} \frac{d}{d\delta} I[u + \delta v],$$

which can be expressed as

$$(\nabla I)(u) = \sum_{i=0}^{\infty} (-1)^i D^i \frac{\partial \rho}{\partial u_{ix}}(u), \quad \text{for} \quad I[u] = \int_{\mathbb{R}} \rho(u)\, dx.$$

The first few conserved densities $\rho_k^{(0)}$ are given by

$$\rho_0^{(0)} = \frac{1}{2}u, \quad \rho_1^{(0)} = \frac{1}{2}u^2, \quad \rho_2^{(0)} = \frac{1}{2}\left(u_x^2 - 2u^3\right),$$

$$\rho_3^{(0)} = \frac{1}{2}\left(u_{2x}^2 - 10uu_x^2 + 5u^4\right).$$

Each density $\rho_k^{(0)}(u)$ can be considered as

$$\rho_k^{(0)} \in \mathscr{P}_{2k+2}[u]/\mathrm{Im}(D).$$

We then define

Definition 4. The set of differential polynomials for the conserved densities are given by

$$\mathscr{F}_k[u] \cong \mathscr{P}_k[u]/\mathrm{Im}(D),$$

where $\mathscr{F}_k[u]$ are defined by

$$\mathscr{F}_k[u] := \mathrm{Span}_{\mathbb{R}} \left\{ u^{l_0} u_x^{l_1} u_{2x}^{l_2} \cdots u_{nx}^{l_n} : \sum_{j=0}^{n} (j+2)\, l_j = k,\ l_n \geq 2 \right\}.$$

Table 5.1 shows the relation between the weight and the dimension of the space $\mathscr{F}_k[u]$:

Table 5.1 The relation between the weight and the dimension of $\mathscr{F}_k[u]$

Weight	2	3	4	5	6	7	8	9	10	11	12	13	14	15	16	17	18	19
Dimension	1	0	1	0	2	0	3	1	4	2	7	3	10	7	14	11	22	17

5.3.2 Approximate Conserved Quantities

Here we discuss *approximate* conserved quantities of the perturbed KdV equation $u_t + K(u; \epsilon) = O(\epsilon^{N+1})$ in (5.1). We look for the conserved quantity in a formal power series,

$$I_l[u; \epsilon] = \int_{\mathbb{R}} \rho_l(u; \epsilon) \, dx = I_l^{(0)}[u] + \epsilon I_l^{(1)}[u] + \cdots + \epsilon^N I_l^{(N)}(u) + O(\epsilon^{N+1}),$$

so that the density $\rho_l(u; \epsilon)$ satisfies

$$V_K \cdot \rho_l(u; \epsilon) = O(\epsilon^{N+1}) \quad (\text{mod Im}(D)), \tag{5.5}$$

$$V_{K_0^{(0)}} \cdot \rho_l^{(m)}(u) = -\sum_{i=1}^{m} V_{K^{(i)}} \cdot \rho_l^{(m-i)}(u), \quad \text{for} \quad m = 1, 2, \cdots. \tag{5.6}$$

Then we have

Lemma 5. *For the linear map* $V_{K^{(0)}}$,

$$V_{K^{(0)}} : \mathscr{F}_{2k}[u] \longrightarrow \mathscr{F}_{2k+3}[u],$$

the kernel of $V_{K^{(0)}}$ *with a fixed weight is a one-dimensional subspace of* $\mathscr{F}_{2k}[u]$ *given by*

$$\text{Ker} V_{K^{(0)}} \cap \mathscr{F}_{2k}[u] = \text{Span}_{\mathbb{R}} \left\{ \rho_{k-1}^{(0)} \right\}.$$

A proof of this lemma can be found in [34]. Namely there is only one conserved density in the form of differential polynomial for each weight. From this lemma, we obtain a sufficient condition for the solvability of $\rho_l^{(m)}$ on the space of differential polynomials $\mathscr{F}_{2(m+l)+2}[u]$,

$$\dim \mathscr{F}_{2(m+l)+5}[u] = \dim \mathscr{F}_{2(m+l)+2}[u] - 1. \tag{5.7}$$

Equation (5.5) is overdetermined in general, and we denote by $N_l^{(m)}$ the number of the constraints for the existence of $\rho_l^{(m)}$, that is,

$$N_l^{(m)} = \dim \mathscr{F}_{2(m+l)+5}[u] - (\dim \mathscr{F}_{2(m+l)+2}[u] - 1).$$

Note here that $N_{l_1}^{(m_1)} = N_{l_2}^{(m_2)}$ if $l_1 + m_1 = l_2 + m_2$. Then from Table 5.1, we obtain

Proposition 6. *For the existence of the higher order corrections of the conserved quantities, we have the following:*

i) There always exist conserved quantities, I_0, I_1, I_2, I_3 up to order ϵ.

ii) At order ϵ^2, there exist I_0, I_1, I_2, and $N_3^{(2)} = 1$, that is, there is one condition $\mu_1^{(2)} = 0$ for the existence of I_3 where

$$
\begin{aligned}
\mu_1^{(2)} := \ & -560a_1^{(2)} + 170a_2^{(2)} - 60a_3^{(2)} - 8a_4^{(2)} \\
& + 24a_5^{(2)} - 9a_6^{(2)} + 3a_7^{(2)} + 24a_8^{(2)} \\
& + \frac{10}{3}a_1^{(1)}\left(240a_1^{(1)} - 67a_2^{(1)} + 6a_4^{(1)}\right) \\
& + \frac{1}{3}a_2^{(1)}\left(4a_2^{(1)} + 30a_3^{(1)} + a_4^{(1)}\right) \\
& - a_3^{(1)}\left(2a_3^{(1)} + a_4^1\right).
\end{aligned}
$$

iii) At order ϵ^3, there always exist I_0, I_1. There are totally three conditions, $\mu_k^{(3)} = 0$, $k = 1, 2, 3$, with $\mu_1^{(3)} = 0$ for the existence of I_2 (i.e., $N_2^{(3)} = 1$), and two conditions $\mu_2^{(3)} = 0$, $\mu_3^{(3)} = 0$ for the existence of I_3 (i.e., $N_3^{(3)} = 2$). (Explicit form of $\mu_k^{(3)}$, $k = 1, 2, 3$, are listed in Appendix.)

iv) At order ϵ^4, there always exist I_0. There are in total seven conditions, $\mu_k^{(4)} = 0$, $k = 1, \cdots, 7$ with $N_1^{(4)} + N_2^{(4)} + N_3^{(4)} = 1 + 2 + 4 = 7$.

One should note here that many physical examples have several conserved quantities, such as the total mass, momentum and energy, which may be assigned as the first three quantities $I_l[u; \epsilon]$ with $l = 0, 1$ and 2. Then the existence of the higher conserved quantity $I_3[u; \epsilon]$ is a key for the integrability of the perturbed equation. The (i) in Proposition 6 suggests the *asymptotic* integrability of the perturbed equation (5.1) up to order ϵ. In fact, in Sect. 5.5, we transform the perturbed equation to an integrable system up to order ϵ and discuss the effect of the nonexistence of $I_3^{(2)}$, i.e., $\mu_1^{(2)} \neq 0$, on the interaction behavior of two solitons.

5.3.3 N-Soliton Solution

One of the most important aspects of the existence of conserved quantities in the form of differential polynomial is to provide several exact solutions of the system. For example, N-soliton solution can be obtained by the variational equation (see for example [34])

$$
\nabla \mathfrak{S}[u; \lambda_1, \ldots, \lambda_N] = 0, \tag{5.8}
$$

where $\mathfrak{S}[u; \lambda_1, \ldots, \lambda_N]$ is the invariants given by

$$\mathfrak{S} = I_{N+1}^{(0)}[u] + \lambda_1 I_N^{(0)}[u] + \cdots + \lambda_N I_1^{(0)}[u].$$

Here λ_k are real constants (the Lagrange multipliers). Thus the N-soliton solution is given by a stationary point of the surface defined by $I_{N+1}^{(0)}[u] = $ constant subject to the constraints $I_k^{(0)}[u] = $ constant for $k = 1, 2, \cdots, N$. This characterization of the N-soliton solution is essential for its spectral stability (see, e.g., [23]).

Equation (5.8) gives a $2N$ order differential equation for $u(x, \cdot)$ and contains one-soliton solution in the form

$$u(x, \cdot) = 2\kappa^2 \text{sech}^2(\kappa(x - x_0(\cdot))), \tag{5.9}$$

with an appropriate choice of the parameters λ_k. In fact, using the bi-Hamiltonian relation (5.4) the variational equation $\nabla \mathfrak{S} = 0$ can be written as

$$\left(\mathfrak{R}^{N-1} + \mu_1 \mathfrak{R}^{N-2} + \cdots + \mu_{N-1}\right) D\left(\nabla I_2^{(0)} + \mu_N \nabla I_1^{(0)}\right) = 0,$$

where μ_ks are some constants related to λ_j. Then by choosing $\mu_N = -4\kappa^2$ this equation admits the one-solitonsolution in the form (5.9). Here the operator $\mathfrak{R}$ is called the recursion operator defined as

$$\mathfrak{R} = \Theta D^{-1} \quad \text{and} \quad \mathfrak{R} D \nabla I_k^{(0)} = D \nabla I_{k+1}^{(0)}. \tag{5.10}$$

The recursion operator plays an important role for the theory of the integrable systems.

5.4 Symmetry and the Perturbed Equation

Here we review the symmetries of the KdV equation and discuss how these symmetries work for the analysis of the nearly integrable systems.

5.4.1 Symmetries of the KdV Equation

Let us define the symmetry for a system

$$u_t = K(u), \quad \text{with} \quad K(u) \in \widehat{\mathscr{P}}[u]. \tag{5.11}$$

Definition 7. A function $S(u) \in \widehat{\mathscr{P}}[u]$ is a symmetry of (5.11) if $S(u)$ satisfies the commutation relation

$$\text{ad}_K \cdot S(u) := [K, S](u) = (V_S \cdot K - V_K \cdot S)(u) = 0,$$

where $\text{ad}_K : \widehat{\mathscr{P}}[u] \to \widehat{\mathscr{P}}[u]$ is the adjoint map of K.

This means that if $S(u)$ is a symmetry of $K(u)$ then the vector fields generated by $K(u)$ and $S(u)$ commute with each other, i.e., $V_K \cdot (V_S \cdot u) - V_S \cdot (V_K \cdot u) = 0$.

The KdV equation has an infinite number of commuting symmetries which are given by the Hamiltonian flows generated by the conserved quantities $I_l^{(0)}[u]$ for $l = 1, 2, \ldots$, that is,

$$K_0^{(n)}(u) := D\nabla I_{n+2}^{(0)}, \quad \text{for} \quad n = -1, 0, 1, 2, \ldots.$$

Then using the recursion operator $\mathfrak{R}$ in (5.10) those symmetries can be constructed as

$$K_0^{(n-1)}(u) = \mathfrak{R}^n K_0^{(-1)}(u), \quad \text{with} \quad K_0^{(-1)}(u) = u_x.$$

The symmetry can also be constructed by the so-called *master symmetry* [10]. The definition of the master symmetries is as follows. A function $M(u)$ is a master symmetry of the evolution equation (5.11) if the Lie bracket defined by M maps symmetries onto symmetries, i.e.,

$$[M, S_i] = S_j,$$

where S_i and S_j are symmetries of (5.11).

The explicit form of the first few commutative symmetries $K_0^{(i)}$, $i \geq -1$, are

$$K_0^{(-1)} = u_x,$$
$$K_0^{(0)} = 6uu_x + u_{3x},$$
$$K_0^{(1)} = u_{5x} + 10u_{3x}u + 20u_{2x}u_x + 30u_x u^2,$$
$$K_0^{(2)} = u_{7x} + 14u_{5x}u + 42u_{4x}u_x + 70u_{3x}u^2 + 70u_{3x}u_{2x} + 280u_{2x}u_x u$$
$$\qquad + 140u_x u^3 + 70u_x^3,$$
$$K_0^{(3)} = u_{9x} + 18u_{7x}u + 72u_{6x}u_x + 168u_{5x}u_{2x} + 126u_{5x}u^2 + 252u_{4x}u_{3x}$$
$$\qquad + 756u_{4x}u_x u + 1260u_{3x}u_{2x}u + 966u_{3x}u_x^2 + 420u_{3x}u^3$$
$$\qquad + 1302u_{2x}^2 u_x + 2520u_{2x}u_x u^2 + 1260u_x^3 u + 630u_x u^4,$$

and the master symmetries $M_0^{(j)}$, $j \geq 0$, are

$$M_0^{(0)} = xK_0^{(-1)} + 2u,$$
$$M_0^{(1)} = xK_0^{(0)} + 8u^2 + 4u_{2x} + 2K_0^{(-1)}D^{-1}(u),$$
$$M_0^{(2)} = xK_0^{(1)} + 32u^3 + 48uu_{2x} + 36u_x^2 + 6u_{4x} + 2K_0^{(0)}D^{-1}(u)$$
$$\qquad + 6K_0^{(-1)}D^{-1}(u^2),$$
$$M_0^{(3)} = xK_2 + 8u_{6x} + 96uu_{4x} + 240u_x u_{3x} + 160u_{2x}^2 + 384u^2 u_{2x} + 576uu_x^2$$
$$\qquad + 128u^4 + 2K_0^{(1)}D^{-1}(u) + 6K_0^{(0)}D^{-1}(u^2) + 10K_0^{(-1)}D^{-1}\left(2u^3 - u_x^2\right).$$

The set of those polynomials $\{K_0^{(n)}, M_0^{(m)}\}$ forms an infinite-dimensional Lie algebra containing a classical Virasoro algebra of the master symmetries, that is,

$$[M_n, K_m] = (2m+3)K_{n+m}, \quad [M_n, M_m] = 2(m-n)M_{n+m}. \tag{5.12}$$

One notes here that the symmetries $K_0^{(n)}(u)$ are the odd weight (differential) polynomials, while the master symmetries are the even weight polynomials. In particular, each $M_0^{(k)}(u) - xK_0^{(k-1)}(u)$ can be considered as an element of the space of even weight polynomials generated by the recursion formula starting from $k = 0$,

$$\begin{aligned} Y^{(k)}(u) = {} & \alpha_1^{(k)} u_{2kx} \\ & + \sum_{\substack{k_1+k_2=k-1 \\ 1 \le i \le N(k_1) \\ 1 \le i \le N(k_2)}} \left(\beta_{ij}^{(k)} (Y_i^{(k_1)} Y_j^{(k_2)})(u) + \gamma_{ij}^{(k)} (X_i^{(k_1-1)} D^{-1} Y_j^{(k_2)})(u) \right), \end{aligned}$$

where $X_i^{(k)}(u)$ is a monomial in $\widehat{\mathscr{P}}_{2k+5}[u]$. This defines the space $\widehat{\mathscr{P}}_{\text{even}}[u]$,

$$\widehat{\mathscr{P}}_{\text{even}}[u] = \bigoplus_{k=0}^{\infty} \widehat{\mathscr{P}}_{\text{even}}^{(k)}[u], \quad \text{with} \quad \widehat{\mathscr{P}}_{\text{even}}^{(k)}[u] = \mathscr{P}_{\text{even}}^{(k)}[u] \oplus \mathscr{Q}_{\text{even}}^{(k)}[u],$$

where $\mathscr{P}_{\text{even}}^{(k)}[u] = \mathscr{P}_{2k+2}[u]$. The space $\widehat{\mathscr{P}}_{\text{even}}^{(k)}[u]$ will be important for the normal form theory discussed in the next section.

Remark 8. The actions of the symmetries and the master symmetries on one-soliton solution (5.9) give a representation of the algebra (5.12) in terms of the soliton parameters κ and $\theta = \kappa x_0$: For the symmetry $K_0^{(n)}$, the action generates the vector field

$$V_n = 4^{n+1} \kappa^{2n+3} \frac{\partial}{\partial \theta}, \quad \text{for} \quad n \ge -1,$$

and for the master symmetry,

$$W_m = 4^n \kappa^{2n+1} \frac{\partial}{\partial \kappa}, \quad \text{for} \quad m \ge 0.$$

This gives a representation of the algebra (5.12), that is,

$$[V_n, V_m] = 0, \quad [W_n, V_m] = (2m+3)V_{n+m}, \quad [W_n, W_m] = 2(n-m)W_{n+m}.$$

5.4.2 Approximate Symmetries

As we discussed about the approximate conserved quantity for the perturbed equation (5.1) in the previous section, one can also discuss the approximate symmetries for (5.1). We say that a differential polynomial $H(u) \in \mathscr{P}[u]$ is an approximate symmetry of order n of the perturbed equation (5.1), if the commutator of $K(u)$ and $H(u)$ gives

$$[K,H](u) = O\left(\epsilon^{n+1}\right).$$

Expanding $H(u)$ in the power series of ϵ, $H = H^{(0)} + \epsilon H^{(1)} + \epsilon^2 H^{(2)} + \cdots$, we have the equation for $H^{(m)}(u)$,

$$\text{ad}_{K_0^{(0)}} \cdot H^{(m)} = -\sum_{j=1}^{m} \left[K^{(j)}, H^{(m-j)}\right], \quad \text{for} \quad m = 1, 2, \cdots,$$

where $K_0^{(0)} = u_{3x} + 6uu_x$, and $K^{(j)}$s are the higher order corrections of the KdV equation (5.1). Then choosing $H^{(0)}$ to be one of the symmetries of the KdV equation, say $H^{(0)} = K_0^{(n)}$, we find the obstacles for the existence of higher order corrections $H^{(m)}$ for the approximate symmetry and obtain the same conditions for the existence as stated in Proposition 6 [31]. In the proof, one needs the following lemma similar to Lemma 5 for the kernel of the adjoint action $\text{ad}_{K_0^{(0)}}$:

Lemma 9. *The kernel of* $\text{ad}_{K_0^{(0)}}$ *on the space of differential polynomials* $\mathscr{P}_{2n+3}[u]$ *is given by*

$$\ker\left(\text{ad}_{K_0^{(0)}}\right) \cap \mathscr{P}_{2n+5}[u] = \text{Span}_{\mathbb{R}}\left\{K^{(n)}\right\}, \quad \text{for} \quad n \geq -1.$$

5.5 Normal Form Theory

The normal form theory has been well developed in the study of finite-dimensional dynamical systems (cf. [3]). The main purpose of the normal form is to classify the vector fields near critical points in terms of the symmetries of the leading order equation. This concept has been applied for the perturbed KdV equation as well as the nonlinear Schrödinger equation in [15, 18, 19].

In this section we review the normal form theory for the perturbed KdV equation (5.1) [19, 20]. The main result is to give an explicit construction of a near-identity transformation which transforms the perturbed equation (5.1) into a normal form. In particular, the normal form up to third order is derived when the perturbed KdV equation has the first three nontrivial integrals, such as mass, momentum and energy. This result shows that the perturbed KdV equations with higher order corrections are *asymptotically equivalent* to the KdV equation up to the first order (see also [8, 9, 28, 29] for a further discussion on the asymptotic equivalence for shallow water waves). As an explicit example, we also discuss the Gardner–Miura transformation (which is an invertible version of the Miura transformation) in terms of the normal form theory and show that the Gardner–Miura transformation is just a normal form transformation.

5.5.1 Normal Form

The basic idea of the normal form for the perturbed KdV equation (5.1) is to re-move all the nonresonant (nonsecular) terms in the higher order corrections using a near-identity transformation given by the Lie transformation. Since the symmetries $K_0^{(n)}(u)$ of the KdV equation give the obvious resonant terms, we write each higher order term $K^{(n)}(u)$ of (5.1) in the form

$$K^{(n)}(u) = a_1^{(n)} K_0^{(n)}(u) + F^{(n)}(u), \quad \text{for} \quad n = 1, 2, \ldots,$$

so that $F^{(n)}(u)$ has no linear term. Then the point of the transformation is to simplify the term by removing the nonresonant terms in $F^{(n)}(u)$. If one succeeds to remove the entire $F^{(n)}(u)$ up to $n = N$, then the perturbed equation is *asymptotically* inte-grable up to the order ϵ^N, and it possesses approximate integrals $I_l^{(n)}[u; \epsilon]$ for all $l \in \mathbb{Z}_{\geq 0}$ and $n = 1, \ldots, N$. However, as we have shown in Proposition 6, there are conditions for the existence of approximate conserved quantities. Thus we expect to see *obstacles* in removing all the nonlinear terms $F^{(n)}(u)$ from the higher order corrections. In fact we have

Theorem 10. *There exists a near-identity transformation, $T_\epsilon : v \mapsto u$,*

$$u = T_\epsilon(v) = v + \epsilon \phi^{(1)}(v) + \cdots, \quad \text{with} \quad \phi^{(n)}(v) \in \widehat{\mathscr{P}}_{\text{even}}^{(n)}[v]$$

such that the perturbed equation (5.1) is transformed to

$$v_t + G(v; \epsilon) = O(\epsilon^{N+1}), \tag{5.13}$$
$$\text{with} \quad G(u; \epsilon) = K_0^{(0)}(v) + \epsilon G^{(1)}(v) + \epsilon^2 G^{(2)}(v) + \cdots + \epsilon^N G^{(N)}(v),$$

where $G^{(n)}(u)$ are given by

$$G^{(n)}(v) = a_1^{(n)} K_0^{(n)}(v) + R^{(n)}(v), \quad \text{with} \quad R^{(n)}(v) = \sum_{i=1}^{\Delta(n)} \mu_i^{(n)} R_i^{(n)}(v).$$

Here the constants $\mu_i^{(n)}$ are given in Proposition 6 for the existence of the approx-imate conserved quantities, $\Delta(n)$ is the total number of the conditions for the exis-tence, and some $R_i^{(n)}(v) \in \widehat{\mathscr{P}}_{2n+5}[v]$.

Proof. We take the Lie (exponential) transform for a near-identity transform, that is,

$$u = T_\epsilon(v) = \exp V_\phi \cdot v,$$

where the generating function ϕ is expanded in the power series of ϵ,

$$\phi = \epsilon \phi^{(1)} + \epsilon^2 \phi^{(2)} + \cdots + \epsilon^N \phi^{(N)} + O\left(\epsilon^{(N+1)}\right).$$

Substituting this into (5.1) leads to

$$V_G = \text{Ad}_{\exp V_\phi} \cdot V_K := \exp V_\phi \cdot V_K \cdot \exp(-V_\phi),$$

which gives

$$G = \sum_{n=0}^{\infty} \frac{1}{n!} \left(\text{ad}_\phi\right)^n \cdot K = K + [\phi, K] + \frac{1}{2!}[\phi, [\phi, K]] + \frac{1}{3!}[\phi, [\phi, [\phi, K]]] + \cdots.$$

Then from each order of ϵ, we obtain

$$\left[K_0^{(0)}, \phi^{(1)}\right] = K^{(1)} - G^{(1)},$$

$$\left[K_0^{(0)}, \phi^{(2)}\right] = K^{(2)} - G^{(2)} + \frac{1}{2}\left[\phi^{(1)}, K^{(1)} + G^{(1)}\right],$$

$$\left[K_0^{(0)}, \phi^{(3)}\right] = K^{(3)} - G^{(3)} + \frac{1}{2}\left[\phi^{(2)}, K^{(1)} + G^{(1)}\right],$$

$$+ \frac{1}{2}\left[\phi^{(1)}, K^{(2)} + G^{(2)}\right] + \frac{1}{12}\left[\phi^{(1)}, \left[\phi^{(1)}, K^{(1)} + G^{(1)}\right]\right],$$

$$\vdots \cdot \vdots \cdot .$$

Thus we have the equation for $\phi^{(n)}$ in the form called the *homological* equation,

$$\text{ad}_{K_0^{(0)}} \cdot \phi^{(n)} := \left[K_0^{(0)}, \phi^{(n)}\right] = \tilde{K}^{(n)} - G^{(n)}, \tag{5.14}$$

where $\tilde{K}^{(n)}$ is successively determined from the previous equations. Since the ad-action $\text{ad}_{K_0^{(0)}}$ raises the weight by three, we have

$$\text{ad}_{K_0^{(0)}} : \widehat{\mathscr{P}}_{2n+2}[v] \longrightarrow \widehat{\mathscr{P}}_{2n+5}[v].$$

Then we assume the following form for $\phi^{(i)} \in \widehat{\mathscr{P}}_{2n+2}[v]$:

$$\phi^{(1)} = \alpha_1^{(1)} v^2 + \alpha_2^{(1)} v_{2x} + \alpha_3^{(1)} v_x D^{-1}(v),$$

$$\phi^{(2)} = \alpha_1^{(2)} v^3 + \alpha_2^{(2)} v v_{2x} + \alpha_3^{(2)} v_x^2 + \alpha_4^{(2)} v_{4x}$$
$$+ \alpha_5^{(2)} K^{(0)} D^{-1}(v) + \alpha_6^{(2)} v_x D^{-1}(v^2),$$

$$\phi^{(3)} = \alpha_1^{(3)} v^{4x} + \alpha_2^{(3)} v v_x^2 + \alpha_3^{(3)} v^2 v_{2x} + \alpha_4^{(3)} v_{2x}^2 + \alpha_5^{(3)} v_x v_{3x}$$
$$+ \alpha_6^{(3)} v v_{4x} + \alpha_7^{(3)} v_{6x} + \alpha_8^{(3)} v_x D^{-1}(v^3) + \alpha_9^{(3)} v_x D^{-1}(v_x^2)$$
$$+ \alpha_{10}^{(3)} K^{(0)} D^{-1}(v^2) + \alpha_{11}^{(3)} K_1 D^{-1}(v) + \alpha_{12}^{(3)} K^{(1)} D^{-1}(v).$$

The homological equation (5.14) gives a linear system of equations for the column vector $\alpha^{(n)} := (\alpha_1^{(n)}, \cdots, \alpha_{N(n)}^{(n)})^T$ with $N(n) = \dim \widehat{\mathscr{P}}_{2n+2}[v]$,

$$A\alpha^{(n)} = b^{(n)}, \tag{5.15}$$

where A is a $(M(n) - 1) \times N(n)$ matrix representation of the linear map $\mathrm{ad}_{K_0^{(0)}}$ on $\widehat{\mathscr{P}}_{2n+2}[v]$, and $b^{(n)}$ represents the coefficients of the monomial in $\tilde{K}^{(n)} - G^{(n)}$ which has $M(n) - 1$ elements with $M(n) = \dim \widehat{\mathscr{P}}_{2n+5}[v]$. The system (5.15) is overdetermined, and the total number of constraints for the consistency of the system is given by

$$N(m) := \dim \widehat{\mathscr{P}}_{2n+5} - 1 - \dim \widehat{\mathscr{P}}_{2n+2}.$$

This number should agree with that of the conditions for the existence of approximate symmetries, that is, the number of $\mu_i^{(n)}$ in Proposition 6.

Let us give an explicit form of $G^{(n)}$ up to $n = 2$:

At order ϵ, the matrix A in (5.15) is given by a 3×3 matrix with rank 3, so that we have $R^{(1)} = 0$, that is, no obstacle. The explicit transformation $\phi^{(1)}(v)$ is given by

$$\alpha_1^{(1)} = \frac{1}{6}\left(20a_1^{(1)} + a_2^{(1)} - a_4^{(1)}\right), \quad \alpha_2^{(1)} = \frac{1}{12}\left(10a_1^{(1)} + a_3^{(1)} - a_4^{(1)}\right),$$

$$\alpha_3^{(1)} = \frac{1}{3}\left(10a_1^{(1)} - a_2^{(1)}\right).$$

At order ϵ^2, the matrix A is a 7×6 matrix with rank 6, and we have one obstacle with the form

$$R^{(2)}(v) = b_2^{(2)}v_{5x}v + b_3^{(2)}v_{4x}v_x + b_4^{(2)}v_{3x}v^2 + b_5^{(2)}v_{3x}v_{2x}$$
$$+ b_6^{(2)}v_{2x}v_xv + b_7^{(2)}v_xv^3 + b_8^{(2)}v_x^3,$$

Here $b_i^{(2)}$, $i = 2, 3, \cdots, 8$, are constants satisfying the condition

$$170b_2^{(2)} - 60b_3^{(2)} - 8b_4^{(2)} + 24b_5^{(2)} - 9b_6^{(2)} + 3b_7^{(2)} + 24b_8^{(2)} = 1.$$

The explicit formula of the $\alpha^{(2)} = (\alpha_1^{(2)}, \cdots, \alpha_6^{(2)})^T$ is given in Appendix.   □

Theorem 10 shows that in particular, the perturbed KdV equation (5.1) can be transformed into the integrable equation of the KdV equation with the fifth-order symmetry. This implies that the weakly dispersive nonlinear wave equations are asymptotically equivalent to the integrable system up to the first-order correction.

The perturbed equation (5.1) may have several approximate conserved quantities based on the original physical setting. Then we consider a particular form of the transformed equation (5.13) whose conserved quantities are given by those of the KdV equation. We call this form of equation the normal form of (5.1), that is, we define

Definition 11. For a subset of integers $\Gamma \subset \mathbb{Z}_{\geq 0}$, suppose that the perturbed KdV equation (5.1) has the approximate conserved quantities $I_l[u, \epsilon]$ for $l \in \Gamma$ up to order ϵ^N. Then the normal form of (5.1) is defined by (5.13)

$$v_t + \sum_{n=0}^{N} \epsilon^n G^{(n)}(v) = O\left(\epsilon^{N+1}\right),$$

whose conserved quantities $J_l[v; \epsilon] := I_l[T_\epsilon(v); \epsilon]$ for $l \in \Gamma$ are expressed in terms of the conserved quantities of the KdV equation $I_l^{(0)}[v]$,

$$J_l[v; \epsilon] = I_l^{(0)}(v) + \epsilon c_l^{(1)} I_{l+1}^{(0)}(v) + \cdots + \epsilon^N c_l^{(N)} I_{N+l}^{(0)}(v) + O\left(\epsilon^{N+1}\right), \qquad (5.16)$$

where $c_l^{(i)}$, $i = 1, 2, \ldots, N$, are some real constants.

In particular, if the set Γ contains the first three numbers, $\Gamma \supseteq \{0, 1, 2\}$, then the normal form admits a solitary wave solution in the form of KdV soliton (5.9). This can be seen by taking the variation

$$\nabla(J_2[v, \epsilon] + \lambda J_1[v. \epsilon]) = 0.$$

Since many physically interesting systems possess those conserved quantities as mass, momentum and energy, we expect to find a solitary wave close to the KdV soliton for such systems. Now we show the existence of such normal form for the case $\Gamma = \{0, 1, 2\}$:

Theorem 12. *Suppose that the perturbed KdV equation (5.1) has the first three approximate conserved quantities $I_l[u, \epsilon]$, $l = 0, 1, 2$ up to order ϵ^3. Then the corresponding normal form takes the form (5.13) with*

$$R^{(1)}(v) = 0,$$
$$R^{(2)}(v) = \mu_1^{(2)} R_1^{(2)},$$
$$R^{(3)}(v) = -\mu_1^{(2)} c_2^{(1)} \mathfrak{R}(R_1^{(2)}) + \mu_2^{(3)} R_1^{(3)} + \mu_3^{(3)} R_2^{(3)},$$

where the conserved quantities $J_l[v, \epsilon]$ for the normal form are given by

$$J_l[v, \epsilon] = I_l[v] + \epsilon c_l^{(1)} I_{l+1}[v] + \cdots + \epsilon^3 c_l^{(3)} I_{l+3}[v] + O(\epsilon^4).$$

The obstacles $R_k^{(n)}$ are expressed as

$$R_1^{(2)} = \frac{-1}{50} \left(v_{5x} v + \frac{3}{2} v_{4x} v_x + 5 v_{3x} v^2 - \frac{5}{2} v_{3x} v_{2x} + 20 v_{2x} v_x v + 10 v_x v^3 + 5 v_x^3 \right),$$

$$R_1^{(3)} = \frac{1}{175} \left(v_{7x} v + \frac{5}{2} v_{6x} v_x + 14 v_{5x} v^2 - \frac{7}{2} v_{4x} v_{3x} + 63 v_{4x} v_x v + 35 v_{3x} v_{2x} v \right.$$
$$+ 56 v_{3x} v_x^2 + 42 v_{3x} v^3 + 42 v_{2x}^2 v_x + 273 v_{2x} v_x v^2 + 168 v_x^3 v$$
$$\left. + \frac{105}{2} v_x v^4 + 21 v_x D^{-1}(v_x^3) \right),$$

$$R_2^{(3)} = \frac{1}{175}\left(4v_{7x}v + 10v_{6x}v_x + \frac{91}{2}v_{5x}v^2 - 14v_{4x}v_{3x} + \frac{441}{2}v_{4x}v_xv\right.$$
$$+ \frac{385}{2}v_{3x}v_{2x}v + 203v_{3x}v_x^2 + 147v_{3x}v^3 + \frac{357}{2}v_{2x}^2v_x + 903v_{2x}v_xv^2$$
$$\left. + 483v_x^3v + 210v_xv^4 + 21v_xD^{-1}(v_x^3)\right)$$

and $\mathfrak{R}$ is the recursion operator.

Proof. Recall that $J_l[v,\epsilon] = \int_{\mathbb{R}} \rho(v;\epsilon)dx$ is an approximate conserved quantity for (5.13) if

$$V_G \cdot \rho(v;\epsilon) + (\epsilon^{N+1}) \in \text{Im}(D).$$

Then using the form (5.13) and $\rho_l = \rho_l^{(0)} + \epsilon c_l^{(1)}\rho_{l+1}^{(0)} + \cdots$, we have, at order ϵ^2,

$$V_{R^{(2)}} \cdot \rho_l^{(0)}(v) \in \text{Im}(D), \quad \text{for} \quad l = 0, 1, 2, \tag{5.17}$$

and at order ϵ^3,

$$V_{R^{(3)}} \cdot \rho_k^{(0)}(v) \in \text{Im}(D), \quad \text{for} \quad k = 0, 1,$$
$$V_{R^{(3)}} \cdot \rho_2^{(0)} + c_2^{(1)}V_{R^{(2)}} \cdot \rho_3^{(0)} \in \text{Im}(D). \tag{5.18}$$

Then from a direct computation with the explicit form of $R^{(2)} \in \mathscr{P}_9[v]$, the conditions (5.17) lead to the required form of $R^{(2)}(v)$.

For (5.18), we first write $R^{(3)}$ in the sum of homogeneous solution $R_h^{(3)}$ and a particular solution $R_p^{(3)}$, that is,

$$V_{R_h^{(3)}} \cdot \rho_l^{(0)} \in \text{Im}(D), \quad \text{for} \quad l = 0, 1, 2,$$
$$V_{R_p^{(3)}} \cdot \rho_2^{(0)} + V_{R^{(2)}} \cdot \rho_3^{(0)} \in \text{Im}(D).$$

Then one can find a particular solution by using the recursion operator $\mathfrak{R} = \Theta D^{-1}$,

$$\begin{aligned}
V_{R^{(2)}} \cdot \rho_3^{(0)} &= R^{(2)}\nabla I_3^{(0)} && \text{mod Im}(D) \\
&= R^{(2)}D^{-1}D\nabla I_3^{(0)} && \text{mod Im}(D) \\
&= R^{(2)}D^{-1}\Theta\nabla I_2^{(0)} && \text{mod Im}(D) \\
&= (\Theta D^{-1}R^{(2)})\nabla I_2^{(0)} && \text{mod Im}(D) \\
&= V_{\mathfrak{R}R^{(2)}} \cdot \rho_2^{(0)} && \text{mod Im}(D)
\end{aligned}$$

from which we have

$$R_p^{(3)} = -\mu_1^{(2)}c_2^{(1)}\mathfrak{R}(R_1^{(2)}).$$

Then from a direct computation we have the homogeneous solution in the desired form,

$$R_h^{(3)} = \mu_2^{(3)}R_1^{(3)} + \mu_3^{(3)}R_2^{(3)}.$$

□

In [5], Bilge also studied the integrability of the perturbed KdV equation (5.1) using the normal form theory and the formal symmetry approach developed in [30]. The main result is to show that the only integrable KdV like seventh-order equations are the KdV, Sawada–Kotera and Kaup equations.

Remark 13. We have the following remarks on the obstacles:

a) In general, the explicit form of the obstacles $R^{(n)}$ may be successively obtained by solving the linear equations

$$V_{R^{(n)}} \cdot \rho_l^{(0)} + \sum_{k=1}^{n-2} c_l^{(k)} V_{R^{(n-k)}} \rho_{l+k}^{(0)} \in \mathrm{Im}(D), \quad \text{for} \quad n = 2, 3, \cdots.$$

Suppose we found $R^{(k)}$ up to $k = n-1$. Then we have a particular solution in the form

$$R_p^{(n)} = -\sum_{k=1}^{n-2} c_l^{(k)} \Re^k R^{(n-k)}.$$

Here one has to check the compatibility among the different $l = 0, 1, 2$.

b) Since the normal form with $\Gamma = \{0, 1, 2\}$ admits the solitary wave in the form of the KdV soliton, one can see that all the obstacles vanish when v assumes the KdV soliton solution

$$R^{(n)}(v) = 0, \quad \text{when} \quad v = 2\kappa^2 \mathrm{sech}^2(\kappa(x - x_0)).$$

5.5.2 The Gardner–Miura Transformation

It was found in [32] that there is an *invertible* transformation between the KdV equation and the KdV equation with a cubic nonlinear term,

$$u_t + 6uu_x + u_{3x} = \epsilon a u^2 u_x, \tag{5.19}$$

where a is an arbitrary constant. The transformation is called the Gardner–Miura transformation which is an invertible version of the Miura transformation,

$$v = u - \alpha \epsilon^{1/2} u_x - \alpha^2 \epsilon u^2, \quad \text{with} \quad \alpha = -\sqrt{a/6}. \tag{5.20}$$

Here we treat (5.19) as an example of the perturbed KdV equation and give an explicit formulation of the Gardner–Miura transformation in terms of the normal form theory. Since the perturbed equation (5.19) has an infinite number of conserved densities and there is no resonant term as the symmetry of the KdV equation in the higher order, the normal form is just the KdV equation. Then we construct the normal form transformation $u = T_\epsilon(v)$ which is the inverse of the Gardner–Miura transformation (5.20).

Since the Gardner–Miura transformation (5.20) is of a Riccati type, the change of the variable

$$\delta u = \frac{1}{\delta}\left(D\ln\varphi + \frac{1}{2\delta}\right), \quad \text{with} \quad \delta = \alpha\epsilon^{1/2}, \tag{5.21}$$

leads to the Schrödinger equation

$$L^2\varphi := (D^2 + v)\varphi = k^2\varphi, \quad \text{with} \quad k = -1/(2\delta).$$

Using the notion of the pseudo-differential operators D^v for $v \in \mathbb{Z}$, one can define $L = (D^2 + v)^{1/2}$ as

$$L = D + q_1 D^{-1} + q_2 D^{-2} + \cdots, \quad q_i \in \mathscr{P}_{i+2}[v].$$

Then writing D in the power series of L, we have

$$D = L + p_1 L^{-1} + p_2 L^{-2} + \cdots, \quad p_i \in \mathscr{P}_{i+2}[v]$$

and using $L\varphi = k\varphi$, we find

$$D\ln\varphi = -\frac{1}{2\delta} + \sum_{i=1}^{\infty}(-2\delta)^i p_i(v). \tag{5.22}$$

Note in particular that we have

$$q_1 = \frac{1}{2}v, \quad p_1 = -\frac{1}{2}v.$$

From (5.21), Eq. (5.22) leads to the inverse of the Gardner–Miura transformation

$$u = \sum_{i=1}^{\infty}(-2)^i\delta^{i-1}p_i(v, v_x, \cdots)$$
$$= v + \delta v_x + \delta^2(v_{2x} - v^2) + \cdots.$$

We now remove the terms of the non-integer powers of ϵ, the odd integer of $\delta = \alpha\epsilon^{1/2}$, in this equation. The first term δv_x can be removed by the translation of x. After removing the term δv_x, we have $K_0^{(0)}(v)$ at the order of $\epsilon^{3/2}$. This can be removed by the translation of t. Continuing this process, we see the symmetries of the KdV equation at the non-integer powers of ϵ. Then shifting the symmetry parameters x_{2n+1}, those can be removed. Here $x = x_1$ and $t = -x_3$. Now we can show

Theorem 14. *The inverse of the Gardner–Miura transformation (5.20) gives a normal form transform*

$$u = \frac{1}{\delta}D^{-1}\sinh\left(\delta\sum_{i=0}^{\infty}\frac{\delta^{2i}}{2i+1}\frac{\partial}{\partial x_{2i+1}}\right)v, \quad \delta = -\sqrt{\frac{\epsilon a}{6}},$$

where the derivative of v with respect to x_{2n+1} defines the symmetry, that is,
$\partial v/\partial x_{2n+1} = K_0^{(n)}(v)$.

Proof. It is well known [13] that the wave function φ of the Schrödinger equation can be expressed by the τ-function

$$\varphi(\mathbf{x},k) = \frac{\tau(\mathbf{x}+2\langle\delta\rangle)}{\tau(\mathbf{x})} e^{kx_1}, \quad \text{with} \quad k = -\frac{1}{2\delta},$$

where we denote $\mathbf{x} := (x_1, x_2, \ldots)$ and

$$\langle\delta\rangle = \left(\delta, \frac{\delta^3}{3}, \frac{\delta^5}{5}, \ldots\right).$$

Expanding the equation $D \ln \varphi$ with this equation in the power of δ, we find

$$v(\mathbf{x}) = 2D^2 \ln \tau(\mathbf{x}). \tag{5.23}$$

Then from (5.21) we have

$$u(\mathbf{x}) = \frac{1}{\delta} D\left(\ln \tau(\mathbf{x}+2\langle\delta\rangle) - \ln \tau(\mathbf{x})\right).$$

Now applying the vertex operator

$$V(\delta) = \exp\left(-\delta \sum_{i=0}^{\infty} \frac{\delta^{2i}}{(2i+1)} \frac{\partial}{\partial x_{2i+1}}\right)$$

and using (5.23), we obtain the result. Note here that both $u(\mathbf{x})$ and $u(\mathbf{x}+\langle\delta\rangle)$ satisfy the same equation (5.19). □

5.6 Interactions of Solitary Waves

As an important application of the normal form theory, we consider the interaction problem of solitary waves and show how the theory enables us to understand the interaction properties under the influence of the higher orders in the perturbed KdV equation (5.1). We assume that the perturbed equation possesses the first three conserved quantities, i.e., $\Gamma = \{0,1,2\}$. Then the first obstacle appears in the order ϵ^2 as $\mu_1^{(2)} \neq 0$, that is, we consider the normal form

$$v_t + K_0^{(0)}(v) + \epsilon a_1^{(1)} K_0^{(1)}(v) + \epsilon^2\left(a_1^{(2)} K_0^{(2)}(v) + \mu_1^{(2)} R_1^{(2)}(v)\right) = 0.$$

Several physical examples of this type will be discussed in Sect. 5.7. We use the method of perturbed inverse scattering transform [16, 17] to analyze the normal form. We start with a brief description of the method.

5.6.1 Inverse Scattering Transform

The key of the inverse scattering transform is based on the one-to-one correspondence for each t between the scattering data $S(t)$ and the potential $v(x,t)$, the solution of the normal form, of the Schrödinger equation (see for example [1]),

$$\frac{\partial^2 \varphi}{\partial x^2} + (v + k^2)\varphi = 0.$$

The correspondence is given by the formula of $v(x,t)$ in terms of the squared eigenfunctions

$$v(x,t) = 4 \sum_{j=1}^{N} \kappa_j(t) C_j(t) \varphi^2(x,t; i\kappa_j) + \frac{2i}{\pi} \int_{-\infty}^{\infty} kr(t;k) \varphi^2(x,t;k)\, dk. \qquad (5.24)$$

Here the scattering data $S(t)$ is defined by

$$S(t) := \left\{ \{\kappa_j(t) > 0, C_j(t)\}_{j=1}^{N},\ r(t;k) \text{ for } k \in \mathbb{R} \right\},$$

and the eigenfunction $\varphi(x,t;k)$ is assumed to satisfy the boundary condition

$$\varphi(x,t;k) \longrightarrow e^{-ikx}, \qquad \text{as } x \longrightarrow -\infty.$$

Note that the squared function $\varphi^2(x,t;k)$ satisfies

$$\Theta \varphi^2 := (D^3 + 2(Dv + vD))\varphi^2 = -4k^2 D\varphi^2,$$

so that the function $D\varphi^2$ is the eigenfunction of the recursion operator $\mathfrak{R}$ with the eigenvalue $-4k^2$.

With the formula of v in (5.24), the conserved quantities $J_i[v; \epsilon]$ can be expressed in terms of the scattering data

$$J_i[v;\epsilon] = I_i^{(0)}[v] + \epsilon c_i^{(1)} I_{i+1}^{(0)}[v] + \epsilon^2 c_i^{(2)} I_{i+2}^{(0)}[v] + O(\epsilon^3), \quad i = 0,1,2, \qquad (5.25)$$

where $I_j^{(0)}$ can be expressed as

$$I_m^{(0)} = \frac{2^{2m+1}}{2m+1}(-1)^{m+1} \left[\sum_{j=1}^{M} \kappa_j^{2m+1} - (-1)^m \Delta_m \right] \qquad (5.26)$$

with the radiation part $\Delta_m(t)$

$$\Delta_m = -\frac{2m+1}{2\pi} \int_0^{\infty} k^{2m} \ln(1 - |r(t;k)|^2)\, dk \geq 0.$$

The existence of such conserved quantities plays a crucial rule for the interaction as in the case of the KdV solitons.

The time evolution of the scattering data is determined by

$$\frac{d}{dt}S(t) = -V_K \cdot S(t), \quad \text{for} \quad v_t + K(v) = 0.$$

In particular, the equation for the reflection coefficient $r(t;k)$ is given by

$$\frac{d}{dt}r(t;k) = \frac{-1}{2ika^2(t;k)} \int_{\mathbb{R}} v_t(x,t)\varphi^2(x,t;k)\, dx,$$

where $a(t;k)$ is the reciprocal transmission coefficient determined by $r(t;k)$ and κ_j. Using the normal form $v_t + K(v) = 0$, we obtain the equation for $r(t;k)$

$$\frac{dr}{dt} = i\omega r + \epsilon^2 \frac{\mu_1^{(2)}}{2ika^2} \int_{-\infty}^{\infty} R_1^{(2)}\varphi^2(x,t;k)dx + O\left(\epsilon^3\right), \tag{5.27}$$

where $\omega = 8k^3(1 - 4\epsilon a_1^{(1)}k^2 + 16a_1^{(2)}\epsilon^2 k^4)$.

Those given above provide enough information for our purpose of studying the interaction of solitary waves.

5.6.2 Solitary Wave Interaction

Recall that the obstacle vanishes for one-soliton solution of the KdV equation, that is, there is no effect of the obstacle on the solitary wave. The higher order terms lead to the shift of the velocity of the soliton solution due to the resonance caused by the symmetries of the KdV equation [24]. In order to see the effect of obstacles, we now consider the interaction of two solitary waves and show the inelasticity in the interaction which can be considered as a nonintegrable effect of the obstacle.

5.6.2.1 Inelasticity in the Interaction

Let us assume that the initial data consist of two well-separated solitary waves with parameters $\kappa_1 > \kappa_2$ in the form of (5.9), traveling with speed,

$$s_j = 4\kappa_j^2\left(1 - 4\epsilon a_1^{(1)}\kappa_j^2 + 16\epsilon^2 a_1^{(2)}\kappa_j^4\right), \quad j = 1,2, \tag{5.28}$$

and they are approaching each other. Then we analyze the interaction by using a perturbation method where the leading order solution is assumed to be the exact two-soliton solution of the KdV equation, that is, for large $x_{02} - x_{01} \gg 1$

$$v(x,0) \simeq 2\kappa_1^2\text{sech}^2(\kappa_1(x - x_{01})) + 2\kappa_2^2\text{sech}^2(\kappa_2(x - x_{02})).$$

We then determine the evolution of the scattering data, $\kappa_j(t)$ and the radiation $\Delta_m(t)$ in (5.29). First we have

Theorem 15. *[19] Due to the interaction of two solitary waves, their parameters κ_j, $j = 1, 2$, are shifted by $\Delta\kappa_j(t)$ which can be expressed in terms of the radiation $\Delta_m(t)$ for any $m = 0, 1, 2$:*

$$\Delta\kappa_1 = \frac{5\kappa_2^2\Delta_1 + 3\Delta_2}{15\kappa_1^2\left(\kappa_1^2 - \kappa_2^2\right)} + O(\epsilon\Delta_m) \geq 0,$$

$$\Delta\kappa_2 = -\frac{5\kappa_1^2\Delta_1 + 3\Delta_2}{15\kappa_2^2\left(\kappa_1^2 - \kappa_2^2\right)} + O(\epsilon\Delta_m) \leq 0.$$

There is also a production of the third soliton with the parameter

$$\Delta\kappa_3 = \frac{15\kappa_1^2\kappa_2^2\Delta_0 + 5\left(\kappa_1^2 + \kappa_2^2\right)\Delta_1 + 3\Delta_2}{15\kappa_1^2\kappa_2^2} + O(\epsilon\Delta_m) \geq 0.$$

Proof. From the conserved quantities $J_l[v; \epsilon] = \text{constant for } l = 0, 1, 2$ up to order ϵ^3, we obtain

$$\Delta I_l^{(0)} + \epsilon c_l^{(0)}\Delta I_{l+1}^{(0)} + \epsilon^2 c_l^{(2)}\Delta I_{l+2}^{(0)} = O\left(\epsilon^3\right),$$

where the variations $\Delta I_n^{(0)}$ are taken over the shifts $\Delta\kappa_j$ and the radiation Δ_m, that is,

$$(2l + 1)\sum_{j=1}^{N} \kappa_j^{2l}\Delta\kappa_j - (-1)^l\Delta_l = O(\epsilon\Delta_l), \quad \text{for} \quad l = 0, 1, 2.$$

Here $N - 2$ is a possible number of new solitary waves. Since the $\Delta\kappa_j$ for $j > 2$ represents a new eigenvalue of the Schrödinger equation, it is nondegenerate and an isolated point on the imaginary axis of the spectral domain . This implies that the number of new eigenvalues should be just one for a sufficiently small ϵ, i.e., $N = 3$.

Because of $\kappa_j = 0$ initially for $j > 2$, one can find the formulae of $\Delta\kappa_j$ for $j = 1, 2$ from those variations for $l = 1, 2$. Also the first variation with $l = 0$,

$$\sum_{j=1}^{3} \Delta\kappa_j - \Delta_0 = O(\epsilon\Delta_0),$$

leads to the formula $\Delta\kappa_3$ for a new solitary wave. □

Thus we find the following:

a) The total mass, $M = \int v\, dx \propto \kappa$, of the larger solitary wave is increased, and contrarily that of the smaller solitary wave is decreased by the interaction.
b) The amount of the energy change, $E = \int v^2 dx \propto \kappa^3$, has the property $1 < |\Delta E_2/\Delta E_1| < (\kappa_1/\kappa_2)^2$, i.e., the energy expense of the smaller solitary wave is more than the energy gain of the larger one.
c) The interaction produces a new solitary wave as well as radiation.

Those results except for a new solitary wave production are consistent with the numerical observation in [6]. It may be difficult to observe the new solitary wave from the numerical calculation, since this solitary wave has long width and small amplitude $(\Delta \kappa_3)^2$ which is of order ϵ^8 (see below).

The function $\Delta_m(t)$ of the radiation can be computed as follows: The reflection coefficient $r(t;k)$ in (5.27) can be expressed as

$$r(t;k) = \frac{\epsilon^2 \mu_1^{(2)}}{2ik} \left(\int_0^t d\tau \frac{e^{-i\omega\tau}}{a(\tau;k)^2} \int_{\mathbb{R}} dx \varphi^2(x,\tau;k) R_1^{(2)}(v) \right) e^{i\omega t}.$$

Then $\Delta_m(t)$ is given by

$$\Delta_m(t) = \frac{2m+1}{2\pi} \int_0^\infty D_m(t;k) \, dk + o(\epsilon^4), \tag{5.29}$$

where

$$D_m(t;k) = k^{2m} |r(t;k)|^2 + o(\epsilon^4)$$
$$= \epsilon^4 \frac{(\mu_1^{(2)})^2 k^{2(m-1)}}{4} \left| \int_0^t d\tau \, e^{-i\omega\tau} \int_{\mathbb{R}} dx \, \varphi^2(x,\tau;k) R_1^{(2)}(v) \right|^2 + o(\epsilon^4).$$

We numerically calculate the formula $\Delta_m(t)$ by means of perturbation, that is, we assume $v(x,t)$ to be a two-soliton solution of the integrable one $v_t + K^{(0)} + \epsilon a_1^{(1)} K_1 + \epsilon^2 a_1^{(2)} K_2 = 0$ and $\varphi(x,t,k)$ is the corresponding eigenfunction. The result will be shown in Sect. 5.7 for the examples of ion acoustic waves and the Boussinesq equation.

5.6.2.2   Additional Phase Shifts of Solitary Wave

As a consequence of the nonlocal terms in the normal form transformation, one can find the additional phase shifts on the solitary waves $u(x,t)$ of the perturbed KdV equation (5.1) through their interaction.

Let us first recall the phase shifts of the two-soliton solution for $v(x,t)$ [2, 11]. The asymptotic form of $v(x,t)$ consists of well-separated one solitons,

$$v(x,t) \approx v_j^\pm(x,t), \quad \text{as} \quad \begin{cases} t \to \pm\infty \\ \kappa_j x \sim \omega_j t \end{cases}$$

$$\text{with} \quad v_j^\pm(x,t) = 2\kappa^2 \text{sech}(\kappa_j x - \omega_j t - \theta_j^\pm),$$

where $\omega_j = \kappa_j s_j$ with the speed s_j in (5.28). Then the phase shifts $\Delta x_j^{(0)} := (\theta_j^+ - \theta_j^-)/\kappa_j$ are given by

$$\Delta x_1^{(0)} = -\frac{1}{\kappa_1} \ln\left(\frac{\kappa_1 - \kappa_2}{\kappa_1 + \kappa_2}\right), \quad \text{and} \quad \Delta x_2^{(0)} = \frac{1}{\kappa_2} \ln\left(\frac{\kappa_1 - \kappa_2}{\kappa_1 + \kappa_2}\right).$$

We now compute the correction to the shift $\Delta x_j^{(0)}$ using the normal form transformation. From the asymptotic form of the two-soliton solution for $t \to -\infty$, we have up to order ϵ

$$u(x,t) \underset{\kappa_1 x \sim \omega_1 t}{\longrightarrow} v_1^- + \epsilon \left(\alpha_1^{(1)}(v_1^-)^2 + \alpha_2^{(1)}(v_1^-)_{2x} + \alpha_3^{(1)}(v_1^-)_x \int_{-\infty}^{x} v_1^- \, dx \right)$$

$$\underset{\kappa_2 x \sim \omega_2 t}{\longrightarrow} v_2^- + \epsilon \left(\alpha_1^{(1)}(v_2^-)^2 + \alpha_2^{(1)}(v_2^-)_{2x} + \alpha_3^{(1)}(v_2^-)_x \int_{-\infty}^{x} v_2^- \, dx \right)$$

$$+ \epsilon \alpha_3 (v_2^-)_x \int_{\mathbb{R}} v_1^- \, dx.$$

One should note here that there is an extra term in the v_2 solitary wave which is the key term for the additional phase shift. Namely the term can be absorbed as a translation of x in v_2. The other terms contribute to modify the shape of the soliton, the dressing part.

Also for $t \to +\infty$, we have

$$u(x,t) \underset{\kappa_1 x \sim \omega_1 t}{\longrightarrow} v_1^+ + \epsilon \left(\alpha_1^{(1)}(v_1^+)^2 + \alpha_2^{(1)}(v_1^+)_{2x} + \alpha_3^{(1)}(v_1^+)_x \int_{-\infty}^{x} v_1^+ \, dx \right)$$

$$+ \epsilon \alpha_3 (v_1^+)_x \int_{\mathbb{R}} v_2^+ \, dx$$

$$\underset{\kappa_2 x \sim \omega_2 t}{\longrightarrow} v_2^+ + \epsilon \left(\alpha_1^{(1)}(v_2^+)^2 + \alpha_2^{(1)}(v_2^+)_{2x} + \alpha_3^{(1)}(v_2^+)_x \int_{-\infty}^{x} v_2^+ \, dx \right).$$

Now the additional shift appears to v_1. This can be extended for the next order where the shift also appears as a translation of t with a term like $K_0^{(0)}(v_1) \int_{\mathbb{R}} v_2 dx$. Thus we have the total phase shift for v_1 solitary wave

$$\Delta x_1 = \Delta x_1^{(0)} + \epsilon \Delta x_1^{(1)} + \epsilon^2 \Delta x_1^{(2)} + O(\epsilon^3), \tag{5.30}$$

where the additional phase shifts are given by

$$\Delta x_1^{(1)} = \alpha_3^{(1)} \int_{\mathbb{R}} v_2 dx = 4\kappa_2 \alpha_3^{(1)},$$

$$\Delta x_1^{(2)} = \alpha_6^{(2)} \int_{\mathbb{R}} (v_2)^2 dx + \frac{\alpha_3^{(1)}\left(\alpha_1^{(1)} - \alpha_3^{(1)}\right)}{2} \int_{\mathbb{R}} (v_2)^2 dx + 4\kappa_1^2 \alpha_5^{(2)} \int_{\mathbb{R}} v_2 dx$$

$$= \left(\frac{16}{3}\alpha_6^{(2)} + \frac{8}{3}\alpha_3^{(1)}\left(\alpha_1^{(1)} - \alpha_3^{(1)}\right)\right)\kappa_2^3 + 16\alpha_5^{(2)}\kappa_1^2\kappa_2.$$

5.7 Examples

In this section, the normal form theory is applied to some explicit models including the ion acoustic wave equation, the Boussinesq equation as a model of the shallow water waves and the regularized long wave equation (sometimes called BBM equation). We also carry out the numerical simulation for those examples and compare the results with the predictions obtained from the normal form theory such as the phase shift (5.30) and the radiation energy (5.29).

We also consider the seventh-order Hirota KdV equation which is known to be nonintegrable even though it admits an exact two solitary wave solution. The main issue is to determine the order of the obstacle of the corresponding normal form, which indicates a nonintegrability of the equation in the asymptotic sense. It turns out that the obstacles appear at order ϵ^4.

5.7.1 Ion Acoustic Waves

An asymptotic property of the ion acoustic waves has been discussed in several papers (see for example [38]). These studies show that the KdV equation is derived as the first approximation of the ion acoustic wave equation under the weakly dispersive limit. The higher order corrections to the KdV soliton solution has also been discussed in [24]. Recently, Li and Sattinger in [26] studied numerically the interaction problem of solitary waves and showed that the amplitude of the radiation after two solitary wave interaction can be observed as small as 10^{-5} order, and they concluded that the KdV equation gives an excellent approximation. Here we explain those observations based on the normal form theory developed in the previous sections.

Following the method in [38], we first derive the KdV equation with the higher order corrections. The ion acoustic wave equation is expressed by the system of three partial differential equations in the dimensionless form

$$
\begin{cases}
(n_i)_T + (n_i v_i)_X = 0, \\
(v_i)_T + \left(\dfrac{v_i^2}{2} + \phi \right)_X = 0, \\
\phi_{2X} - \exp \phi + n_i = 0,
\end{cases}
\tag{5.31}
$$

where n_i, v_i and ϕ are the normalized variables for ion density, ion velocity and electric potential. Electron density n_e is related with ϕ as $n_e = \exp(\phi)$. Assuming the weak nonlinearity and the weak dispersion, we introduce the scaled variables

$$
v_i = \epsilon v, \quad n_e = 1 + \epsilon n \quad \text{and} \quad x = \epsilon^{1/2} X, \quad t = \epsilon^{1/2} T.
$$

Then we write (5.31) in the following form for (n, v):

$$\begin{cases} \left(1-\epsilon D^2 \cdot \dfrac{1}{1+\epsilon n}\right) n_t + v_x + \epsilon(nv)_x - \epsilon(v\ln(1+\epsilon n))_x = 0, \\[2ex] v_t + vv_x + \dfrac{1}{1+\epsilon n} n_x = 0. \end{cases}$$

Then inverting the operator in front of n_t, we have the matrix equation

$$\frac{\partial}{\partial t}U + A_0 \frac{\partial}{\partial x}U + \epsilon\frac{\partial}{\partial x}B(U) = 0, \quad \text{with} \quad U = \begin{pmatrix} n \\ v \end{pmatrix}, \tag{5.32}$$

where A_0 is the constant matrix $A_0 = \begin{pmatrix} 0 & 1 \\ 1 & 0 \end{pmatrix}$, and the vector function $B(U)$ is given by the expansion

$$B = B^{(1)} + \epsilon B^{(2)} + \epsilon^2 B^{(3)} + \cdots,$$

$$\text{with} \begin{cases} B^{(1)} = \begin{pmatrix} v_{2x} + nv \\ \frac{1}{2}(v^2 - n^2) \end{pmatrix}, \quad B^{(2)} = \begin{pmatrix} v_{4x} + n_x v_x \\ \frac{1}{3}n^3 \end{pmatrix}, \\[3ex] B^{(3)} = \begin{pmatrix} v_{6x} + (n_x v_x)_{2x} - (nv_{3x})_x - vnn_x \\ -\frac{1}{4}n^4 \end{pmatrix}. \end{cases}$$

Thus for the case with $\epsilon = 0$, we have two simple linear waves propagating with the speeds $\lambda_\pm = \pm 1$ given by the eigenvalues of the matrix A_0. We then look for an asymptotic wave along with the speed $\lambda_+ = 1$, so that we introduce the scaled variables on this moving frame,

$$x' = x - t, \quad t' = \epsilon t,$$

which gives, after dropping the prime on the new variable,

$$(A_0 - I)\frac{\partial U}{\partial x} + \epsilon\left(\frac{\partial U}{\partial t} + \frac{\partial B(U)}{\partial x}\right) = 0, \tag{5.33}$$

where I is the 2×2 identity matrix. Let us decompose U in the form

$$U(x,t) = u(x,t)R_+ + f(x,t)R_-,$$

where $R_\pm$ are the eigenvectors corresponding to the eigenvalues $\lambda_\pm = \pm 1$. Then taking the projections of (5.33) on the $R_\pm$-directions, we have

$$\begin{cases} u_t + (L_+ B)_x = 0, \\[1.5ex] -2f_x + \epsilon(f_t + (L_- B)_x) = 0, \end{cases}$$

where $L_\pm$ are the left eigenvectors with the normalization $L_\pm R_\pm = 1$, $L_\pm R_\mp = 0$. Then we can see that $f(x,t)$ can be expressed in an expansion form

$$f(x,t) = \epsilon f^{(1)}(u) + \epsilon^2 f^{(2)}(u) + \epsilon^3 f^{(3)}(u) + \cdots, \quad \text{with} \quad f^{(k)}(u) \in \widehat{\mathscr{P}}^{(k)}_{\text{even}}[u],$$

where $f^{(n)}$ are determined iteratively from

$$f = \frac{\epsilon}{2}(L_- B) + \frac{\epsilon}{2} D^{-1} f_t = \frac{\epsilon}{2}\left(L_- B^{(1)}\right) + \frac{\epsilon^2}{2}\left(L_- B^{(2)}\right) + \frac{\epsilon}{2} f_t + \cdots.$$

Thus we obtain the perturbed KdV equation which takes the following form up to order ϵ^2 after an appropriate normalization

$$u_t + 6uu_x + u_{3x} + \epsilon\left(\frac{1}{4}u_{5x} - u_{3x}u - \frac{3}{2}u_x u^2\right) + \epsilon^2\left(\frac{5}{72}u_{7x} - \frac{1}{2}u_{5x}u\right.$$
$$\left. -\frac{17}{24}u_{4x}u_x + \frac{1}{6}u_{3x}u_{2x} + \frac{3}{4}u_{3x}u^2 - \frac{1}{4}u_{2x}u_x u - \frac{7}{8}u_x^3 + \frac{1}{4}u_x u^3\right) = O(\epsilon^3),$$

from which we have

$$\mu_1^{(2)} = -\frac{11}{9} \neq 0.$$

Thus the normal form of the ion acoustic wave equation has the obstacle $R_1^{(2)}$. The normal form transformation is given by

$$\begin{cases} \alpha_1^{(1)} = \dfrac{11}{12}, \quad \alpha_2^{(1)} = \dfrac{1}{3}, \quad \alpha_3^{(1)} = \dfrac{7}{6}, \\[2mm] \alpha_1^{(2)} = \dfrac{731}{3600}, \quad \alpha_2^{(2)} = \dfrac{89}{600}, \quad \alpha_3^{(2)} = \dfrac{433}{3600}, \\[2mm] \alpha_4^{(2)} = \dfrac{51}{800}, \quad \alpha_5^{(2)} = \dfrac{23}{1800}, \quad \alpha_6^{(2)} = \dfrac{87}{400}. \end{cases}$$

Remark 16. The physical variables n and v are expressed as

$$n = u + f, \quad v = u - f, \quad \text{with} \quad R_\pm = \begin{pmatrix} 1 \\ \pm 1 \end{pmatrix}.$$

Since f has an expansion of the power series of ϵ and each coefficient $f^{(n)}$ is an element in $\widehat{\mathscr{P}}^{(n)}_{\text{even}}$, the expressions of n, v have the same form as of the normal form transformation. In the asymptotic sense, all the physical variables are expressed by one function u as $U = uR_+$. Then the choice of the higher order terms f has a freedom. Then the main purpose of the normal form is to use this freedom to classify near-integrable systems in the asymptotic sense.

We now compare the results in Sect. 5.6 with numerical results. We used the spectral method [40] for the numerical computation. For convenience, we consider (5.31) in a moving frame with a speed c:

$$\begin{cases} (n)_t - (6+c)n_x + 6v_x + 6(nv)_x = 0, \\ (v)_t - (6+c)v_x + 6\left(\dfrac{v^2}{2} + \phi\right)_x = 0, \\ \phi_{2x} - 3\exp\phi + 3(n+1) = 0. \end{cases} \qquad (5.34)$$

The computation is done with the 2^{12} number of Fourier modes and the time step $dt = 0.008$. In general, the spectral method yields aliasing errors from the high-frequency modes in the nonlinear terms, so we try to get rid of their errors by adopting the 3/2 rule [36].

Let us first compare a solitary wave solution of the ion acoustic wave equation with the KdV soliton solution. Since the one-soliton solution (5.9) is a kernel of the obstacle $R_1^{(2)}$, one can construct a solitary wave for the variable u by the normal form transformation, i.e.,

$$u = v + \epsilon\phi^{(1)} + \epsilon^2\left(\phi^{(2)} + \frac{1}{2}V_{\phi^{(1)}} \cdot \phi^{(1)}\right) + O(\epsilon^3),$$

$$\text{with} \quad v = 2\kappa^2\text{sech}^2(\kappa(x - x_0)), \qquad (5.35)$$

where x_0 is a center position of the soliton at $t = 0$. The solitary wave for the variable u is constructed by the core v with the dressing terms $\epsilon\phi^{(1)} + \epsilon^2(\phi^{(2)} + \frac{1}{2}V_{\phi^{(1)}} \cdot \phi^{(1)}) + O(\epsilon^3)$. Now let us see the effect of the dressing terms, which is considered as nonresonant higher harmonics in a periodic solution of finite-dimensional problem.

Figure 5.1 shows the difference of the amplitude of radiation emitted during the time evolution for each initial wave. The initial solitary wave is given by $u = v$ in the left figure, $u = v + \epsilon\phi^{(1)}$ in the central figure and $u = v + \epsilon\phi^{(1)} + \epsilon^2(\phi^{(2)} + \frac{1}{2}V_{\phi^{(1)}} \cdot \phi^{(1)})$ in the right figure. The core part is given by (5.35) for $\kappa = 0.2$. These figures show that the generation of the radiation can be suppressed by adding the higher order dressing terms into the core.

Now let us discuss the interaction of the two solitary waves. In Fig. 5.2, we show the result of the phase shift Δx_1 of (5.30) and the radiation $\Delta_1(t)$ of (5.29)

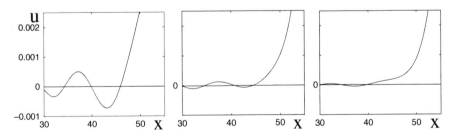

Fig. 5.1 Effect of the dressing term for ion acoustic solitary wave. The figures plot the solutions of (5.34) for the initial function to be either (**a**) $u = v$ (*left*), (**b**) $u = v + \epsilon\phi^{(1)}(v)$ (*middle*) or $u = v + \epsilon^{(1)}(v) + \epsilon^2\psi^{(2)}(v)$ (*right*) where $\psi^{(2)} = \phi^{(2)} + \frac{1}{2}V_{\phi^{(1)}} \cdot \phi^{(1)}$

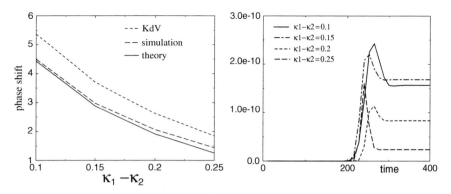

Fig. 5.2 Phase shifts Δx_1 of (5.30) and the time evolution of the radiation $\bar{\Delta}_1(t) := \Delta_1(t)/(\epsilon^4 (\mu_1^{(2)})^2)$ of (5.29) for the ion acoustic waves

generated by the interaction of the solitary waves with the parameter $\kappa_1(> \kappa_2)$. We fix $\kappa_1 + \kappa_2 = 0.5$ and $\epsilon = (\kappa_1^2 + \kappa_2^2)/2 \simeq 0.07$. For the phase shift (the left figure), the solid line is calculated from the formula (5.30), and the broken line is obtained by the numerical simulation. The shift of the KdV soliton is also shown as the dotted line.

As we can see, the phase shift formula gives a good agreement to the numerical results. In the right figure, the energy of the radiation $\Delta_1(t)$ emitted after the interaction is calculated from (5.29) with $\Delta_1(t) = \bar{\Delta}_1(t)/(\epsilon^4 (\mu_1^{(2)})^2)$. This shows that the radiation energy is of order 10^{-5} which also agrees with the result in [26]. Thus the normal form theory provides an accurate description of the deviation from the KdV equation.

5.7.2 Boussinesq Equation

The Boussinesq equation as an approximate equation for the shallow water waves is given by

$$\begin{cases} \eta_T + v_X + (\eta v)_X = 0, \\ v_T + \dfrac{1}{2}(v^2)_X + \eta_X - \dfrac{1}{3} v_{XXT} = 0, \end{cases}$$

where η and v are the normalized variables which represent the amplitude and the velocity [39]. Since this equation is truncated at the first order from the shallow water wave equation, the normal form may not provide a structure of the asymptotic integrability at the second order. However, the Boussinesq equation itself has an interesting mathematical structure, such as the regularization at the higher dispersion regime, and it may be interesting to study its own asymptotic integrability. The normal form for the shallow water wave equation has been studied in [21].

Following the similar process as in the case of ion acoustic waves, we obtain the perturbed KdV equation up to order ϵ^2:

$$u_t + 6uu_x + u_{3x} + \epsilon \left(\frac{3}{8}u_{5x} + \frac{3}{2}u_{3x}u + 4u_{2x}x - \frac{3}{4}u_x u^2 \right)$$

$$+ \epsilon^2 \left(\frac{5}{32}u_{7x} + \frac{11}{16}u_{5x}u + \frac{99}{32}u_{4x}u_x + \frac{95}{16}u_{3x}u_{2x} - \frac{3}{16}u_{3x}u^2 \right.$$

$$\left. - \frac{33}{16}u_{2x}u_x u - \frac{27}{32}u_x^3 + \frac{3}{16}u_x u^3 \right) = O(\epsilon^3).$$

Due to the non-zero integrability condition ($\mu_1^{(2)} = 3/2$), the obstacle $R_1^{(2)}$ does not disappear in the second-order correction. The generating functions in the Lie transformation are given by

$$
\begin{cases}
\alpha_1^{(1)} = \dfrac{13}{8}, & \alpha_2^{(1)} = \dfrac{17}{24}, & \alpha_3^{(1)} = \dfrac{3}{4}, \\[2mm]
\alpha_1^{(2)} = \dfrac{2591}{14400}, & \alpha_2^{(2)} = \dfrac{71}{100}, & \alpha_3^{(2)} = \dfrac{2747}{3600}, \\[2mm]
\alpha_4^{(2)} = \dfrac{1583}{9600}, & \alpha_5^{(2)} = \dfrac{17}{800}, & \alpha_6^{(2)} = \dfrac{571}{4800}.
\end{cases}
$$

Now let us consider the phase shift during the two solitary wave interactions in the same way as the ion acoustic wave equation. As shown in Fig. 5.3, the numerical data of the phase shift can be well explained by the phase shift formula. In the right figure, we show the time evolution of $\Delta_1(t)$. This figure shows that the energy of the radiation after the interaction is about the same as that for the ion acoustic wave equation.

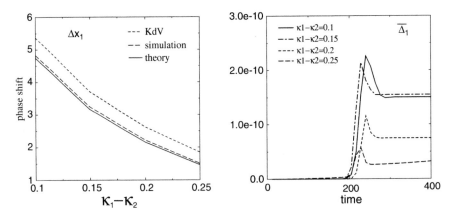

Fig. 5.3 Phase shifts Δx_1 of (5.30) and the time evolution of the radiation $\bar{\Delta}_1(t) = \Delta_1(t)/(\epsilon^4(\mu_1^{(2)})^2)$ of (5.29) for the Boussinesq equation

5.7.3 Regularized Long Wave Equation

Since the KdV equation is derived under the assumption of weak dispersion, it is not valid for the wave phenomena involving short waves. As a model of shallow water waves for a large range of wavelength, Benjamin et al. in [4, 37] proposed the equation

$$w_T + w_X + 6ww_X - w_{XXT} = 0.$$

This equation is called the regularized long wave equation (RLW) or the BBM equation. It has been shown that this equation has only three nontrivial independent conserved quantities [35] indicating its nonintegrability. The BBM equation admits a solitary wave solution

$$w = \frac{2\kappa^2}{1 - 4\kappa^2} \operatorname{sech}^2 \left(\kappa X - \frac{\kappa}{1 - 4\kappa^2} T \right).$$

The numerical simulations by Bona et al. [6] showed that the interaction of the solitary waves is inelastic and generates radiation. Their study also found the shifts of the amplitudes of two solitary waves after the collision. The normal form theory has been applied to this equation in [19]. Here we review the work [19] and add the phase shift results. We also compare the BBM equation with the ion acoustic wave equation and the Boussinesq equation.

As in the previous cases, we introduce the scaled variables

$$x = \epsilon^{\frac{1}{2}}(X - T), \quad t = \epsilon^{\frac{3}{2}} T, \quad w = \epsilon u,$$

which yield

$$(1 - \epsilon D^2) u_t + 6u u_x + u_{3x} = 0.$$

Then inverting the operator in front of u_t, we obtain the perturbed KdV equation

$$u_t + K_0^{(0)}(u) + \sum_{k=1}^{\infty} \epsilon^k D^{2k} K_0^{(0)}(u) = 0.$$

Then the obstacle $R^{(2)}(u)$ appears with $\mu_1^{(2)} = 40$. The comparison of the phase shift between the numerical results and those by the formula (5.30) is shown in Fig. 5.4.

Because of the large value of $\mu_1^{(2)}$, the agreement between the numerical results and the results from (5.30) is poor. In the computation, the value of $\epsilon^2 \mu_1^{(2)}$ is about order one, and thus one needs to consider much higher corrections to get a better agreement. However, the numerical observation of the radiation agrees with the formula (5.29) which is of order 10^{-2}–10^{-3}, and the shifts in the parameters observed in [6] also agree with the result from the normal form theory. In [27], Marchant discussed the solitary wave interaction for the BBM equation and obtained the similar results presented here.

Fig. 5.4 Phase shifts Δx_1 of (5.30) for the regularized long wave equation

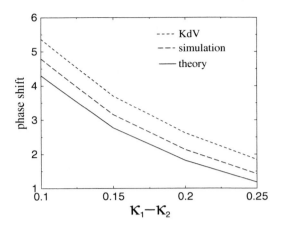

5.7.4 Seventh-Order Hirota KdV Equation

Here we consider

$$w_T + w_{7X} + 28ww_{5X} + 28w_X w_{4X} + 70w_{2X}w_{3X}$$

$$+ 210w^2 w_{3X} + 420w^3 w_X = 0, \tag{5.36}$$

which has a Hirota bilinear form,

$$D_X(D_X^7 + D_T)\tau \cdot \tau = 0,$$

where D_X and D_T mean the Hirota derivative, and w is given by $w = 2D^2 \ln \tau$. If the order of the derivative is either *three* or *five*, the equation becomes the KdV equation or the Sawada–Kotera equation which are both integrable. However, (5.7.4) is known to be nonintegrable [33], but it admits both one and two solitary wave solutions in the same form as the KdV solitons except their time evolution. Since there is an exact two solitary wave solution, several conditions $\mu_k^{(n)} = 0$ should be satisfied. Now the question is to determine the order in which the condition $\mu_k^{(n)} = 0$ breaks.

In order to apply the normal form theory, we first introduce

$$w = \epsilon u + c, \quad t = 210\epsilon^{3/2}c^2 T, \quad x = \epsilon^{1/2}\left(X - 420c^3 T\right),$$

which puts (5.36) in the form (5.1) of the perturbed KdV equation

$$u_t + K_0^{(0)}(u) + \epsilon \frac{2}{15c}(u_{5x} + 15u_{3x}u + 15u_{2x}u_x + 45u_x u^2)$$

$$+ \frac{\epsilon^2}{210c^2}(u_{7x} + 28u_{5x}u + 28u_{4x}u_x + 70u_{3x}u_{2x}$$

$$+ 210u_{3x}u^2 + 420u_{2x}u_x u + 420u_x u^3) = 0,$$

where c is an arbitrary non-zero constant. The direct calculation shows

$$\begin{cases} \mu_1^{(2)} = 0, \\ \mu_i^{(3)} = 0, \ i = 1,2,3, \\ \mu_i^{(4)} \neq 0, \ i = 1,2,\ldots,7. \end{cases}$$

Thus the seventh-order Hirota KdV equation passes the asymptotic integrability conditions not only at order ϵ^2 but also at order ϵ^3, and the first obstacles appear at order ϵ^4. Then the normal form up to order ϵ^4 of the seventh-order Hirota KdV equation takes the form

$$v_t + K_0^{(0)} + \epsilon a_1^{(1)} K_0^{(1)} + \epsilon^2 a_1^{(2)} K_0^{(2)} + \epsilon^3 a_1^{(3)} K_0^{(3)}$$
$$+ \ \epsilon^4 \left(a_1^{(4)} K_0^{(4)} + R^{(4)} \right) = O\left(\epsilon^5 \right), \tag{5.37}$$

where $R^{(4)}$ consists of seven obstacles. Since the seventh-order Hirota KdV equation admits the two-soliton solution, the kernel of $R^{(4)}$ should include not only one-soliton solution but also two-soliton solution. Thus we have

Corollary 17. *If the normal form of a perturbed KdV equation has the form (5.37), then there exists two-soliton solution up to order ϵ^4. Equivalently, if $\mu_1^{(2)}$, $\mu_1^{(3)}$, $\mu_2^{(3)}$ and $\mu_3^{(3)}$ are all zero, then the perturbed KdV equation can admit a two-soliton solution up to order ϵ^4.*

Appendix

Here we give the explicit formulae for the coefficients $\alpha_k^{(2)}$ of the generating function $\phi^{(2)}$ in Theorem 10 and the conditions $\mu_k^{(3)}$ for the asymptotic integrability in Proposition 6.

The coefficients $\alpha_k^{(2)}$ in $\phi^{(2)}$

$$\alpha_1^{(2)} = -\frac{100}{9} \left(a_1^{(1)} \right)^2 - \frac{10}{27} a_1^{(1)} a_2^{(1)} - \frac{5}{108} \left(a_2^{(1)} \right)^2 - \frac{1}{54} a_2^{(1)} a_4^{(1)} + \frac{1}{36} \left(a_4^{(1)} \right)^2$$
$$+ \frac{112}{9} a_1^{(2)} + \frac{2}{9} a_2^{(2)} + \frac{1}{9} a_4^{(2)} - \frac{1}{6} a_7^{(2)} - \frac{4}{225} \mu_1^{(2)},$$

$$\alpha_2^{(2)} = -\frac{175}{9} \left(a_1^{(1)} \right)^2 + \frac{10}{9} a_1^{(1)} a_2^{(1)} + \frac{5}{18} a_1^{(1)} a_3^{(1)} - \frac{1}{36} a_2^{(1)} a_3^{(1)} - \frac{5}{18} a_1^{(1)} a_4^{(1)}$$

$$-\frac{1}{18}a_2^{(1)}a_4^{(1)} - \frac{1}{36}a_3^{(1)}a_4^{(1)} + \frac{1}{18}\left(a_4^{(1)}\right)^2 + \frac{56}{3}a_1^{(2)} - \frac{1}{2}a_2^{(2)} + \frac{1}{12}a_6^{(2)}$$

$$-\frac{1}{4}a_7^{(2)} - \frac{2}{75}\mu_1^{(2)},$$

$$\alpha_3^{(2)} = -\frac{25}{3}\left(a_1^{(1)}\right)^2 - \frac{85}{108}a_1^{(1)}a_2^{(1)} + \frac{1}{108}\left(a_2^{(1)}\right)^2 + \frac{5}{18}a_1^{(1)}a_3^{(1)} + \frac{1}{72}a_2^{(1)}a_3^{(1)}$$

$$-\frac{5}{36}a_1^{(1)}a_4^{(1)} - \frac{1}{216}a_2^{(1)}a_4^{(1)} - \frac{1}{24}a_3^{(1)}a_4^{(1)} + \frac{1}{24}\left(a_4^{(1)}\right)^2 + \frac{91}{9}a_1^{(2)}$$

$$+\frac{7}{18}a_2^{(2)} - \frac{1}{18}a_4^{(2)} - \frac{1}{6}a_7^{(2)} + \frac{1}{6}a_8^{(2)} - \frac{13}{900}\mu_1^{(2)},$$

$$\alpha_4^{(2)} = -\frac{25}{36}\left(a_1^{(1)}\right)^2 + \frac{5}{48}a_1^{(1)}a_2^{(1)} - \frac{1}{144}a_2^{(1)}a_3^{(1)} - \frac{5}{72}a_1^{(1)}a_4^{(1)} - \frac{1}{288}a_2^{(1)}a_4^{(1)}$$

$$-\frac{1}{288}a_3^{(1)}a_4^{(1)} + \frac{1}{144}\left(a_4^{(1)}\right)^2 + \frac{7}{6}a_1^{(2)} - \frac{5}{48}a_2^{(2)} + \frac{1}{24}a_3^{(2)} + \frac{1}{96}a_6^{(2)}$$

$$-\frac{1}{32}a_7^{(2)} - \frac{7}{2400}\mu_1^{(2)},$$

$$\alpha_5^{(2)} = -\frac{50}{9}\left(a_1^{(1)}\right)^2 + \frac{5}{9}a_1^{(1)}a_2^{(1)} + \frac{14}{3}a_1^{(2)} - \frac{1}{3}a_2^{(2)} - \frac{1}{150}\mu_1^{(2)},$$

$$\alpha_6^{(2)} = \frac{50}{3}\left(a_1^{(1)}\right)^2 - \frac{85}{18}a_1^{(1)}a_2^{(1)} + \frac{5}{36}\left(a_2^{(1)}\right)^2 + \frac{5}{18}a_1^{(1)}a_4^{(1)} + \frac{1}{36}a_2^{(1)}a_4^{(1)}$$

$$-\frac{28}{3}a_1^{(2)} + \frac{7}{3}a_2^{(2)} - \frac{1}{3}a_4^{(2)} + \frac{1}{75}\mu_1^{(2)}.$$

The conditions $\mu_k^{(3)} = 0$ for the asymptotic integrability

$$\mu_1^{(3)} = \frac{280}{3}a_2^{(3)} - \frac{280}{3}a_3^{(3)} + 60a_4^{(3)} - 20a_6^{(3)} - \frac{5}{3}a_7^{(3)} - a_8^{(3)} + 8a_9^{(3)} - a_{10}^{(3)}$$

$$-4a_{11}^{(3)} + \frac{2}{3}a_{12}^{(3)} - a_{13}^{(3)} - 2a_{15}^{(3)} + \frac{800}{3}\left(a_1^{(1)}\right)^3 + \frac{130}{9}\left(a_1^{(1)}\right)^2 a_2^{(1)}$$

$$+\frac{62}{27}a_1^{(1)}\left(a_2^{(1)}\right)^2 - \frac{2}{27}\left(a_2^{(1)}\right)^3 - \frac{400}{9}\left(a_1^{(1)}\right)^2 a_3^{(1)} - \frac{175}{27}a_1^{(1)}a_2^{(1)}a_3^{(1)}$$

$$+\frac{1}{9}\left(a_2^{(1)}\right)^2 a_3^{(1)} + \frac{34}{9}a_1^{(1)}\left(a_3^{(1)}\right)^2 + \frac{1}{9}a_2^{(1)}\left(a_3^{(1)}\right)^2 - \frac{2}{27}\left(a_3^{(1)}\right)^3$$

$$+\frac{20}{3}\left(a_1^{(1)}\right)^2 a_4^{(1)} - \frac{7}{27}a_1^{(1)}a_2^{(1)}a_4^{(1)} - \frac{4}{27}a_1^{(1)}a_3^{(1)}a_4^{(1)} - \frac{560}{3}a_1^{(1)}a_1^{(2)}$$

$$-\frac{560}{9}a_2^{(1)}a_1^{(2)}+\frac{280}{9}a_3^{(1)}a_1^{(2)}-\frac{230}{3}a_1^{(1)}a_2^{(2)}+\frac{10}{9}a_2^{(1)}a_2^{(2)}+\frac{55}{9}a_3^{(1)}a_2^{(2)}$$

$$+\frac{340}{3}a_1^{(1)}a_3^{(2)}-\frac{10}{9}a_2^{(1)}a_3^{(2)}-\frac{20}{3}a_3^{(1)}a_3^{(2)}+\frac{5}{9}a_4^{(1)}a_3^{(2)}-\frac{64}{9}a_1^{(1)}a_4^{(2)}$$

$$+\frac{1}{3}a_2^{(1)}a_4^{(2)}-\frac{1}{3}a_3^{(1)}a_4^{(2)}-\frac{136}{3}a_1^{(1)}a_5^{(2)}+2a_2^{(1)}a_5^{(2)}+2a_3^{(1)}a_5^{(2)}$$

$$-\frac{1}{3}a_4^{(1)}a_5^{(2)}+\frac{28}{9}a_1^{(1)}a_6^{(2)}-\frac{1}{6}a_2^{(1)}a_6^{(2)}+a_1^{(1)}a_7^{(2)}-12a_1^{(1)}a_8^{(2)}-\frac{1}{3}a_2^{(1)}a_8^{(2)}$$

$$+\frac{2}{3}a_3^{(1)}a_8^{(2)}+\frac{1}{3}\mu_1^{(2)},$$

$$\mu_2^{(3)}=-4200a_1^{(3)}+770a_2^{(3)}-\frac{280}{3}a_3^{(3)}-40a_4^{(3)}-\frac{130}{3}a_5^{(3)}+40a_6^{(3)}+\frac{10}{3}a_7^{(3)}$$

$$-8a_8^{(3)}-6a_9^{(3)}+7a_{10}^{(3)}+8a_{11}^{(3)}-\frac{4}{3}a_{12}^{(3)}+2a_{13}^{(3)}+9a_{15}^{(3)}-\frac{95000}{9}\left(a_1^{(1)}\right)^3$$

$$+\frac{73450}{27}\left(a_1^{(1)}\right)^2a_2^{(1)}-\frac{325}{3}a_1^{(1)}\left(a_2^{(1)}\right)^2+\frac{14}{27}\left(a_2^{(1)}\right)^3+\frac{800}{9}\left(a_1^{(1)}\right)^2a_3^{(1)}$$

$$-\frac{1720}{27}a_1^{(1)}a_2^{(1)}a_3^{(1)}+\frac{8}{9}\left(a_2^{(1)}\right)^2a_3^{(1)}+\frac{70}{9}a_1^{(1)}\left(a_3^{(1)}\right)^2-\frac{2}{9}a_2^{(1)}\left(a_3^{(1)}\right)^2$$

$$+\frac{4}{27}\left(a_3^{(1)}\right)^3-\frac{500}{3}\left(a_1^{(1)}\right)^2a_4^{(1)}+10a_1^{(1)}a_2^{(1)}a_4^{(1)}+\frac{65}{27}a_1^{(1)}a_3^{(1)}a_4^{(1)}$$

$$+\frac{123200}{9}a_1^{(1)}a_1^{(2)}-\frac{12250}{9}a_2^{(1)}a_1^{(2)}-\frac{560}{9}a_3^{(1)}a_1^{(2)}+\frac{280}{3}a_4^{(1)}a_1^{(2)}$$

$$-\frac{15950}{9}a_1^{(1)}a_2^{(2)}+\frac{685}{9}a_2^{(1)}a_2^{(2)}+\frac{310}{9}a_3^{(1)}a_2^{(2)}-\frac{55}{9}a_4^{(1)}a_2^{(2)}$$

$$+\frac{700}{3}a_1^{(1)}a_3^{(2)}-\frac{40}{9}a_2^{(1)}a_3^{(2)}-\frac{10}{9}a_4^{(1)}a_3^{(2)}+\frac{310}{3}a_1^{(1)}a_4^{(2)}-\frac{7}{3}a_2^{(1)}a_4^{(2)}$$

$$-a_3^{(1)}a_4^{(2)}-\frac{280}{3}a_1^{(1)}a_5^{(2)}+\frac{28}{3}a_2^{(1)}a_5^{(2)}-4a_3^{(1)}a_5^{(2)}+\frac{2}{3}a_4^{(1)}a_5^{(2)}$$

$$+\frac{115}{9}a_1^{(1)}a_6^{(2)}-\frac{1}{2}a_2^{(1)}a_6^{(2)}-\frac{25}{3}a_1^{(1)}a_7^{(2)}-\frac{80}{3}a_1^{(1)}a_8^{(2)}+\frac{7}{3}a_2^{(1)}a_8^{(2)}$$

$$-\frac{4}{3}a_3^{(1)}a_8^{(2)}-\frac{7}{3}\mu_1^{(2)},$$

$$\mu_3^{(3)}=560a_2^{(3)}-\frac{1400}{3}a_3^{(3)}+300a_4^{(3)}-\frac{170}{3}a_5^{(3)}-100a_6^{(3)}+\frac{5}{3}a_7^{(3)}-9a_8^{(3)}$$

$$+40a_9^{(3)}-a_{10}^{(3)}-20a_{11}^{(3)}+\frac{19}{3}a_{12}^{(3)}-9a_{13}^{(3)}-2a_{14}^{(3)}-10a_{15}^{(3)}$$

$$+ \frac{32000}{9} \left(a_1^{(1)}\right)^3 - \frac{1600}{27} \left(a_1^{(1)}\right)^2 a_2^{(1)} - \frac{2270}{27} a_1^{(1)} \left(a_2^{(1)}\right)^2$$

$$- \frac{2800}{9} \left(a_1^{(1)}\right)^2 a_3^{(1)} + \frac{545}{27} a_1^{(1)} a_2^{(1)} a_3^{(1)} + \frac{86}{27} \left(a_2^{(1)}\right)^2 a_3^{(1)} + \frac{40}{3} a_1^{(1)} \left(a_3^{(1)}\right)^2$$

$$- \frac{10}{9} a_2^{(1)} \left(a_3^{(1)}\right)^2 - \frac{4}{27} \left(a_3^{(1)}\right)^3 - \frac{400}{9} \left(a_1^{(1)}\right)^2 a_4^{(1)} + 30 a_1^{(1)} a_2^{(1)} a_4^{(1)}$$

$$- \frac{1}{27} \left(a_2^{(1)}\right)^2 a_4^{(1)} - \frac{155}{27} a_1^{(1)} a_3^{(1)} a_4^{(1)} - \frac{19}{27} a_2^{(1)} a_3^{(1)} a_4^{(1)} + \frac{2}{9} \left(a_3^{(1)}\right)^2 a_4^{(1)}$$

$$- \frac{10}{9} a_1^{(1)} \left(a_4^{(1)}\right)^2 - \frac{22400}{9} a_1^{(1)} a_1^{(2)} - 560 a_2^{(1)} a_1^{(2)} + \frac{1960}{9} a_3^{(1)} a_1^{(2)}$$

$$+ \frac{280}{3} a_4^{(1)} a_1^{(2)} - \frac{1600}{9} a_1^{(1)} a_2^{(2)} + 90 a_2^{(1)} a_2^{(2)} + \frac{35}{3} a_3^{(1)} a_2^{(2)} - \frac{200}{9} a_4^{(1)} a_2^{(2)}$$

$$+ 400 a_1^{(1)} a_3^{(2)} - \frac{200}{9} a_2^{(1)} a_3^{(2)} - \frac{80}{3} a_3^{(1)} a_3^{(2)} + \frac{85}{9} a_4^{(1)} a_3^{(2)} + \frac{50}{3} a_1^{(1)} a_4^{(2)}$$

$$- \frac{1}{9} a_2^{(1)} a_4^{(2)} - \frac{37}{9} a_3^{(1)} a_4^{(2)} + \frac{1}{3} a_4^{(1)} a_4^{(2)} - 160 a_1^{(1)} a_5^{(2)} + \frac{50}{3} a_2^{(1)} a_5^{(2)}$$

$$+ \frac{22}{3} a_3^{(1)} a_5^{(2)} - \frac{13}{3} a_4^{(1)} a_5^{(2)} - \frac{85}{9} a_1^{(1)} a_6^{(2)} - \frac{7}{2} a_2^{(1)} a_6^{(2)} + \frac{5}{3} a_3^{(1)} a_6^{(2)}$$

$$+ \frac{2}{3} a_4^{(1)} a_6^{(2)} + \frac{10}{3} a_1^{(1)} a_7^{(2)} - \frac{1}{3} a_2^{(1)} a_7^{(2)} + \frac{1}{6} a_3^{(1)} a_7^{(2)} + \frac{20}{3} a_1^{(1)} a_8^{(2)}$$

$$+ a_2^{(1)} a_8^{(2)} + \frac{2}{3} a_3^{(1)} a_8^{(2)} - \frac{4}{3} a_4^{(1)} a_8^{(2)} + \frac{7}{3} \mu_1^{(2)}.$$

Acknowledgments Y.K. would like to thank the organizers for the invitation to attend this program and to give the series of lectures. He also appreciates the financial support from the Newton Institute. This work was supported in part by an NSF Grants #DMS0404931 and 0806219.

References

1. M.J. Ablowitz and H. Segur, Solitons and the Inverse Scattering Transform, SIAM Publication, Philadelphia, 1981.
2. M.J. Ablowitz and Y. Kodama, Note on asymptotic solutions of the Kortewegde Vries equation with solitons, Stud. Appl. Math. 66, 159, 1982.
3. V.I. Arnold, Geometrical Methods in the Theory of Ordinary Differential Equations, Springer-Verlag, Berlin, 1983.
4. T.B. Benjamin, J.L. Bonna, and J.J. Mahony, Model equations for long waves in nonlinear dispersive systems, Plilos. Trans. R. Soc. A. 272, 47, 1972.
5. A.H. Bilge, Integrability of seventh-order KdV-like evolution equations using formal symmetries and perturbations of conserved densities, J. Math. Phys. 33, 3025, 1992.
6. J.L. Bona, W.G. Pritchard, and L.R. Scott, Solitary-wave interaction, Phys. Fluids. 23, 438, 1980.

7. W. Craig, An existence theory for water waves and the Boussinesq and Korteweg-de Vries scaling limits, Comm. PDEs 10, 787, 1985.
8. H.R. Dullin, G.A. Gottwald, and D.D. Holm, Camassa-Holm, Korteweg-de Vries-5 and other asymptotically equivalent equations for shallow water waves, Fluid Dynamics Res. 33, 73, 2003.
9. H.R. Dullin, G.A. Gottwald, and D.D. Holm, On asymptotically equivalent shallow water wave equations, Physica D, 190, 1, 2004.
10. B. Fucshteiner, Mastersymmetries, higher-order time-dependent symmetries and conserved densities of nonlinear evolution equations, Prog. Theor. Phys. 70, 1508, 1983.
11. C.S. Gardner, J.M. Green, M.D. Kruskal, and R.M. Miura, Korteweg-de Vries equation and generalizations. VI. Methods for exact solution, Comm. Pure Apple. Math. 27, 97, 1974.
12. Y. Hiraoka, Studies on nearly integrable systems in terms of normal form theory, Master Thesis, Department of Informatics and Mathematical Science, Osaka University, February, 2002.
13. M. Jimbo and T. Miwa, Soliton and infinite dimensional Lie algebras, Publ. RIMS, Kyoto Univ. 19, 943, 1983.
14. T. Kano and T. Nishida, A mathematical justification for Korteweg-de Vries equation and Boussinesq equation of water surface waves, Osaka J. Math. 23, 289, 1986.
15. T. Kano, Normal form of nonlinear Schrödinger equation, J. Phys. Soc. Japan 58, 4322, 1989.
16. D.J. Kaup, A perturbation expansion for the Zakharov-Shabat inverse scattering transform, SIAM J. Appl. Math. 31, 121, 1976.
17. D.J. Kaup and A.C. Newell, Solitons as particles, oscillators, and in slowly changing media: a singular perturbation theory, Proc. R. Soc. Lond. A361, 413, 1978.
18. Y. Kodama, Normal forms for weakly dispersive wave equations, Phys. Lett. A 112, 193, 1985.
19. Y. Kodama, On solitary-wave interaction, Phys. Lett. A 123, 276, 1987.
20. Y. Kodama, Normal form and solitons, in: Topics in Soliton Theory and Exactly Solvable Nonlinear Equations, M.J. Ablowitz et al. (eds.), World Scientific, 319–340, 1987.
21. Y. Kodama Normal form and solitons in the shallow water waves, in: Nonlinear Water Waves, K. Horikawa and H. Maruo (eds.) IUTAM Symposium Tokyo/Japan 1987, Springer-Verlag, Berlin Heidelberg, 85–91, 1988.
22. Y. Kodama and A.V. Mikhailov, Obstacles to asymptotic integrability, in: Algebraic Aspects of Integrable Systems, in Memory of Irene Dorfman, Progress in Nonlinear Differential Equations, Vol. 26, A.S. Fokas and I.M. Gel'fand, eds., Birkhäuser, Boston, 173–204, 1996.
23. Y. Kodama and D. Pelinovsky, Spectral stability and time evolution of N-solitons in the KdV hierarchy, J. Phys. A: Math. Gen. 38, 6129, 2005.
24. Y. Kodama and T. Taniuti, Higher order approximation in the reductive perturbation method. I. The weakly dispersive system, J. Phys. Soc. Japan 45, 298, 1978.
25. R.A. Kraenkel, M. Senthilvelan, and A.I. Zenchuk, On the integrable perturbations of the Camassa-Holm equation, J. Math. Phys. 41, 3160, 2000.
26. Yi. Li and D.H. Sattinger, Soliton collisions in the ion acoustic plasma equation, J. Math. Fluid. Mech. 1, 117, 1999.
27. T.R. Marchant, Solitary wave interaction for the extended BBM equation, Proc. R. Soc. Lond. A, 456, 433, 2000.
28. T.R. Marchant, High-order interaction of solitary waves on shallow water, Stud. Appl. Math. 109, 1, 2002.
29. T.R. Marchant, Asymptotic solitons for a third-order Korteweg-de Vries equation, Chaos Sol. Frac. 22, 261, 2004.
30. A.V. Mikhilov, A.B. Shabat and V.V. Sokolov, The symmetry approach to the classification of integrable equations, in What is integrability? Springer-Verlag, New York, 1990.
31. A.V. Mikhailov, talk given at NEEDS-91 conference, Gallipoli, Italy (1991) and at NATO Advanced Reseach Workshop: Singular Limits of Dispersive Waves, Lyon, France, 1991, (unpublished).
32. R.M. Miura, C.S. Gardner and M.D. Kruskal, Korteweg-de Vries equation and generalizations. II. Existence of conservation laws and constants of motion, J. Math. Phys. 9, 1204, 1968.
33. A.C. Newell, Solitons in Mathematics and Physics CBMS 48, SIAM, Philadelphis, 1985.

34. S. Novikov, S.V. Manakov, L.P. Pitaevski and V.E. Zakharov, Theory of Solitons Contemporary Soviet Mathematics, Consultants Bureau, Plenum, NY, 1984.
35. P.J. Olver, Euler operators and conservation laws of the BBM equation, Math. Proc. Camb. Phil. Soc. 85, 143, 1979.
36. S.A. Orszag, Numerical simulation of incompressible flows within simple boundaries. I. Galerkin (Spectral) representations, Stud. Appl. Math. 50, 293, 1971.
37. D.H. Peregrine, Calculations of the development of an undular bore, J. Fluid. Mech. 25, 321, 1964.
38. T. Taniuti and K. Nishihara, Nonlinear Waves, Monograph and Studies inMathematics, 15, Pitman, Boston MA, 1983.
39. G.B. Whitham, Linear and Nonlinear Waves, John Wiley, New York, 1074.
40. S.B. Wineberg, J.F. McGrath, E.F. Gabl, L.R. Scott, and C.E. Southwell, Implicit spectral methods for wave propagation problems, J. Comp. Phys. 97, 311, 1991.

Chapter 6
Multiscale Expansion and Integrability of Dispersive Wave Equations

A. Degasperis

6.1 Introduction

The propagation of *nonlinear dispersive* waves is of great interest and relevance in a variety of physical situations for which model equations, as infinite-dimensional dynamical systems, have been investigated from various perspectives and to different purposes. In the ideal case in which waves propagate in a one-dimensional medium (no diffraction) without losses and sources, quite a number of special models, so-called *integrable* models, have been discovered together with the mathematical tools to investigate them. This important progress has provided important contributions to such matters as dispersionless propagation (solitons), wave collisions, wave decay, long-time asymptotics among others. On the mathematical side, such progress on integrable models has considerably contributed also to our present (admittedly not concise) answer to the question "What is integrability?", which can be found in [1], and a partial guide to the vast literature on the theory of solitons is given in [2].

It is plain that integrable models, though both useful and fascinating, remain exceptional: nonlinear partial differential equations (PDEs) in 1+1 variables (space+time) are generically not integrable. The aim of these notes is to show how an algorithmic technique, based on *multiscale* analysis and *perturbation* theory, may be devised as a tool to establish how "far" is a given PDE from being integrable. This method basically associates to a given PDE one or more, generally simpler, PDEs with respect to rescaled space and time variables. This approach [3] has been known in applicative contexts [4–8] since several decades as it provides approximate solutions when only one, or a few, monochromatic "carrier waves" propagate in a strongly dispersive and weakly nonlinear medium. More recently [9] it has proved to be also a simple way to obtain *necessary conditions* which a given PDE has to satisfy in order to be integrable, and to discover integrable PDEs as well [10].

A. Degasperis (✉)

Dipartimento di Fisica, Universitá di Roma "La Sapienza", P.le A.Moro 2, 00185 Roma, Italy and Istituto Nazionale di Fisica Nucleare, Sezione di Roma
antonio.degasperis@roma1.infn.it

Degasperis, A.: *Multiscale Expansion and Integrability of Dispersive Wave Equations*. Lect. Notes Phys. **767**, 215–244 (2009)
DOI 10.1007/978-3-540-88111-7_6

The basic philosophy of this approach is to derive from a nonlinear PDE one or more PDEs whose integrability properties are either already known or easily found. In this respect, a general remark on this method is the following. Integrability is not a precise notion, and different degrees of integrability can be attributed to a PDE within a certain class of solutions and boundary conditions, according to the technique of solving it. For instance, C-integrable are those nonlinear equations which can be transformed into linear equations via a change of variables [10], and S-integrable are those equations which can be linearized (within a certain class of solutions) by the method of the spectral (or scattering) transform (see, f.i., [11, 12]). Examples of C-integrability are the equations ($u_t = \partial u / \partial t$, $u_x = \partial u / \partial x$, etc.)

$$u_t + a_1 u_x - a_3 u_{xxx} = a_3 \left(3uu_x + u^3 \right)_x, \qquad\qquad u = u(x,t), \qquad (6.1a)$$

$$u_t + a_1 u_x - a_3 u_{xxx} = 3a_3 c \left(u^2 u_{xx} + 3uu_x^2 \right) + 3a_3 c^2 u^4 u_x, \quad u = u(x,t), \qquad (6.2a)$$

which are both mapped to their linearized version (a_1, a_3, c are constant coefficients)

$$v_t + a_1 v_x - a_3 v_{xxx} = 0, \qquad\qquad v = v(x,t), \qquad (6.3)$$

the first one, (6.1a), by the (Cole–Hopf) transformation

$$u = v_x / v \qquad\qquad (6.1b)$$

and the second one, (6.2a), by the transformation [10]

$$u = v / (1 + 2cw)^{1/2}, \quad w_x = v^2. \qquad (6.2b)$$

Well-known examples of S-integrable equations are the modified Korteweg–de Vries (mKdV) equation

$$u_t + a_1 u_x - a_3 u_{xxx} = 6a_3 c u^2 u_x, \quad u = u(x,t), \qquad (6.4a)$$

and the nonlinear Schrödinger (NLS) equation (a_1, a_2, a_3, c are real constant coefficients)

$$u_t - ia_2 u_{xx} = 2ia_2 c |u|^2 u, \qquad\qquad u = u(x,t), \qquad (6.5a)$$

whose method of solution is based on the eigenvalue problem

$$\psi_x + ik\sigma\psi = Q\psi, \qquad \psi = \psi(x,k,t), \qquad (6.6)$$

where ψ is a 2-dim vector, σ is the diagonal matrix $\mathrm{diag}(1,-1)$ and $Q(x,t)$ is the off-diagonal matrix

$$Q = \begin{pmatrix} 0 & u \\ -cu & 0 \end{pmatrix}, \qquad\qquad (6.4b)$$

where u is real for the mKdV equation (6.4a) and (the asterisk indicates complex conjugation)

$$Q = \begin{pmatrix} 0 & u \\ -cu^* & 0 \end{pmatrix}, \tag{6.5b}$$

where u is complex for the NLS equation (6.5a). Here k is the spectral variable. In any case, whatever type of integrability is involved, we adopt in our treatment the "first principle" (*axiom*) that integrability is preserved by the multiscale method. Though in some specific cases, where integrability can be formulated as a precise mathematical property, one can give this principle a rigorous status, we prefer to maintain it throughout our treatment as a robust assumption. Its use, according to contexts, may lead to interesting consequences. One is that it provides a way to obtain other (possibly new) integrable equations. On the other hand, if a PDE, which has been obtained by this method from a given PDE, is proved to be nonintegrable, then from our first principle it there follows that the given PDE cannot be integrable, and this implication leads to conditions of integrability. Some of these conditions are found to be simple and, therefore, of ready practical use. Others conditions are instead the results of lengthy algebraic manipulations which require a rather heavy computer assistance. Finally, this way of reasoning leads to the following observation, which has been pointed out in [10]. Suppose the same PDE is obtained by multiscale reduction from any member of a fairly large family of PDEs; so we can call it a "model PDE". Then the principle stated above explains why a model PDE may be at the same time widely applicable (because it derives from a large class of different PDEs) and integrable (because it suffices that just one member equation of that large family of PDEs be integrable). The most widely known example of such case is the NLS equation (6.5a) which is certainly a model equation (as shown below) with many applications (f.i. nonlinear optics and fluid dynamics [4–8]), and whose integrability has been discovered in 1971 [13] but it could have been found even earlier by multiscale reduction from the KdV equation $u_t + u_{xxx} = 6uu_x$ (the way to infer the S-integrability of the NLS equation from the S-integrability of the KdV equation has been first pointed out in [14]), whose integrability has been unveiled in 1967 [15].

The method of multiscale reduction which we now introduce is a perturbation technique based on three main ingredients : (i) Fourier expansion in harmonics, (ii) power expansion in a small parameter ϵ, (iii) dependence on a (finite or infinite) number of "slow" space and time variables, which are first introduced via an ϵ-dependent rescaling of x and t and are then treated as independent variables. This last feature explains why this approach is also referred to as multiscale perturbation method or multiscale reduction.

In order to briefly illustrate how these basic ingredients naturally come into play in the simpler context of ordinary differential equations (ODEs), let us consider the well-known Poincaré–Lindstedt perturbation scheme to construct small amplitude oscillations of an anharmonic oscillator around a stable equilibrium position. Let our one-degree dynamical system be given by the nonlinear equation ($\dot{q} \equiv dq/dt$)

$$\ddot{q} + \omega_o^2 q = c_2 q^2 + c_3 q^3 + \dots , \qquad q = q(t, \epsilon) \tag{6.7a}$$

where the small perturbative parameter ϵ is here introduced as the initial amplitude,

$$q(0,\epsilon) = \epsilon, \qquad \dot{q}(0,\epsilon) = 0 . \tag{6.7b}$$

The equation of motion (6.7a) is autonomous as all coefficients $\omega_0, c_2, c_3, \ldots$, are time independent, and it has been written with its linear part in the lhs and its non-linear (polynomial or, more generally, analytic) part in the rhs. In this elementary context, the model equation which is associated with this family of dynamical systems, is of course the harmonic oscillator equation, $\ddot{q} + \omega_0^2 q = 0$, which is obtained when the amplitude ϵ is so small that all nonlinear terms can be neglected. In fact, the purpose of the Poincaré–Lindstedt approach is to capture the deviations from the harmonic motion which are due to the nonlinear terms in the rhs of (6.7a). Since, for sufficiently small ϵ, the motion is periodic, namely

$$q(t,\epsilon) = q(t + \frac{2\pi}{\omega(\epsilon)}, \epsilon) , \tag{6.8}$$

it is natural to change the time variable t into the phase variable θ,

$$\theta = \omega(\epsilon)t , \qquad q(t,\epsilon) = f(\theta,\epsilon) , \tag{6.9}$$

even if the frequency $\omega(\epsilon)$ is not known as it is expected to depend on the initial amplitude ϵ. Then Eqs. (6.7a, b) now read ($f' \equiv df/d\theta$)

$$\omega^2(\epsilon)f'' + \omega_0^2 f = c_2 f^2 + c_3 f^3 + \ldots , \quad f(0,\epsilon) = \epsilon, \ f'(0,\epsilon) = 0, \tag{6.10}$$

and we look for approximate solutions via the power expansions

$$\omega^2(\epsilon) = \omega_0^2 + \gamma_1 \epsilon + \gamma_2 \epsilon^2 + \ldots , \tag{6.11}$$

$$f(\theta,\epsilon) = \epsilon f_1(\theta) + \epsilon^2 f_2(\theta) + \ldots . \tag{6.12}$$

We note that the periodicity condition $f(\theta) = f(\theta + 2\pi)$ implies that $\omega(0) = \omega_0$; inserting the expansions (6.11) and (6.12) in the differential equation (6.10) and equating the lhs coefficients with the rhs coefficients of each power of ϵ yields an infinite system of differential equations, the first one, at $O(\epsilon)$, is homogeneous, while all others, at $O(\epsilon^n)$ with $n > 1$, are nonhomogeneous, i.e.

$$O(\epsilon): \quad f_1'' + f_1 = 0, \quad f_1(0) = 1, \ f_1'(0) = 0, \tag{6.13}$$

$$O(\epsilon^n): \quad f_n'' + f_n = \{-n, -n+1, \ldots, -1, 0, 1, \ldots, n-1, n\}, f_n(0) = 0, f_n'(0) = 0. \tag{6.14}$$

The notation in this last equation refers to harmonic expansion with the following meaning. Since the functions $f_n(\theta)$ are periodic in the interval $(0, 2\pi)$, one can Fourier-expand them; however, because of the differential equation they satisfy, only a finite number of the Fourier exponentials $\exp(i\alpha\theta)$, α being an integer, enters in their representation. This is easily seen by recursion: $f_1(\theta) = \frac{1}{2}(\exp(i\theta) + \exp(-i\theta))$, and since $f_n(\theta)$, for $n > 1$, satisfies the forced harmonic oscillator equation where the forcing term in the rhs of (6.14) is an appropriate polynomial of $f_1, f_2, \ldots, f_{n-1}$, its expansion can only contain the harmonics $\exp(i\alpha\theta)$ with $|\alpha| \leq n$. Thus, the integers in the curly bracket in the rhs of (6.14) indicate the harmonics which enter in the Fourier expansion of the forcing term, and this implies that $f_n(\theta)$ itself has the Fourier expansion

$$f_n(\theta) = \sum_{\alpha=-n}^{n} f_n^{(\alpha)} \exp(i\alpha\theta), \quad n \geq 1, \tag{6.15}$$

where the complex numbers $f_n^{(\alpha)}$ have to be recursively computed. To this aim, it is required that also the coefficients γ_n in the expansion (6.11) be computed, and the way to do it is to use the periodicity condition $f_n(\theta) = f_n(\theta + 2\pi)$, or, equivalently, the condition that the ϵ-expansion (6.12) be uniformly asymptotic (note that we do not address here the problem of convergence of the series (6.12) but we limit ourselves to establish uniform asymptoticity). The point is that, for each $n \geq 2$, the forcing term in (6.14) contains the fundamental harmonics $\exp(i\theta)$ and $\exp(-i\theta)$ which are solutions of the lhs equation (i.e. of the homogeneous equation), and are therefore secular, namely at resonance.

At this point, and for future use, we observe that, in a more general setting, if

$$v'(\theta) - Av(\theta) = w(\theta) + u(\theta) \tag{6.16}$$

is the equation of the motion of a vector $v(\theta)$ in a linear (finite or infinite dimensional) space and A is a linear operator, then, if the vector $w(\theta)$ solves the homogeneous equation,

$$w'(\theta) - Aw(\theta) = 0, \tag{6.17}$$

then the forcing term $w(\theta)$ in (6.16) is secular. This is apparent from the θ-dependence of the general solution of (6.16), which reads

$$v(\theta) = \tilde{v}(\theta) + \theta w(\theta), \tag{6.18}$$

where $\tilde{v}(\theta)$ is the general solution of the equation $\tilde{v}'(\theta) - A\tilde{v}(\theta) = u(\theta)$.

In our present case, the occurrence of the harmonics $\exp(i\theta)$ and $\exp(-i\theta)$ in the forcing term in the rhs of (6.14) forces the solution $f_n(\theta)$ to have a nonperiodic dependence on θ, and therefore the condition that the coefficients of $\exp(i\theta)$ and $\exp(-i\theta)$ must vanish is a crucial ingredient of our computational scheme. In fact, this condition fixes the value of the coefficient γ_{n-1} and this completes the recurrent procedure of computing, at each order in ϵ, both the frequency

$$\omega(\epsilon) = \omega_0 + \omega_1 \epsilon + \omega_2 \epsilon^2 + \ldots, \tag{6.19}$$

and the solution $f(\theta, \epsilon)$, see (6.12). As an instructive exercise, we suggest the reader to compute the frequency $\omega(\epsilon)$ up to $O(\epsilon^2)$ (answer: $\omega_1 = 0$, $\omega_2 = -(10c_2^2 + 9\omega_0^2 c_3)/24\omega_0^3)$.

This approach has been often used in applications with the aim of computing approximate solutions; in that context the properties of the series (6.11) and (6.12) of being convergent, or asymptotic, and also uniformly so in t, is of crucial importance (see, f.i., [16] and the references quoted there), particularly when one is interested also in the large time behaviour. Our emphasis here is instead in the formal use of the double expansion (see (6.12) and (6.15))

$$q(t,\epsilon) = \sum_{n=1}^{n} \sum_{\alpha=-n} \epsilon^n \exp(i\alpha\theta) f_n^{(\alpha)}, \qquad (6.20)$$

where $\theta = \omega_0 t + \omega_1 \epsilon t + \omega_2 \epsilon^2 t + \ldots$ and therefore here and in the following we drop any question related to convergence and approximation.

Let us consider now the propagation of nonlinear waves, and let us apply the Poincaré–Lindstedt method to PDEs. For the sake of simplicity, here and also throughout these notes, we focus our attention on the following family of equations which are first order in the variable time

$$Du = F[u, u_x, u_{xx}, \ldots] \quad, \quad u = u(x,t), \qquad (6.21)$$

with the assumptions that this equation be real, that the linear differential operator D in the lhs have the expression

$$D = \partial/\partial t + i\omega(-i\partial/\partial x) \quad, \qquad (6.22)$$

where $\omega(k)$ is a real odd analytic function,

$$\omega(k) = \sum_{m=0} a_{2m+1} k^{2m+1} \quad, \qquad (6.23)$$

and that F in the rhs be a nonlinear real analytic function of u and its x - derivatives. For instance, the subfamily

$$\omega(k) = a_1 k + a_3 k^3 \quad, \quad F = cu_x^3 + \left(c_2 u^2 + c_3 u^3 + \ldots\right)_x \quad, \qquad (6.24)$$

contains three S-integrable equations, i.e. the KdV equation ($c = 0, c_n = 0$ for $n \geq 3$), the mKdV equation (6.4a) and the equation [17]

$$u_t + a_1 u_x - a_3 u_{xxx} = -a_3[\alpha \sinh u + \beta(\cosh u - 1) + u_x^2/8]u_x. \qquad (6.25)$$

Since the linearized version of the PDE (6.21), $Du = 0$, has the harmonic wave solution

$$u = \exp[i(k_0 x - \tilde{\omega}_0 t)] \quad, \quad \tilde{\omega}_0 = \omega(k_0) \quad, \qquad (6.26)$$

one way to extend the Poincaré–Lindstedt approach to the PDE (6.21) is to look for solutions, if they exist, which are periodic plane waves,

$$u(x,t) = f(\theta,\epsilon), \quad \theta = k(\epsilon)x - \tilde{\omega}(\epsilon)t, \quad f(\theta,\epsilon) = f(\theta + 2\pi, \epsilon) \quad, \qquad (6.27)$$

together with the power expansions

$$f(\theta,\epsilon) = \epsilon f_1(\theta) + \epsilon^2 f_2(\theta) + \dots,$$
$$k(\epsilon) = k_0 + k_1\epsilon + k_2\epsilon^2 + \dots, \quad \tilde{\omega}(\epsilon) = \tilde{\omega}_0 + \tilde{\omega}_1\epsilon^2 + \tilde{\omega}_2\epsilon^2 + \dots \qquad (6.28)$$

This approach can be easily carried out as for the anharmonic oscillator since the function $f(\theta,\epsilon)$ does now satisfies the real ODE

$$-\tilde{\omega}(\epsilon)f^{(1)}(\theta,\epsilon) + i\omega(-ikd/d\theta)f(\theta,\epsilon) = F\left[f, kf^{(1)}, k^2 f^{(2)}, \dots\right], k = k(\epsilon), \qquad (6.29)$$

where $f^{(j)} \equiv d^j f(\theta,\epsilon)/d\theta^j$. Periodic plane waves in fluid dynamics have been investigated along these lines and, though exact solutions are known for instance for water waves models (such as the KdV equation) in terms of Jacobian elliptic functions (cnoidal waves), approximate expressions have been found more than a century ago (Stokes approximation) [18].

The class of periodic plane-wave solutions (if they exist) is too restrictive to our purpose. In fact their construction requires going from the PDE (6.21) to the ODE (6.29), a step which implies loss of information about the PDE itself. Therefore we now turn our attention to the class of solutions of the wave equation (6.21) whose leading term in the perturbative expansion is a quasi-monochromatic wave, namely a wave-packet whose Fourier spectrum is not one point but is well localized in a small interval of the wave number axis, $(k - \Delta k, k + \Delta k)$, where k is a fixed real number and $\Delta k/k$ is small,

$$u(x,t) \simeq \Delta k \int_{-\infty}^{+\infty} d\eta A(\eta) \exp\{i[x(k + \eta\Delta k) - t\omega(k + \eta\Delta k)]\} + c.c.; \qquad (6.30)$$

here the amplitude $A(\eta)$ is sharply peaked at $\eta = 0$, and the additional complex conjugated term is required by the condition (which we maintain here and in the following) that $u(x,t)$ is real, $u = u^*$.

The perturbation formalism which is suited to deal with this class of solutions is still close to the Poincaré–Lindstedt approach to the anharmonic oscillator. In fact, let us go back to the two-index series (6.20) and substitute θ with the expansion $\theta = \omega_0 t + \omega_1 t_1 + \omega_2 t_2 + \dots$, where we have formally introduced the rescaled "slow" times $t_n = \epsilon^n t$; then the formal expansion (6.20) reads

$$q(t,\epsilon) = \sum_{n=1}^{n} \sum_{\alpha=-n}^{n} \epsilon^n E^\alpha q_n^{(\alpha)}(t_1, t_2, \dots) , E \equiv \exp(i\omega_0 t) , \qquad (6.31)$$

where the functions $q_n^{(\alpha)}$ depend only on the slow-time variables t_n. The scheme of computation based on the expansion (6.31) is equivalent to that shown above, and it goes with inserting the expansion (6.31) into the eq. (6.7a) and by treating the time variables t_n as independent variables. In particular the derivative operator d/dt takes the ϵ-expansion

$$d(E^\alpha q_n^{(\alpha)})/dt = E^\alpha(i\alpha\omega_0 + \epsilon\partial/\partial t_1 + \epsilon^2\partial/\partial t_2 + \dots)q_n^{(\alpha)}, \qquad (6.32)$$

and similarly expanding the lhs and rhs of (6.7a) in powers of ϵ and of E finally yields a system of PDEs whose solution (after eliminating secular terms) gives the same result as the (much simpler) frequency-renormalization method based on (6.9) and (6.11). In this case the service of the multiscale technique is merely to display the three ingredients of the approach we use below for PDEs, i.e. the power expansion in a small parameter ϵ, the expansion in harmonics and the dependence on slow variables.

Let us now proceed with applying the multiscale perturbation approach to solutions of the PDE (6.21) along the line discussed above. As a preliminary observation, in the case the PDE (6.21) is linear, i.e. $F = 0$, the expression (6.30) is exact as it yields the Fourier representation of the solution. If we introduce the harmonic solution

$$E(x,t) \equiv \exp[i(kx - \omega t)], \quad \omega = \omega(k), \tag{6.33}$$

the small parameter $\epsilon \equiv \Delta k/k$ and the slow variables $\xi \equiv \epsilon x, t_n \equiv \epsilon^n t$ for $n \geq 1$, the Fourier integral takes the expression of a "carrier wave" whose small amplitude is modulated by a slowly varying envelope (no higher harmonics are generated in the linear case)

$$u(x,t) = \epsilon E(x,t) u^{(1)}(\xi, t_1, t_2, \ldots) + c.c.. \tag{6.34}$$

Since the envelope function is (see (6.30))

$$u^{(1)}(\xi, t_1, t_2, \ldots) = k \int_{-\infty}^{+\infty} d\eta A(\eta) \exp[i(k\eta\xi - k\omega_1 \eta t_1 - k^2 \omega_2 \eta^2 t_2 - \ldots)], \tag{6.35}$$

it satisfies the set of PDEs

$$\partial_{t_n} u^{(1)} = (-i)^{n+1} \omega_n \partial_\xi^n u^{(1)}, \quad n = 1, 2, \ldots \tag{6.36}$$

In order to write down these equations, we have assumed that the dispersion function $\omega(k)$ is analytic at k, so that its Taylor series

$$\omega(k + \epsilon \eta k) = \sum_{n=0}^{\infty} \omega_n \eta^n k^n \epsilon^n, \quad \omega_n(k) = \frac{1}{n!} \frac{d^n}{dk^n} \omega(k), \tag{6.37}$$

is convergent. This shows that one has to ask that $u^{(1)}$ depends on as many rescaled times t_n as the number of nonvanishing coefficients ω_n in the expansion (6.37); f.i. if $\omega(k)$ is a polynomial of degree N, the multiscale method requires the introduction of at most N new independent time variables, this being a rule which holds also in the nonlinear case. More interestingly, we note that in the *linear* case, because of the hierarchy of compatible evolution equations (6.36) with respect to the slow times, the commutativity property $[\partial_{t_n}, \partial_{t_m}] = 0$ is trivially satisfied, whereas, in the *nonlinear* case this commutativity condition is of paramount importance and is strictly related to integrability in more than one way. Indeed, the purpose of Sect. 6.3 is to show that the picture we have outlined in the linear case can be extended to the nonlinear case under appropriate conditions. The main consequence of nonlinearity is the generation of harmonics which are different from the fundamental one (6.33), together with the occurrence of undesired secular terms which force the amplitudes to grow with time. Killing the secular terms to keep the amplitudes bounded for all times is the basic way to derive a number of evolution equations. An old result in this direction, first derived in nonlinear optics and in fluid dynamics [4–8], is the dependence of the leading order amplitude $u_1^{(1)}(\xi, t_1, t_2)$ of the fundamental harmonic on the first two slow times t_1 and t_2, namely $u_1^{(1)}$ translates with respect to t_1 with the group velocity ω_1 and evolves with respect to t_2 according to the NLS equation. Thus, at this order, the solution $u(x,t)$ of the PDE (6.21) is approximated by the expression

$$u(x,t) = \epsilon v(\xi - \omega_1 t_1, t_2)E(x,t) + c.c. + O\left(\epsilon^2\right) , \tag{6.38}$$

where

$$v_{t_2} = i\omega_2 \left(v_{\xi\xi} - 2c|v|^2 v\right) \equiv K_2(v) . \tag{6.39}$$

In order to proceed further, the natural point to start from is the harmonic expansion of the solution $u(x,t)$,

$$u(x,t) = \sum_{\alpha=-\infty}^{+\infty} u^{(\alpha)}(\xi, t_1, t_2, \ldots)E^{\alpha}(x,t) , \tag{6.40}$$

where $E(x,t)$ si defined by (6.33) and, since u is real, $u = u^*$, the coefficients $u^{(\alpha)}$ satisfy the reality condition

$$u^{(\alpha)*} = u^{(-\alpha)} . \tag{6.41}$$

As for the slow variables, and guided by the approximate expression (6.30) where we set $\Delta k = \epsilon^p k$, with $p > 0$, we define

$$\xi = \epsilon^p x, \ t_n = \epsilon^{np} t , \ p > 0, \ n = 1, 2, \ldots. \tag{6.42}$$

As a consequence, the differential operators ∂_t and ∂_x, as acting on the expansion (6.40), are replaced by the power expansions

$$\partial_x \to \partial_x + \epsilon^p \partial_\xi , \ \partial_t \to \partial_t + \epsilon^p \partial_{t_1} + \epsilon^{2p} \partial_{t_2} + \ldots. \tag{6.43}$$

Inserting these expansions in the linear operator D, see (6.22), yields the formula

$$D\left[u^{(\alpha)}E^{\alpha}\right] = E^{\alpha}D^{(\alpha)}u^{(\alpha)}, \tag{6.44}$$

which defines the differential operator $D^{(\alpha)}$ acting *only* on the slow variables (6.42). Moreover, like the operators (6.43), also the differential operator $D^{(\alpha)}$ has a power expansion in ϵ,

$$D^{(\alpha)} = D_0^{(\alpha)} + \epsilon^p D_1^{(\alpha)} + \epsilon^{2p} D_2^{(\alpha)} + \ldots, \tag{6.45}$$

the first term being just the multiplication by the constant

$$D_0^{(\alpha)} = i[\omega(\alpha k) - \alpha\omega(k)], \tag{6.46}$$

since $DE^{\alpha} = D_0^{(\alpha)}E^{\alpha}$.

Let us consider now the nonlinear part, namely the rhs of the PDE (6.21). Since F is supposed to be an analytic function, its decomposition in harmonics,

$$F[u, u_x, u_{xx}, \ldots] = \sum_{\alpha=-\infty}^{+\infty} F^{(\alpha)} \left[u^{(\beta)}, u_\xi^{(\beta)}, u_{\xi\xi}^{(\beta)}, \ldots \right] E^\alpha, \qquad (6.47)$$

which is implied by the expansion (6.40), defines the functions $F^{(\alpha)}$ of the amplitudes $u^{(0)}, u^{(\pm 1)}, u^{(\pm 2)}, \ldots$ and their derivatives with respect to ξ. For future reference, we note that the functions $F^{(\alpha)}$ have the gauge property of transformation

$$F^{(\alpha)} \to \exp(i\alpha\theta) F^{(\alpha)}$$

when the amplitude $u^{(\alpha)}$ in its arguments is replaced by $\exp(i\alpha\theta) u^{(\alpha)}$, where θ is an arbitrary constant.

Combining now the expansion (6.40), and the definition (6.44), with the expansion (6.47) shows that the PDE (6.21) is equivalent to the (infinite) set of equations

$$D^{(\alpha)} u^{(\alpha)} = F^{(\alpha)}, \qquad (6.48)$$

which, since also $F^{(\alpha)}$ obviously satisfies the reality condition

$$F^{(\alpha)*} = F^{(-\alpha)}, \qquad (6.49)$$

needs to be considered only for nonnegative α, i.e. for $\alpha \geq 0$.

In the following sections, Eq. (6.48) will be investigated after expanding the amplitudes $u^{(\alpha)}$ in power of ϵ. In this respect, it should be pointed out that the approximate expression (6.30) of the solution $u(x, t)$ clearly shows that the smallness of u may originate in two ways, one from $\Delta k/k$ and the other from the amplitude A. In fact, we find it convenient to define ϵ by requiring that u itself be $O(\epsilon)$, and this explains why we have introduced the so far arbitrary parameter p in the rescaling (6.42) which defines the slow variables.

In Sect. 6.2, since we will look at Eq. (6.48) at the lowest order in ϵ, only few harmonics will be considered. This analysis, when carried out in a systematic way, eventually yields a certain number of model PDEs in the slow variables, whose integrability properties, if known, lead to the formulation of necessary conditions of integrability for the original PDE (6.21).

In the third section we tackle instead the problem of pushing the investigation of (6.48) to higher orders in the ϵ-expansion. This analysis displays interesting connections with integrability and it gives a way to set up an entire hierarchy of necessary conditions of integrability.

We end this introduction with few remarks. First, for pedagogical reasons, we have constrained the family of PDEs considered here to satisfy appropriate conditions in order to simplify the formalism. These limitations are mainly technical and do not play an essential role. For instance, extensions of the family of PDEs (6.21) may include differential equations of higher order in t for complex vector, or matrix, solutions in higher spatial dimensions.

Second, we have confined our interest to the multiscale technique which yields model equations of nonlinear Schrödinger type. Similar arguments, however, do apply also to the weakly dispersive regime where the prototypical model equation is instead the KdV equation [19], or to the resonant, or nonresonant, interaction of N waves [10].

Finally, a different approach which similarly yields necessary conditions for integrability, and has common features with the one described in Sect. 6.3, has been introduced by Kodama and Mikhailov [20]. There the perturbation expansion is combined with the property of integrable systems of possessing symmetries, and the order-by-order construction of such symmetries is the core of the method. Other ways to relate integrability to perturbative expansions in a small parameter have been investigated within different mathematical settings. The interested reader may refer to Zakharov and Schulman [21] for the Hamiltonian formalism. Also the use of normal form theory has been designed to this purpose in various contexts, see f.i. [22–24].

6.2 Nonlinear Schrödinger-Type Model Equations and Integrability

In this section we investigate the basic equations (6.48) which have been obtained via the harmonic expansion (6.40) of a quasi-monochromatic solution of the PDE (6.21). Here we consider only the lowest significant order in the small parameter ϵ, but before illustrating our computational scheme, which is mainly based on Refs. [25, 26] that the interested reader should consult for details and generalizations, we point out first the main ideas and aims of our approach.

Consider first that once the ϵ-expansion is introduced into the Eq. (6.48), the linear operator $D^{(\alpha)}$ takes the expression (6.45) whose coefficients, in addition to the first one (6.46), are easily found to be

$$D_n^{(\alpha)} = \partial_{t_n} - (-i)^{n+1} \omega_n(\alpha k) \partial_\xi^n , \quad n \geq 1 , \tag{6.50}$$

where the function $\omega_n(k)$ is defined by (6.37). Then, at the lowest order in ϵ, the operator $D^{(\alpha)}$ in (6.48) should be replaced by the coefficient $D_0^{(\alpha)} = i[\omega(\alpha k) - \alpha\omega(k)]$; therefore, if $D_0^{(\alpha)}$ is not vanishing, Eq. (6.48) for $u^{(\alpha)}$ becomes merely an algebraic equation whose solution is readily obtained. Because of this simple property, we term "slave harmonics" those harmonics such that, for their corresponding integer α, the quantity $D_0^{(\alpha)}$ does not vanishes, i.e.

$$\omega(\alpha k) - \alpha\omega(k) \neq 0. \tag{6.51}$$

If instead α is such that $D_0^{(\alpha)} = 0$, then we say that its corresponding harmonic is at resonance or, shortly, that α is a "resonance". The important feature of resonant harmonics is that their amplitude satisfies a differential equation in the slow variables (see (6.50)) rather than an algebraic equation as for slave harmonics. Of course, the harmonics $\alpha = 0, \pm 1$ are always (i.e. for any wave number k) at resonance (recall that $\omega(k)$ is on odd function, $\omega(-k) = -\omega(k)$). However, it may well happen that $D_0^{(\alpha)} = 0$ for $|\alpha| \neq 0, 1$ for a particular value of k; in this case also their corresponding harmonics are *accidentally* (i.e. not for all values of k) at resonance and their amplitudes are expected to satisfy differential equations which may be coupled to the equations for the fundamental harmonics amplitude.

The repeated application of this argument to the next term of the expansion of $D^{(\alpha)}$ will be shown below to lead to the introduction of weak and strong resonances, and the systematic investigation of all resonant cases does finally produce a list of *ten* model PDEs of nonlinear Schrödinger type. These evolution equations are reported and discussed below in this section, together with the implication of these findings with respect to integrability.

The starting ansatz is the ϵ-dependence at the leading order of the amplitude $u^{(\alpha)}$ in (6.40):

$$u^{(\alpha)} = \epsilon^{1+\gamma_\alpha} \psi_\alpha \quad \alpha = 0, \pm 1, \pm 2, \ldots, \tag{6.52}$$

where the parameters γ_α are nonnegative, $\gamma_\alpha \geq 0$, and, of course, even, $\gamma_{-\alpha} = \gamma_\alpha$, with the condition

$$\gamma_1 = 0, \tag{6.53}$$

which fixes the small parameter ϵ.

Looking only at the lowest order in ϵ greatly simplifies our analysis in two ways: it restricts our attention only to the first harmonics $|\alpha| = 0, 1, 2$ and, second, it allows the amplitudes ψ_α, see (6.52), to be considered as functions only of the slow variables ξ, t_1 and t_2. Moreover, since ξ and t_1 are of the same order in ϵ (see (6.42)), it turns out to be convenient to replace the slow space coordinate ξ with the new coordinate

$$\xi = \epsilon^p (x - Vt) \tag{6.54}$$

in the frame moving with the group velocity,

$$V = d\omega(k)/dk = \omega_1(k), \tag{6.55}$$

of the fundamental harmonics ($|\alpha| = 1$), so that the amplitudes ψ_α depend throughout this section only on two variables,

$$\psi_\alpha = \psi_\alpha(\xi, \tau) , \quad \tau \equiv t_2 = \epsilon^{2p} t. \tag{6.56}$$

As an additional remark, the following treatment suggests that it is convenient to take advantage of the fact that the nonlinear function in the rhs of the PDE (6.21) under investigation could be an x-derivative of a (polynomial or analytic) function, namely that it could be written as $\partial_x^h F(u, u_x, u_{xx}, \ldots)$, where it is advisable to choose for the integer h its highest possible value. This is only a technical point as the final results can be also derived, though more painfully, by starting with a lower value of h or by setting tout court $h = 0$, as in (6.21). Thus we rewrite the PDE (6.21)

$$Du = (\partial/\partial x)^h F[u, u_x, u_{xx}, \ldots],$$ (6.57)

where

$$F[u, u_x, u_{xx}, \ldots] = \sum_{m=2}^{\infty} \sum_{j_1=0}^{\infty} \sum_{j_2=j_1}^{\infty} \cdots \sum_{j_m=j_{m-1}}^{\infty} c_{j_1,\ldots,j_m}^{(m)} u^{(j_1)} u^{(j_2)} \ldots u^{(j_m)},$$ (6.58)

with $u^{(j)} \equiv (\partial/\partial x)^j u(x,t)$. Thus the family of PDEs we consider below is fully characterized by the following parameters: the real coefficients a_{2m+1} which define the dispersion function $\omega(k)$, see (6.22) and (6.23), the integer h (see (6.57)) and the real coefficients $c_{j_1,\ldots,j_m}^{(m)}$, see (6.58). The method described here provides necessary conditions which these parameters have to satisfy in order that the PDE (6.57) be integrable.

By taking into account the x-derivative in the rhs of (6.57) together with the ansatz (6.52), we first rewrite Eq. (6.48) in the form

$$\epsilon^{1+\gamma_\alpha} D^{(\alpha)} \psi_\alpha = (i\alpha k + \epsilon^p \partial_\xi)^h F^{(\alpha)}.$$ (6.59)

We obtain thereby nontrivial evolution equations for the quantities $\psi_\alpha(\xi, \tau)$ by first taking the limit $\epsilon \to 0$ (after having made an appropriate choice for the exponents γ_α and p) and then by performing some algebraic calculations and also some "cosmetic rescalings" on the dependent and independent variables, so as to present the results in neater form.

Let us first treat the linear part, namely the lhs of (6.57). Clearly we get

$$D^{(\alpha)} = \epsilon^{2p} \partial/\partial\tau + i \sum_{m=0}^{M} \epsilon^{pm} A_\alpha^{(m)}(k)(-i\partial/\partial\xi)^m$$ (6.60)

and

$$A_\alpha^{(0)}(k) = \omega(\alpha k) - \alpha\omega(k) ,$$ (6.61a)

$$A_\alpha^{(1)}(k) = \omega_1(\alpha k) - \omega_1(k) ,$$ (6.61b)

$$A_\alpha^{(s)}(k) = \frac{1}{s!} \frac{d^s}{dq^s} \omega(q)|_{q=\alpha k}, \quad s \geq 2.$$ (6.61c)

Here the coefficients $A_\alpha^{(s)}(k)$ with $s = 0, 1$ have been singled out because of the special role they play in the following. Note that by definition

$$A_1^{(0)} = A_1^{(1)} = 0;$$ (6.62)

this corresponds to the pivotal role of the component $\psi_1(\xi, \tau)$ which is the amplitude of the fundamental harmonic. It is indeed clear from (6.59) and (6.60) that the value of γ_α which is determined by the requirement to match the dominant terms as $\epsilon \to 0$ of the quantities in the rhs of (6.59), tends to be smaller if $A_\alpha^{(0)}$ vanishes and even smaller if in addition also $A_\alpha^{(1)}$ vanishes and so on. Of course the smaller is the value of γ_α, the larger is the role that the component $\psi_\alpha(\xi, \tau)$ plays in the regime of weak nonlinearity (small ϵ). This qualitative notion is given quantitative substance below; but already at this stage it indicates that the different possibilities discussed below emerge from various different assumptions about the vanishing of some of the quantities $A_\alpha^{(s)}(k)$; a vanishing which might occur for all values of k, as it were for *structural reasons*, or it might happen only for some special value of k, on which attention may then be focussed.

For these reasons, in the following the harmonic α is called *weak resonance* if $A_\alpha^{(0)}(k)$, but not $A_\alpha^{(1)}(k)$, vanishes,

$$A_\alpha^{(0)}(k) = 0 \ , \ A_\alpha^{(1)}(k) \neq 0, \tag{6.63}$$

while we say that the harmonic α is a *strong resonance* if, in addition to $A_\alpha^{(0)}(k)$, also $A_\alpha^{(1)}(k)$ vanishes,

$$A_\alpha^{(0)}(k) = A_\alpha^{(1)}(k) = 0. \tag{6.64}$$

Of course, one could consider also the case of even stronger resonances by requiring that, in addition to (6.64), also the condition $A_\alpha^{(2)}(k) = 0$ be satisfied. However, these cases are obviously less generic, and they will not be treated here.

Let us now consider the nonlinear rhs of (6.59). Inserting the ansatz (6.52) in the rhs of (6.58) yields the expression

$$F^{(\alpha)} = \sum_{m=2}^{\mu} \epsilon^{m-1} f_\alpha^{(m)} + O(\epsilon^\mu), \tag{6.65}$$

with

$$f_\alpha^{(m)} = \sum_{\{\alpha_1 \leq \alpha_2 \leq \ldots \leq \alpha_m; \sum_{j=1}^{m} \alpha_j = \alpha\}} \epsilon^\Gamma \{g(\alpha_1, \alpha_2, \ldots, \alpha_m) \psi_{\alpha_1} \ldots \psi_{\alpha_m} + O(\epsilon^p)\}; \tag{6.66}$$

here

$$\Gamma \equiv \gamma_{\alpha_1} + \gamma_{\alpha_2} + \ldots + \gamma_{\alpha_m} \ , \tag{6.67}$$

and for the constants g we get

$$g(\alpha_1, \ldots, \alpha_m) = \sum_{\{0 \leq j_1 \leq \ldots \leq j_m\}} (ik)^J c_{j_1, \ldots, j_m}^{(m)} \left[\sum_{P(\alpha_1, \ldots, \alpha_m)} \Pi_{p=1}^m (\alpha_p)^{j_p} \right] , \tag{6.68}$$

where $J = j_1 + j_2 + \ldots + j_m$, and the notation $\sum_{P(\alpha_1, \ldots, \alpha_m)}$ indicates the sum over all permutations of the indices $\alpha_1, \ldots, \alpha_m$ having different values.

Additional, drastic simplifications occur when further steps are taken towards implementing the $\epsilon \to 0$ limit; indeed in this context we shall generally need to consider only the quadratic and cubic terms of F in (6.57), because the contribution of all other terms turns out to be negligible. Hence (6.59) can now be written, in more explicit form, as follows:

$$\epsilon^{2p}\left[\psi_{1\tau}-iA_1^{(2)}\psi_{1\xi\xi}\right]=(ik)^h$$

$$\cdot\left[\epsilon^{1+\gamma_0}g(0,1)\psi_0\psi_1+\epsilon^{1+\gamma_2}g(-1,2)\psi_1^*\psi_2+\epsilon^2g(1,1,-1)|\psi_1|^2\psi_1\right],\qquad(6.69a)$$

$$\epsilon^{\gamma_0+p}\left[A_0^{(1)}\psi_{0\xi}+\epsilon^p\psi_{0\tau}\right]=(\partial/\partial\xi)^h.$$

$$\cdot\epsilon^{hp}\left[\epsilon^{1+2\gamma_0}g(0,0)\psi_0^2+\epsilon g(-1,1)|\psi_1|^2+\epsilon^{1+2\gamma_2}g(-2,2)|\psi_2|^2\right],\qquad(6.69b)$$

$$\epsilon^{\gamma_2}\{iA_2^{(0)}\psi_2+\epsilon^pA_2^{(1)}\psi_{2\xi}+\epsilon^{2p}\left[\psi_{2\tau}-iA_2^{(2)}\psi_{2\xi\xi}\right]\}=(2ik)^h.$$

$$\cdot\left[\epsilon g(1,1)\psi_1^2+\epsilon^{1+\gamma_0+\gamma_2}g(0,2)\psi_0\psi_2\right].\qquad(6.69c)$$

The coefficients g which appear in these PDEs are found, via the formula (6.68), to have the expressions

$$g(0,0)=c_{0,0}^{(2)},\qquad(6.70a)$$

$$g(0,n)=2c_{0,0}^{(2)}+\sum_{j=1}^{\infty}(-1)^j(nk)^{2j}c_{0,2j}^{(2)}+i\sum_{j=0}^{\infty}(-1)^j(nk)^{2j+1}c_{0,2j+1}^{(2)},\ n\neq0,$$

$$(6.70b)$$

$$g(n_1,n_2)=\left(1-\frac{1}{2}\delta_{n_1n_2}\right)\left[\sum_{j=0}^{\infty}(-1)^jk^{2j}\sum_{j'=0}^{j}c_{j',2j-j'}^{(2)}\left(n_1^{j'}n_2^{2j-j'}+n_1^{2j-j'}n_2^{j'}\right)\right.$$

$$+i\sum_{j=0}^{\infty}(-1)^jk^{2j+1}\sum_{j'=0}^{j}c_{j',2j+1-j'}^{(2)}\left(n_1^{j'}n_2^{2j+1-j'}+n_1^{2j+1-j'}n_2^{j'}\right)\Bigg],\ n_1\neq0,\ n_2\neq0.$$

$$(6.70c)$$

Equations (6.69a,b,c) contain terms of different order in the small parameter ε, and this requires some explaining.

In the first place, many other terms which might have been present have been omitted because they are of higher order in ε than terms which are present. This is for instance the case for *cubic* terms in the rhs of (6.69a) involving ψ_0, ψ_2, which are of higher order than *quadratic* terms which are present. Of course this argument and analogous ones below are applicable only if the relevant dominant terms are indeed present, namely provided they are not absent. Note that such an absence might happen for some "accidental" reason (possibly only for some special value of k) or for a "structural" reason, for instance if the original equation (6.57) contains nonlinear terms only of cubic order and higher, but no quadratic terms.

The second point that must be emphasized about (6.69a,b,c) is that these equations generally contain contributions of different orders in ε, and only those of lowest order are relevant. The identification of these depends of course on the assignments of specific numerical values to p (of course $p>0$) and to the parameters γ_α (of course $\gamma_\alpha\geq0,\alpha=0,1,2$). These assignments are dictated by the structure of these equations (6.69a,b,c), and by assumptions which have to be made about the vanishing or nonvanishing of the quantities $A_\alpha^{(m)}(k)$, $m = 0, 1, 2, \alpha = 0, 1, 2$, appearing in the lhs of (6.69b,c); hence one must consider many subcases, according to which resonances are present. Let us reemphasize that, in this treatment which yields the results reported here, the assumption is made that *all* nonlinear terms which might be present at the lowest order in ε are indeed present, namely that no nonlinear terms are missing due to "accidental" cancellations or "structural" causes. Whenever this hypothesis turns out not to hold, the analysis leading to the assignment of the exponents p and γ_α must be performed anew by taking into account higher order terms in ε. This analysis can be based on Eqs. (6.69a,b,c) only if all the relevant higher order terms are already present in the rhs of these equations, otherwise account of additional terms in the ε-expansion is necessary. Explicit instances of this phenomenon are reported in [25].

We finally display the model equations which are obtained from (6.69a,b,c) in the notation $\psi_0 = \theta, \psi_1 = \varphi, \psi_2 = \chi, \xi = x$ and $\tau = t$. There are 10 such equations:

$$i\varphi_t + v\varphi_{xx} = \lambda |\varphi|^2 \varphi ; \tag{6.71}$$

$$\begin{cases} i\varphi_t + v\varphi_{xx} = \lambda^{(1)} \theta \varphi , \\ \theta_x = \lambda^{(2)} |\varphi|^2 ; \end{cases} \tag{6.72}$$

$$\begin{cases} i\varphi_t + v\varphi_{xx} = \lambda^{(1)} \chi \varphi^* , \\ \chi_x = \lambda^{(2)} \varphi^2 ; \end{cases} \tag{6.73}$$

$$\begin{cases} i\varphi_t + v\varphi_{xx} = \lambda^{(1)} \theta \varphi + \lambda^{(2)} \chi \varphi^* , \\ \theta_x = \lambda^{(3)} |\varphi|^2 , \\ \chi_x = \lambda^{(4)} \varphi^2 ; \end{cases} \tag{6.74}$$

$$\begin{cases} i\varphi_t + v\varphi_{xx} = \lambda^{(1)} \theta \varphi , \\ \theta_t = \lambda^{(2)} \theta^2 + \lambda^{(3)} |\varphi|^2 ; \end{cases} \tag{6.75}$$

$$\begin{cases} i\varphi_t + v\varphi_{xx} = \lambda^{(1)} \theta \varphi , \\ \theta_t = \lambda^{(2)} (|\varphi|^2)_x ; \end{cases} \tag{6.76}$$

$$\begin{cases} i\varphi_t + v\varphi_{xx} = \lambda^{(1)} |\varphi|^2 \varphi + \lambda^{(2)} \theta \varphi , \\ \theta_t = \lambda^{(3)} (|\varphi|^2)_{xx} ; \end{cases} \tag{6.77}$$

$$\begin{cases} i\varphi_t + v\varphi_{xx} = \lambda^{(1)} \theta \varphi + \lambda^{(2)} \chi \varphi^* , \\ \theta_t = \lambda^{(3)} (|\varphi|^2)_x , \\ \chi_x = \lambda^{(4)} \varphi^2 ; \end{cases} \tag{6.78}$$

$$\begin{cases} i\varphi_t + v^{(1)}\varphi_{xx} = \lambda^{(1)}\chi\varphi^* , \\ i\chi_t + v^{(2)}\chi_{xx} = \lambda^{(2)}\varphi^2 ; \end{cases} \tag{6.79}$$

$$\begin{cases} i\varphi_t + v^{(1)}\varphi_{xx} = \lambda^{(1)}\theta\varphi + \lambda^{(2)}\chi\varphi^* , \\ \theta_t = \lambda^{(3)}\theta^2 + \lambda^{(4)}|\varphi|^2 + \lambda^{(5)}|\chi|^2 , \\ i\chi_t + v^{(2)}\chi_{xx} = \lambda^{(6)}\theta\chi + \lambda^{(7)}\varphi^2 . \end{cases} \tag{6.80}$$

Let us emphasize that the coefficients v and λ appearing in different equations are *different* quantities, even if they have the same symbol. Note moreover that the equations featuring in the lhs of the zeroth harmonic $\psi_0 = \theta$ are *real*, hence *all* coefficients (both v and λ) appearing in them are real; while for the other equations the coefficients v are *real*, the coefficients λ are generally *complex*. It should be also clear that the structure of these equations reflects the existence of structural and/or accidental resonances. In fact, since the fundamental harmonic $\alpha = 1$ is, by definition, strongly at resonance, its amplitude φ always satisfies a PDE which is first order in time and second order in space; on the other hand, the zeroth harmonic is always weakly resonating and either it does not appear at all when $h \geq 1$ (because the first-order differential equation it satisfies can be explicitly integrated) or, when $h = 0$, it couples to the other resonating harmonics through a first-order differential equation which can be either in x or in t depending on whether it is weakly or, respectively, strongly resonating. Similarly for the amplitude χ of the second harmonic: if this harmonic is slave, it does not appear in the model equation, otherwise it satisfies a coupled differential equation which is first order in x if it is only weakly resonating, and is first order in t and second order in x if it is also strongly at resonance.

The derivation by reduction of these ten nonlinear Schrödinger-type model equations is the starting point to make contact with integrability. Indeed, from the knowledge that a model equation is not integrable we deduce that that particular original PDE in the class (6.57), from which the model equation follows by reduction, cannot be integrable. To the aim of illustrating the way to convert this general statement in concrete results we select out of the ten equations (6.71), (6.72), (6.73), (6.74), (6.75), (6.76), (6.77), (6.78), (6.79), (6.80) the following four PDEs, whose integrability properties are already known (for more details and examples, see [26]). *Equation (6.71)*: this is the NLS equation which is obtained if

$$A_0^{(1)}(k) \neq 0, A_1^{(2)}(k) \neq 0, A_2^{(0)}(k) \neq 0$$

and $h \geq 1$, with $v = A_1^{(2)}(k)$ and, if $h = 1$,

$$\lambda = -k\left[A_0^{(2)}(k)g(0,1)g(-1,1) + 2kA_0^{(1)}(k)g(-1,2)g(1,1)\right.$$
$$\left. + A_0^{(1)}(k)A_0^{(2)}(k)g(-1,1,1)\right]/A_0^{(1)}(k)A_0^{(2)}(k); \tag{6.81}$$

this equation is known to be S-integrable if

$$\text{Im}(\lambda) = 0. \tag{6.82}$$

Equation (6.72): it corresponds to $h = 0$, and $A_{(0)}^{(1)}(k) \neq 0$ and $A_1^{(2)}(k) \neq 0$; in this case $v = A_1^{(2)}(k)$, and

$$\lambda^{(1)} = g(0,1), \ \lambda^{(2)} = g(-1,1)/A_0^{(1)}(k); \tag{6.83}$$

this system of equations has been found (A. Ramani, Private Communication) to pass the Painlevé-type test only if

$$\lambda^{(1)}\lambda^{(2)} = 0, \tag{6.84}$$

namely, if it effectively linearizes.
Equation (6.73): this obtains if $h \geq 1$ and if, for some real nonvanishing value $k = \tilde{k}, A_2^{(0)}(\tilde{k}) = 0, A_1^{(2)}(\tilde{k}) \neq 0$ and $A_2^{(1)}(\tilde{k}) \neq 0$. In this case $v = A_1^{(2)}(\tilde{k})$ and, if $h = 1$,

$$\lambda^{(1)} = -\tilde{k}g(-1,2), \ \lambda^{(2)} = 2i\tilde{k}g(1,1)/A_2^{(1)}(\tilde{k}), \tag{6.85}$$

where, of course, the coefficients $g(-1,2)$ and $g(1,1)$ are valued here at $k = \tilde{k}$. Also this equation has been found (R. Conte, Private Communication) to pass the Painlevé-type test only if (6.84) holds.
Equation (6.76): this is the case if $h = 1$, and if, for some real nonvanishing value $k = \tilde{k}, A_0^{(1)}(\tilde{k}) = 0$ and $A_1^{(2)}(\tilde{k}) \neq 0$. Then $v = A_1^{(2)}(\tilde{k})$ and

$$\lambda^{(1)} = i\tilde{k}g(0,1), \ \lambda^{(2)} = g(-1,1), \tag{6.86}$$

where $g(0,1)$ and $g(-1,1)$ are evaluated at $k = \tilde{k}$. This system has been proved to be S-integrable [27] only if

$$\text{Im}\lambda^{(1)} = \text{Im}\lambda^{(2)} = 0. \tag{6.87}$$

With this information in our hands we are now in the position to formulate necessary conditions of integrability. For a systematic exploration of the various cases in which such conditions arise and apply, the reader is refereed to [26], while we limit ourselves to give here only few instances of our method, and of its potentialities.

We first observe that the integrability conditions for the four equations we have selected, i.e. (6.71), (6.72), (6.73) and (6.76), involve both the linear part (through the coefficients $A_\alpha^{(n)}$, see (6.61a,b,c)) and the nonlinear part (through the coefficients g, see (6.70a,b,c) and (6.58)) of the PDE (6.57) we wish to test, and that both the coefficients $A_\alpha^{(n)}$ and g are functions of the real parameter k. It is then clear that the integrability conditions (such as (6.82) and (6.84)) which hold for an arbitrary value of k produce a number of necessary conditions for the PDE (6.57) which is larger than the number of necessary conditions which originates from expressions such as (6.85) and (6.86) since these hold only for special values (if any) of k (say $\tilde{k}$).

Let us first assume that the PDE (6.57) we are going to test by our method is in the class with $h = 0$, namely its nonlinear term is not a derivative. Then, if the appropriate reduced equation is (6.72), the requirement that $g(0,1)$ or $g(-1,1)$ vanish for *all* real values of k entails, via (6.70b) and (6.70c), quite explicit restrictions only on the nonlinear part of (6.57). This is made explicit by the following:

Lemma 1. *A necessary condition for the integrability of a nonlinear evolution PDE of type* (6.57) *with $h = 0$ is that either*

$$c_{0n}^{(2)} = 0 , \; n = 0,1,2,\ldots, \tag{6.88}$$

or

$$\sum_{j=0}^{n} (-1)^j c_{j2n-j}^{(2)} = 0, \; n = 0,1,2,\ldots, \tag{6.89a}$$

namely

$$c_{00}^{(2)} = 0, \; c_{02}^{(2)} - c_{11}^{(2)} = 0, \; c_{04}^{(2)} - c_{13}^{(2)} + c_{22}^{(2)} = 0 \tag{6.89b}$$

and so on. Clearly the condition (6.88) *comes from the requirement that $g(0,1)$ vanish, while* (6.53) *comes from the requirement that $g(-1,1)$ vanish, see* (6.84) *and* (6.83). *Since they both require that $c_{00}^{(2)}$ vanish we obtain the following remarkably neat result.*

Lemma 2. *Every nonlinear PDE of type* (6.57) *with $h = 0$ featuring in its nonlinear part a term $c_{00}^{(2)} u^2$ is not integrable.*

Consider now the class of PDEs (6.57) with $h = 1$, and assume that the appropriate reduced model equation is the NLS equation (6.71). The requirement (6.82) with (6.81) for S-integrability involves both quantities related to the linear and nonlinear parts of the original equation (6.57), but in many cases it amounts to the requirements that (i) the quantity $g(0,1)$ be real (note that $g(-1,1)$ is always real, see (6.70c)); (ii) the quantities $g(-1,2)$ and $g(1,1)$ be both real or both imaginary; (iii) the quantity $g(-1,1,1)$ be real. Given the arbitrariness of k, the first of these three conditions clearly entails the vanishing of all the coefficients $c_{0n}^{(2)}$ with n odd; the second condition entails the vanishing of $c_{12}^{(2)}, c_{14}^{(2)}$ and $c_{23}^{(2)}$ and many other relations for the coefficients $c_{nm}^{(2)}$ with $n + m$ odd; the third condition entails the vanishing of $c_{001}^{(3)}$ and many other relations for the coefficients $c_{nmj}^{(3)}$ with $n + m + j$ odd. These are very stringent, and quite explicit, conditions on the nonlinear part of (6.57) (the case in which $h > 1$ can be similarly treated [26]).

Assume now that the original PDE (6.57), with $h = 1$, has passed the test based on the conditions specified above, namely that all conditions entailed by the requirement (6.82), with (6.81), are satisfied. Since these conditions are only necessary, not much information is gained, a part from a definite hint that our PDE may indeed turn out to be integrable. However, we can still push our method to look for additional conditions to be satisfied. This is in fact the case if a special value of $k, k = \tilde{k}$, exists such that either the condition $A_2^{(0)}(\tilde{k}) = 0$ holds, this being appropriate to obtain the model equation (6.73), or the condition $A_0^{(1)}(\tilde{k}) = 0$ holds, this being the case for the model equation (6.76). In the first case, a necessary condition for the integrability of a PDE of type (6.57) with $h = 1$ is that, for such special value for $k, k = \tilde{k}$, at least one of the two quantities $g(-1, 2), g(1, 1)$ vanish, see (6.84) with (6.85). The applicability and potency of this result is of course somewhat reduced relative to the conditions previously found, due to the requirement to restrict consideration to only those real values $\tilde{k}$ of k (if any) which satisfy the appropriate equality and inequalities specified above. Yet there clearly is a large class of nonlinear evolution PDEs to which these necessary conditions are applicable [26].

In the second case, namely that in which the model equation is (6.76), a necessary condition for the integrability of a PDE (6.57) with $h = 1$ is that, for the appropriate special value of k, i.e. $k = \tilde{k}$ such that the zeroth harmonic is strongly resonating, the quantity $g(0, 1)$ be imaginary (or vanish),

$$\text{Re}[g(0, 1)] = 0 \, , \, k = \tilde{k}. \tag{6.90}$$

This requirement follows from (6.87), (6.86) and from the property of $g(-1, 1)$ to be always real. This result is analogous to the previous one as it requires focussing on special values $\tilde{k}$ of k.

Let us state again that we have presented here only some of the necessary conditions which can be established by the multiscale reduction method and that more instances and applications are discussed in [26] where a distinction between necessary conditions for C-integrability and for S-integrability is also made. We also observe that various extensions are possible and worth of further research; for example, different classes of PDEs other than (6.57) can be investigated, say for vector or matrix solutions as well as with more spatial variables; and/or different model equations, other than the four equations considered here, can be taken as starting points for the derivation of other necessary conditions for integrability.

6.3 Higher Order Terms and Integrability

In this section our perturbative analysis of the original PDE (6.21) is extended to terms of higher order in ϵ. This extension is based on the expansion in powers of ϵ of the amplitude $u^{(\alpha)}$ in Eq. (6.48), with the implication that computations become rather heavy. To the aim of simplifying the formalism by avoiding unessential complications, we add two assumptions which we maintain throughout this section. First we ask that the nonlinear part of our equation (6.21), namely its rhs F, be an odd function of u,

$$F \to -F \text{ if } u \to -u. \tag{6.91}$$

As it is easily verified, this parity property allows us to consistently assume that the amplitudes of all even harmonics be vanishing,

$$u^{(2\alpha)} = 0 , \ |\alpha| \geq 0. \tag{6.92}$$

Therefore, from now on, we will have to deal only with the amplitudes $u^{(2\alpha+1)}$ of the odd harmonics. For instance, this condition on F is satisfied by the mKdV equation (6.4a), the C-integrable equation (6.2a) and by the class of PDEs (6.21) with (6.24) if $c_{2n} = 0$.

Our second assumption is that, in contrast with the analysis carried out in the previous section, no resonance occurs besides the fundamental harmonics $\alpha = \pm 1$. In other words, the resonance condition $D_0^{(\alpha)} = 0$, see (6.46), should hold only in the trivial case $|\alpha| = 1$.

These assumptions imply that all harmonics $\pm(2\alpha+1)$ with $\alpha > 0$ are slave and that the coefficients $u^{(\alpha)}(n)$ of their ϵ-expansion,

$$u^{(\alpha)} = \sum_{n=1} \epsilon^n u^{(\alpha)}(n), \ |\alpha| > 1, \tag{6.93}$$

are therefore expressed as differential polynomials of the coefficients $u(n)$ of the expansion of the fundamental harmonic ($\alpha = 1$)

$$u^{(1)} \equiv u = \epsilon u(1) + \epsilon^2 u(2) + \ldots = \sum_{n=1} \epsilon^n u(n). \tag{6.94}$$

Here, and also in the following, we drop the harmonic upper index in the coefficients of this expansion because of the very special role played by the function $u^{(1)}$ in this scheme (it is the only amplitude which satisfies a differential equation). Moreover, as additional implication which can be easily retrieved from the basic equation (6.48), the leading order of each harmonic amplitude comes from the rule

$$u^{(\alpha)}(n) = 0 , \ \text{for} \ n < |\alpha|, \tag{6.95}$$

which is equivalent to setting $\gamma_{2\alpha+1} = 2\alpha$ for $\alpha \geq 0$ in the notation (6.52); the slow variables ξ and t_n are here defined as in (6.42) with $p = 1$, i.e.

$$\xi = \epsilon x, \ t_n = \epsilon^n t \ , n = 1, 2, \ldots \tag{6.96}$$

In order to perform all operations required by our approach the functions $u(n), n = 1, 2, \ldots$, are required to be smooth in the real variable ξ, namely they are differentiable to any order in the whole ξ-axis.

The first step is inserting in Eq. (6.48) with $\alpha = 1$ the appropriate ϵ-expansions, namely that of the linear operator $D^{(1)} \equiv D$, see (6.45) with $\alpha = 1$ and $p = 1$,

$$D = \epsilon D_1 + \epsilon^2 D_2 + \ldots \ , \tag{6.97}$$

that of the amplitude $u^{(1)} \equiv u$, see (6.94), and finally the expansion of the nonlinear term,

$$F^{(1)} \equiv F = \epsilon^3 F_3 + \epsilon^4 F_4 + \dots ; \tag{6.98}$$

let us reemphasize here that since the differential operators D_n, see (6.50) with $\alpha = 1$, have the expression

$$D_n = \partial_{t_n} - (-i)^{n+1} \omega_n(k) \partial_\xi^n, \ n \geq 1, \tag{6.99}$$

there is no need to introduce the slow time t_n if it happens that $\omega_n(k) = 0$. Thus, if the dispersion relation $\omega(k)$ is a polynomial of degree $N > 1$, the expansion (6.97) turns out to be a polynomial in ϵ of degree N with the implication that only N slow times enter into play. We also note that, because of the parity condition (6.91), the expansion (6.98) of the nonlinear term starts from the third order. In conclusion, the basic equation (6.48) with $\alpha = 1$, i.e. $D^{(1)} u^{(1)} = F^{(1)}$ or, in the present notation

$$Du = F, \tag{6.100}$$

obviously yields the triangular system of convolution type

$$D_1 u(n) + D_2 u(n-1) + \dots + D_n u(1) = F_{n+1} . \tag{6.101}$$

Here, and in the following treatment, it is convenient to consider F_n as an element of the finite-dimensional vector space $\mathscr{P}_n$ defined as the set of all nonlinear differential polynomials in the functions $u(m)$ and $u^*(m)$ of order n and gauge 1. The meaning of this terminology is rather obvious: each monomial appearing in an element of $\mathscr{P}_n$ is a product of some $u(m), u^*(k)$ and their ξ-derivatives with the understanding that

$$\text{order}(u_j(m)) = \text{order}(u_j^*(m)) = m + j, \tag{6.102}$$

where we use the short-hand notation

$$u_j(m) \equiv \partial_\xi^j u(m). \tag{6.103}$$

On the other hand, by requiring that each polynomial in $\mathscr{P}_n$ be of gauge 1 we understand that such polynomials, say F_n, possess the transformation property

$$F_n \to e^{i\theta} F_n \ \text{if} \ u(m) \to e^{i\theta} u(m), \tag{6.104}$$

θ being an arbitrary real constant. By following these rules, the reader may easily verify that $\mathscr{P}_2$ is empty, dim $(\mathscr{P}_3) = 1$, the basis of $\mathscr{P}_3$ being the single monomial $|u(1)|^2 u(1)$, while dim $(\mathscr{P}_4) = 4$ where its basis may be given by the following four monomials: $|u(1)|^2 u(2), u(1)^2 u^*(2), |u(1)|^2 u_1(1), u(1)^2 u_1^*(1)$.

Therefore, each nonlinear term F_{n+1} in the rhs of (6.101) is a linear combination of the basis vectors (f.i. monomials) of the vector space $\mathscr{P}_{n+1}$, where the complex coefficients of such combination are determined by the nonlinear function in the rhs of our original PDE (6.21) (see the expansion (6.58) with m running only on the odd integers).

The next step aims to eliminating all secular terms which may enter in the system (6.101). Our analysis is briefly described below, and the reader who is interested in a detailed investigation of this point is referred to [28].

Consider first Eq. (6.101) for $n = 1$, i.e. $D_1 u(1) = 0$ since $F_2 = 0$ (see (6.98)); because of the expression (6.99), $D_1 = \partial_{t_1} + \omega_1 \partial_\xi$, the function $u(1)$ depends on t_1 through the variable $\xi - \omega_1 t_1$. The next equation, say (6.101) with $n = 2$, reads (see (6.99))

$$D_1 u(2) = -\left[\left(\partial_{t_2} - i\omega_2 \partial_\xi^2 \right) u(1) - F_3 \right], \qquad (6.105)$$

where its rhs plays the role of the nonhomogeneous (forcing) term with respect to the t_1-evolution. On the other hand, this term depends on t_1 through the variable $\xi - \omega_1 t_1$ (recall that $F_3 \in \mathscr{P}_3$) and it satisfies therefore the homogeneous equation $D_1 f = 0$. This implies that the rhs of (6.105) is secular and its elimination requires that $u(1)$ satisfies, with respect to t_2, the evolution equation $(\partial_{t_2} - i\omega_2 \partial_\xi^2)u(1) = F_3$, namely just the NLS equation, which has been derived in the previous section. Killing the secular term in (6.105) implies that also $u(2)$ as $u(1)$ depends on t_1 through the variable $\xi - \omega_1 t_1$. This argument can be easily repeated for each integer n in (6.101) and, together with taking into account the structure of the differential polynomial F_{n+1}, it recursively leads to conclude that the coefficients $u(n)$ all satisfy with respect to the time t_1 the same (trivial) equation

$$D_1 u(n) = (\partial_{t_1} + \omega_1 \partial_\xi)u(n) = 0, \ n \geq 1. \qquad (6.106)$$

The time t_1 plays no essential role and the system (6.101) reduces to

$$D_2 u(n-1) + D_3 u(n-2) + \ldots + D_n u(1) = F_{n+1}, \ n \geq 2, \qquad (6.107)$$

whose first equation (i.e. for $n = 2$) is the NLS equation

$$\partial_{t_2} u(1) = i\omega_2 \left(\partial_\xi^2 u(1) - 2c|u(1)|^2 u(1) \right) \equiv K_2[u(1)]; \qquad (6.108)$$

the rhs of this equation defines the nonlinear operator K_2 and we have set $F_3 = -2i\omega_2 c|u(1)|^2 u(1)$.

Next we consider Eq. (6.107) for $n = 3$, and we look at the evolution with respect to the time t_2. To this aim it is convenient to introduce the linear operator

$$M_2 = \partial_{t_2} - K_2'[u(1)], \qquad (6.109)$$

where $K_2'[u(1)]$ is the Fréchet derivative of $K_2[u(1)]$, see (6.108), that is

$$\frac{d}{ds}K_2[u(1) + sv]|_{s=0} = K_2'[u(1)]v, \qquad (6.110)$$

namely

$$M_2 v = v_{t_2} - i\omega_2 \left(v_{\xi\xi} - 4c|u(1)|^2 v - 2cu^2(1)v^* \right); \qquad (6.111)$$

in fact, with this notation, the $n = 3$ equation (6.107) reads

$$M_2 u(2) + D_3 u(1) = \tilde{F}_4, \tag{6.112}$$

where $\tilde{F}_4 = F_4 + 2i\omega_2 c(2|u(1)|^2 u(2) + u^2(1)u^*(2))\epsilon \mathscr{P}_4$. Again one has to face the problem of secularities for this equation. First we observe that the term $\partial_{t_3} u(1)$ in $D_3 u(1)$ is secular since, obviously, $M_2(\partial_{t_3} u(1)) = 0$ as $M_2 \sigma = 0$ is satisfied by any symmetry σ of the NLS equation (6.108). Second, we note that also the other term $\partial_\xi^3 u(1)$ in $D_3 u(1)$ is secular in the following sense. The requirement that the ϵ-expansion (6.94) of u is uniformly asymptotic in time implies that the coefficients $u(n)$ remain bounded as $t \to \infty$. In particular one should ask that the forcing term $\tilde{F}_4 - D_3 u(1)$ in (6.112) vanishes, as $t_2 \to \infty$, faster than $t_2^{-1/2}$ while the variable ξ/t_2 is kept fixed [28]. This restriction is equivalent to asking that $D_3 u(1)\varepsilon \mathscr{P}_4$, while, at the same time, the t_3-flow for $u(1)$ should also be compatible with the t_2-flow given by the NLS equation. The existence of such evolution of $u(1)$ with respect to t_3 is a fine consequence of the integrability of the NLS equation, provided the parameter c in (6.108) is real, $c = c^*$. Indeed, it is well known that a whole hierarchy of flows,

$$\partial_{t_n} u(1) = K_n[u(1)], \; n = 1, 2, \ldots \tag{6.113}$$

exist which are all compatible (i.e. commuting) with each other; in the present context, these evolution equations may be conveniently rewritten as

$$D_n u(1) = (-i)^{n+1} \omega_n c V_{n+1}, \; n = 1, 2, \ldots, \tag{6.114}$$

where V_n is a special element of $\mathscr{P}_n$ which depends only on $u(1), u(1)^*$ and their ξ-derivatives. The expression of the first few of these polynomials are

$$V_2 = 0, \; V_3 = -2q_0 u(1), \; V_4 = -6q_0 u_1(1),$$

$$V_5 = 2(3q_1 + 3cq_0^2 - q_{0\xi\xi})u(1) - 6(q_0 u_1(1))_\xi,$$

$$V_6 = 10(q_1 + 3cq_0^2 - q_{0\xi\xi})u_1(1) - 6(q_0 u_2(1))_\xi, \tag{6.115}$$

where we use the notation (6.103) together with the definition

$$q_n = |u_n(1)|^2, \; n = 0, 1, 2, \ldots . \tag{6.116}$$

Thus, the requirement that the solution $u(2)$ of (6.112) remains bounded as $t_2 \to \infty$ is that $D_3 u(1) = \omega_3 c V_3$ or, equivalently (see (6.113)), that $u(1)$ satisfies the complex mKdV equation

$$\partial_{t_3} u(1) = K_3[u(1)], \tag{6.117}$$

with the implication that Eq. (6.112) rereads

$$M_2 u(2) = G_4, \tag{6.118}$$

with

$$G_4 = \tilde{F}_4 - \omega_3 c V_4 = F_4 + 2i\omega_2 c(2|u(1)|^2 u(2) + u^2(1)u^*(2)) + 6\omega_3 c|u(1)|^2 u_1(1)\varepsilon \mathscr{P}_4.$$

The way to arrive at Eqs. (6.117) and (6.118) from (6.112) we have sketched here can be repeated for Eq. (6.107) for all n, through a careful analysis of the asymptotic behaviour of the functions $u(n)$ as $t_2 \to \infty$ [28]. The upshot of this analysis is that the system of PDEs (6.107) splits into the NLS hierarchy (6.113) for the first coefficient $u(1)$ and the secularity-free system

$$M_2 u(n) + M_3 u(n-1) + \ldots + M_n u(2) = G_{n+2}, n = 2, 3, \ldots, \qquad (6.119)$$

where G_n is an element of the vector space $\mathscr{P}_n$ and M_n is the linear operator

$$M_n = \partial_{t_n} - K'_n[u(1)], \qquad (6.120)$$

where, again, $K'_n[u(1)]$ is the Fréchet derivative of the nonlinear operator $K_n[u(1)]$ in the rhs of (6.113).

Let us point out here that the derivation of the triangular system of nonlinear PDEs (6.119) requires only that the lowest order nonlinear model equation (in this case the NLS equation) is integrable (i.e. in this case, the condition is that c be real, see (6.108)) so as to guarantee the existence of an infinite hierarchy of independent mutually commuting symmetries (such as (6.113)). However, if no further information on the original PDE (6.21) is at hand, one is left with the (hard) task of integrating the PDEs of the triangular system (6.119). Thus, at this point, the natural question to ask is whether the special property of the original PDE (6.21) of being (C- or S-) integrable reflects itself in a special property of the triangular system (6.119). Here below we briefly show that, indeed, the answer to this question leads to a hierarchy of necessary conditions of integrability which leads to test a given PDE (see also [29]).

First we observe that in the obviously integrable case in which the PDE (6.21) is linear, say $F = 0$, the operator M_n (6.120) reduces to D_n, see (6.99), and the system (6.119) with $G_n = 0$ separates into the hierarchy $D_n u(m) = 0, n = 1, 2, \ldots$, i.e., the same hierarchy for each function $u(m)$. In this case the consistency condition $[D_n, D_m] = 0$ is certainly plain but essential. The basic observation [29] now is that, if the original PDE (6.21) is C-integrable or S-integrable, then, similar to the first coefficient $u(1)$ which satisfies the hierarchy of PDEs (6.113), each function $u(m)$, for $m \geq 2$, satisfies the hierarchy of PDEs

$$M_n u(m) = f_n(m), \ n \geq 2, \ m \geq 2, \qquad (6.121)$$

where the nonhomogeneous nonlinear term $f_n(m)$ in the rhs is a differential polynomial in $\mathscr{P}_{n+m}$. More precisely, one can show that

$$f_n(m) \in \mathscr{P}_{n+m}(m-1), \qquad (6.122)$$

where $\mathscr{P}_n(j)$ is defined as the subspace of $\mathscr{P}_n$ whose elements are the differential polynomials of the functions $u(m)$ and $u^*(m)$ where the index m goes only up to j, say $1 \le m \le j$. Of course, since the functions $u(m)$ are also solutions of the system (6.119), the rhs terms of the hierarchy (3.31) has to be related to the rhs of (6.119) by the triangular condition

$$f_2(n) + f_3(n-1) + \ldots + f_n(2) = G_{n+2}, \ n \ge 2. \tag{6.123}$$

In order for the system of PDEs (6.119) to split into separate PDEs, namely Eq. (6.121), certain compatibility conditions must be met. In fact, since the linear operators M_n given by (6.120) commute with each other,

$$[M_n, M_m] = 0 \ , \ n \ge 1, \ m \ge 1, \tag{6.124}$$

as a straight consequence of the commutativity of the flows of the NLS hierarchy (6.113), then the hierarchy (6.121), for each $m \ge 2$, must satisfy the compatibility condition

$$M_j f_n(m) = M_n f_j(m). \tag{6.125}$$

Eliminating the time-derivatives by using the evolution equation (6.121) leads to rewrite the compatibility equation (6.125) as an algebraic condition which the differential polynomials $f_n(m)$ have to satisfy. In fact, this condition ultimately reads as a set of constraints on the components of $f_n(m)$ on the basis of the vector space $\mathscr{P}_{n+m}$.

The way to prove this interesting property of integrable PDEs is not reported here; it goes via the change of variable which linearizes the PDE (6.21) in the case of C-integrability (see, f.i., the transformation (6.2b)) or it makes use of the multiscale expansion of the spectral equation of the Lax pair in the case of S-integrability (see, f.i., the ODE (6.6) with (6.6b)). We note here that this result opens the way to establish an integrability test as it yields necessary conditions that the PDE (6.21) has to satisfy in order to be integrable. Indeed, if one can prove that no differential polynomials $f_n(m)$ exist such that (6.121) holds together with the relation (6.123), where G_n is given by the multiscale technique, see Eq. (6.119), then the original PDE (6.21) cannot be integrable.

The following two propositions are instrumental in setting up our test.

Proposition 3. *the homogeneous equation $M_n f = 0$ has no solution f in the vector space $\mathscr{P}_m$, namely*

$$\mathrm{Ker}(M_n) \cap \mathscr{P}_m = \phi. \tag{6.126}$$

Proposition 4. *if, for each $n \ge 2$, the equation*

$$M_2 f_3(n) = M_3 f_2(n) \tag{6.127}$$

is satisfied with $f_2(n)$ and $f_3(n)$ given in the appropriate space, see (6.122), then differential polynomials $f_m(n)$, with $m \ge 4$ and (6.122), exist unique such that the flows $M_m u(n) = f_m(n)$ commute with each other for $m \ge 2$. Our method is then better *illustrated by first showing the nonhomogeneous terms of the hierarchies (6.121) in the following table (note that in $f_n(m)$ the index n labels the nth member of the hierarchy of evolution equations, namely it refers to the time t_n, while m indicates the m-th coefficient of u in its ϵ-expansion):*

This table is arranged in such a way that summing up the entries along the vertical lines reproduces the condition (6.123), while the arrow which connects $f_2(n)$ with $f_3(n)$ represents the compatibility equation (6.127). Note also that the pattern pictured in Table 6.1 looks like a ladder if only a finite number of slow times need to be introduced (the simplest picture is obtained when only t_2 and t_3 are present as for the dispersion relation $\omega(k) = a_3 k^3$). Let us now proceed with our test. First one has to compute the differential polynomial G_4; this is obtained from the cubic terms of the nonlinear part F of the PDE (6.21) to be tested (of course, the preliminary step of computing G_3 and, therefore, the real constant c which enters in the operators M_n has been already made). Then, because of (6.127) with $n = 2$ and the equality $f_2(2) = G_4$, one has to verify that the vector $M_3 G_4$ is in the image $M_2(\mathscr{P}_5(1))$ of the operator M_2. In order to envisage the actual computational task, one has to realize that the differential operator M_2 which maps vectors in $\mathscr{P}_n(m)$ onto vectors in the bigger space $\mathscr{P}_{n+2}(m)$ is represented in such spaces as a rectangular matrix, with the implication that its image is a proper subspace of $\mathscr{P}_{n+2}(m)$. If it turns out that $M_3 G_4$ is not in $M_2(\mathscr{P}_5(1))$, then the original PDE (1.21) cannot be integrable and computations stop here. If, instead, $M_3 G_4$ belongs to the image of M_2, one can compute the vector $f_3(2)$ which solves the algebraic equation (6.127), and, because of Proposition 3, see (6.126), the solution $f_3(2)$ is unique. Proceeding to the next step requires first the computation of $f_2(3)$ by subtraction (see Table 6.1),

$$f_2(3) = G_5 - f_3(2) \; , \tag{6.128}$$

where G_5 is obtained directly from the original PDE (6.21), and then the verification that $M_3 f_2(3)$ be in the image $M_2(\mathscr{P}_6(2))$. If this is not the case, this test leads to the conclusion that the original PDE (6.21) is not integrable, otherwise the test goes on with the next step in a similar way, namely one computes $f_3(3)$ by solving (6.127) for $n = 3$. Because of Proposition 4 (see above), the polynomial $f_4(2)$ can be computed and, by subtraction (see Table 1),

Table 6.1 Triangular structure of the terms $f_n(m)$. Vertical summation represents equation (6.123). The arrow indicates the construction of $f_3(n)$ from $f_2(n)$ via equation (6.127)

$f_2(2) \rightarrow$	$f_3(2) ,$	$f_4(2) ,$	$f_5(2) ,$
$---$	$+$	$+$	$+$
G_4	$f_2(3) \rightarrow$	$f_3(3) ,$	$f_4(3) ,$
	$---$	$+$	$+$
	G_5	$f_2(4) \rightarrow$	$f_3(4) ,$
		$---$	$+$
		G_6	$f_2(5) \rightarrow$
			$---$
			G_7

$$f_2(4) = G_6 - f_3(3) - f_4(2) , \qquad (6.129)$$

the polynomial $f_2(4)$ is obtained, this being the starting point for the next order.

Thus this procedure may go on order by order, starting with G_4 at the order $n = 2$. Assume now that one has been able to iterate this computational scheme we have just illustrated up to the calculation of $f_2(n+1)$, and that the polynomial $M_3 f_2(n+1)$ turns out not to belong to the image $M_2(\mathscr{P}_{n+4}(n))$, then we have found an "obstruction" at order $n + 1$ since this procedure cannot be carried on any further. Of course, the higher is n where the obstruction occurs, the *more integrable* is the original PDE (6.21). The specification of this property deserves a notation, so we say that the PDE (6.21) is A_n-*integrable*, meaning *asymptotically integrable up to order n*, if no *obstruction* occurs up to order n and if the *obstruction* (if any) occurs at order $m \geq n + 1$. For instance, the PDE (6.21) is A_1-integrable if the constant c in the NLS equation (6.108) is real. It is also A_2-integrable if

$$M_3 f_2(2) = M_3(a|u(1)|^2 u_1(1) + bu(1)^2 u_1^*(1)) \qquad (6.130)$$

is in the image of M_2, i.e. in $M_2(\mathscr{P}_5(1))$, and recall that the coefficients a and b are directly computed from the PDE (6.21) since $f_2(2) = G_4$. By a straight, but tedious, computation one obtains that $M_3 f_2(2)$ is in $M_2(\mathscr{P}_5(1))$ if and only if a and b are real, $a = a^*$ and $b = b^*$, otherwise one has the obstruction. If a and b are real, one can go further at $n = 3$. In this case $f_2(3)$ is a 12-dim complex vector and the condition that $M_3 f_2(3)$ be in $M_2(\mathscr{P}_2(6))$ turns out to yield 15 real conditions so that the general solution $f_2(3)$ depends on $2 \times 12 - 15 = 9$ real constants. As it is already clear from these first instances, the computational burden rapidly increases with n and a computer code is needed even for the first few orders. An idea on how easily a PC can run out of memory already at $n = 4$ or $n = 5$ is given by the dimensionality of the vector spaces involved. In the notation $\mathscr{P}_n(m) \rightarrow \dim(\mathscr{P}_n(m))$, we have $\mathscr{P}_3(1) \rightarrow 1, \mathscr{P}_4(1) \rightarrow 2, \mathscr{P}_5(1) \rightarrow 5, \mathscr{P}_6(1) \rightarrow 8, \mathscr{P}_4(2) \rightarrow 4, \mathscr{P}_5(2) \rightarrow 12, \mathscr{P}_6(2) \rightarrow 26, \mathscr{P}_5(3) \rightarrow 14, \mathscr{P}_6(3) \rightarrow 34, \mathscr{P}_6(4) \rightarrow 36$.

In conclusion, this test is based on an infinite sequence of necessary conditions of integrability, one at each order of the ϵ-expansion of the amplitude of the fundamental harmonic. Formulated as it is here, several mathematical problems related to this method remain open for future investigations. Among others, natural generalizations of the family of PDEs (6.21) we have considered here are feasible. For instance, one can consider PDEs with more than one dispersion branch, as for PDEs of higher order of the time-derivative or systems of PDEs with vector or matrix solutions, and/or PDEs in more than $1 + 1$ independent variables. As an instance, we have applied [29] this test to the following family of third-order PDEs

$$u_t + c_0 u_x + \gamma u_{xxx} - \alpha^2 u_{xxt} = \left(c_1 u^2 + c_2 u_x^2 + c_3 u u_{xx}\right)_x , \qquad (6.131)$$

which is not in the class (6.21). With the assistance of Mathematica, we have found that only three members of the family (6.131) are A_3-integrable, namely the KdV equation ($\alpha = c_2 = c_3 = 0$), the Camassa–Holm [30] equation ($c_1 = -\frac{3}{2}c_3/\alpha^2$, $c_2 = c_3/2$) and one new equation ($c_1 = -2c_3/\alpha^2$, $c_2 = c_3$) which can be transformed, by a change of variables, to the form

$$m_t + m_x u + 3mu_x = 0 \ , \quad m = u - u_{xx} \ . \tag{6.132}$$

Since the nonlinearity of this equation is quadratic and it passes our test up to order 3 (we could not push the test to higher order because of the heavy algebraic computations involved), we conjectured that this equation be integrable, but with no proof as our conditions are only necessary. Only in a subsequent investigation of (6.132), related in particular with the existence of special solutions known as *peakons*, it has been finally shown that the PDE (6.132) is S-integrable by explicitly displaying the associated Lax pair and conservation laws [31] together with multisoliton solutions [32].

Finally, since the conditions of integrability presented here are only necessary, once they are met, one may try *the daisy petals method*:

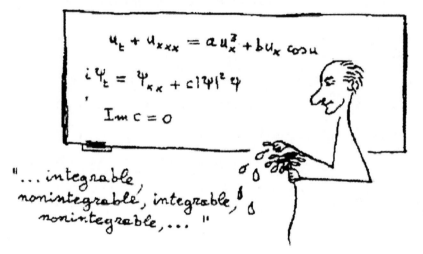

References

1. V.E. Zakharov (ed.), What is Integrability?, Springer, Heidelberg, 1990.
2. A. Degasperis, Resource Letter Sol-1: Solitons, Am. J. Phys. 66, 486–497, 1998.
3. T. Taniuti and N. Yajima, Perturbation method for a nonlinear wave modulation. I, J. Math. Phys. 10, 1369–1372, 1969; Suppl. Prog. Theor. Phys. 55, 1974 (special issue devoted to the Reductive Perturbation Method for Nonlinear Wave Propagation).
4. P.L. Kelley, Self-Focusing of Optical Beams, Phys Rev. Lett. 15, 1005, 1965.
5. D.J. Benney and A.C. Newell, J. Math. and Phys. (now Stud. Appl. Math.) 46, 133–139, 1967.
6. V.E. Zakharov, Instability of self-focusing of light Soviet Phys. JETP 26, 994–998, 1968.
7. H. Hasimoto and H. Ono, Nonlinear Modulation of Gravity Waves, J. Phys. Soc. Japan 33, 805, 1972.
8. A. Hasegawa and F. Tappert, Transmission of stationary nonlinear optical pulses in dispersive dielectric fibers. I. Anomalous dispersion, Appl. Phys. Lett. 23, 142, 1973.

9. F. Calogero and W. Eckhaus, Necessary conditions for integrability, Inverse Probl. 3, L27–L32, 1987.

10. F. Calogero, Why are certain nonlinear PDEs both widely applicable and integrable?. In [1], pp. 1–62.

11. F. Calogero and A. Degasperis, Spectral Transform and Solitons. Vol. I, North Holland, Amsterdam, 1982.

12. M.J. Ablowitz and P.A. Clarkson, Solitons, Nonlinear Evolution Equations and Inverse Scattering, Cambridge U.P., Cambridge, 1991.

13. V.E. Zakharov and A.B. Shabat, Exact theory of two-dimensional self-focusing and one-dimensional self-modulation of waves in nonlinear media, Soviet Phys. JETP 34, 62–69, 1972 [Russian original: Zh. Eksp. Teor. Fiz. 61, 118–134, 1971].

14. V.E. Zalharov and E.A. Kuznetsov, Multiscale Expansion in the Theory of Systems Integrable by the Inverse Scattering Transform, Physica 18D, 455–463, 1986.

15. C.S. Gardner, J.M. Greene, M.D. Kruskal, and R.M. Miura, Method for solving the Korteweg-de Vries equation, Phys. Rev. Lett. 19, 1095–1097, 1967.

16. J.A. Murdock, Perturbations, John Wiley & Sons, Inc., New York, 1991.

17. F. Calogero and A. Degasperis, Reduction technique for matrix nonlinear evolution equations solvable by the spectral transform, J. Math.Phys. 22, 23–31, 1981.

18. F.B. Whitham, Linear and Nonlinear Waves, John Wiley & Sons, Inc., New York, 1974.

19. R.A. Kraenkel, M.A. Manna, and J.G. Pereira, The Korteweg-de Vries hierarchy and long water-waves, J. Math. Phys. 36, 307, 1995.

20. Y. Kodama and A.V. Mikhailov, Obstacle to asymptotic integrability In: Algebraic Aspects of Integrable Systems: in memory of Irene Dorfman, A.S. Fokas and I.M. Gel'fand (eds.), Birkhauser, Boston, 173–204, 1996.

21. V.E. Zakharov and E.I. Schulman, Integrability of Nonlinear Systems and Perturbation Theory. In [1], pp. 185–250.

22. D. Bambusi, Birkhoff normal form for some nonlinear PDEs, Comm. Math. Phys. 234, 253–285, 2003.

23. B. Dubrovin and Y. Zhang, Normal forms of hierarchies of integrable PDEs, Frobenius manifolds and Gromov – Witten invariants, arXiv:math/0108160v1 [math.DG].

24. Y. Hiraoka and Y. Kodama, Normal form and solitons, Chap. 5 of this book and arXiv:nlin.SI/020602.

25. F. Calogero, A. Degasperis, and J. Xiaoda, Nonlinear Schrödinger-type equations from multiscale reduction of PDEs. I. Systematic derivation, J. Math. Phys. 41, 6399–6443, 2000.

26. F. Calogero, A. Desgasperis, and J. Xiaoda, Nonlinear Schrödinger-type equations from multiscale reduction of PDEs. II. Necessary conditions of integrability for real PDEs, J. Math. Phys. 42, 2635–2652, 2001.

27. N. Yajima and M. Oikawa, Formation and Interaction of Sonic-Langmuir Solitons, Prog. Theor. Phys. 56, 1719–1739, 1976.

28. A. Degasperis, S.V. Manakov, and P.M. Santini, Multiple-scale perturbation beyond the Nonlinear Schrödinger Equation. I, Physica D100, 187–211, 1997.

29. A. Degasperis and M. Procesi, Asymptotic Integrability. In: Symmetry and Perturbation Theory, A. Degasperis and G. Gaeta (eds.), World Scientific, Singapore, 23–37, 1999.

30. R. Camassa and D.D. Holm, An integrable shallow water equation with peaked solitons, Phys. Rev. Lett. 71, 1661–1664, 1993.

31. A. Degasperis, D.D. Holm, and A.N.W. Hone, A new integrable equation with peakon solutions, Theor. Math. Phys. 133, 1463–1474, 2002.

32. Y. Matsuno, Multisoliton solutions of the Degasperis–Procesi equation and their peakon limit, Inverse Problems 21, 1553–1570, 2005.

Chapter 7
Painlevé Tests, Singularity Structure and Integrability

A.N.W. Hone

Abstract After a brief introduction to the Painlevé property for ordinary differential equations, we present a concise review of the various methods of singularity analysis which are commonly referred to as Painlevé tests. The tests are applied to several different examples, and the connection between singularity structure and integrability of ordinary and partial differential equations is discussed.

7.1 Introduction

The connection between the integrability of differential equations and the singularity structure of their solutions was first discovered in the pioneering work of Kowalewski [70, 71], who considered the equations for the motion under gravity of a rigid body about a fixed point, namely

$$\frac{d\underline{\ell}}{dt} = \underline{\ell} \times \underline{\omega} + \underline{c} \times \underline{g},$$
$$\frac{d\underline{g}}{dt} = \underline{g} \times \underline{\omega}; \qquad \underline{\ell} = \mathbf{I}\,\underline{\omega}. \qquad (7.1)$$

In the above, $\underline{\ell}$ and $\underline{\omega}$ are, respectively, the angular momentum and angular velocity of the body, $\underline{g}$ is the gravity vector with respect to a moving frame, and the centre of mass vector $\underline{c}$ and inertia tensor $\mathbf{I}$ are both constant. The remarkable insight of Kowalewski was that the system of Eq. (7.1) could be solved explicitly whenever the dependent variables $\underline{\ell}$ and g are *meromorphic* functions of time t extended to the complex plane, $t \in \mathbb{C}$. By requiring that the solutions should admit Laurent expansions around singular points, she found constraints on the constants $\underline{c}$ and $\mathbf{I} = \mathrm{diag}(I_1, I_2, I_3)$ (diagonalized in a suitable frame). Her method isolated the two solvable cases previously known to Euler ($\underline{c} = 0$) and Lagrange ($I_1 = I_2$ with $\underline{c}$ defining the axis of symmetry), as well as a new case having principal moments of

A.N.W. Hone (✉)
Institute of Mathematics & Statistics, University of Kent, Canterbury CT2 7NF, UK,
anwh@kent.ac.uk

Hone, A.N.W.: *Painlevé Tests, Singularity Structure and Integrability.* Lect. Notes Phys. **767**, 245–277 (2009)
DOI 10.1007/978-3-540-88111-7_7   © Springer-Verlag Berlin Heidelberg 2009

inertia $I_1 = I_2 = 2I_3$ and $\underline{c}$ perpendicular to the axis of symmetry. The latter case is now known as the Kowalewski top, and Kowalewski was further able to integrate it explicitly in terms of theta-functions associated with an hyperelliptic curve of genus 2, thereby proving directly that the solutions are meromorphic functions of t. A modern discussion can be found in [8] or [75], for instance.

An important feature of Eq. (7.1) from the point of view of singularity analysis is that they are *nonlinear*. For a linear differential equation

$$\frac{d^n y}{dz^n} + a_{n-1}(z)\frac{d^{n-1} y}{dz^{n-1}} + \ldots + a_1(z)\frac{dy}{dz} + a_0(z)y = 0$$

of arbitrary order n it is well known [45, 58] that the general solution can have only *fixed* singularities at the points in the complex z-plane where the coefficient functions $a_j(z)$ are singular. However, for nonlinear differential equations, as well as the fixed singularities which are determined by the equation itself, the solutions can have *movable* singularities which vary with the initial conditions. For example, the first-order nonlinear differential equation

$$\frac{dy}{dz} + y^2 = 0$$

has the general solution

$$y = \frac{1}{z - z_0}, \qquad z_0 \quad \text{arbitrary},$$

with a movable simple pole at $z = z_0$. If the initial data $y = y_0$ is specified at the point $z = 0$, then the position of the simple pole varies according to

$$z_0 = -\frac{1}{y_0}.$$

The classification of ordinary differential equations (ODEs) in terms of their singularity structure was initiated in the work of Painlevé [84, 85]. The main property that Painlevé sought for ODEs was that their solutions should be single valued around movable singular points. Nowadays this property is usually formulated thus:

Definition 1. The Painlevé property for ODEs: An ODE has the Painlevé property if all movable singularities of all solutions are poles.

Painlevé proved that for first-order ODEs of the general form

$$y' = \frac{\mathscr{P}(y,z)}{\mathscr{Q}(y,z)},$$

where $\mathscr{P}$ and $\mathscr{Q}$ are polynomial functions of y and analytic functions of z (and the prime $'$ denotes d/dz), the only movable singularities that can arise are poles and algebraic branch points. The latter are excluded by Definition 1, and he further showed that the most general first-order equation with the Painlevé property is the Riccati equation

$$y' = a_2(z)y^2 + a_1(z)y + a_0(z),$$

where the coefficients a_0, a_1, a_2 are analytic functions of z.

For second-order ODEs, life is more complicated because movable essential singularities can arise (see Chap. 3 in [6] for an example). Painlevé initiated the classification of second-order ODEs of the form

$$y'' = \mathscr{F}\left(y', y, z\right), \tag{7.2}$$

with $\mathscr{F}$ being a rational function of y' and y and analytic in z. Painlevé and his contemporaries succeeded in classifying all ODEs of the type (7.2) which fulfil the requirements of Definition 1. The complete result is in the form of a list of approximately 50 representative equations, unique up to Möbius transformations, which are detailed in Chap. 14 of Ince's book [58]. It was found that (after suitable changes of variables) all of these ODEs have general solutions in terms of classical special functions (defined by linear equations) or elliptic functions, except for six special equations which are now known as Painlevé I–VI (or just PI–VI).

As an example, consider the second-order ODE

$$y'' = 6y^2 - \frac{1}{2}g_2. \tag{7.3}$$

This can be immediately integrated once, because the equation is autonomous (the right-hand side is independent of z), which yields

$$\left(y'\right)^2 = 4y^3 - g_2 y - g_3, \tag{7.4}$$

with g_3 being an integration constant. The general solution of the first-order ODE (7.4) is given by the Weierstrass elliptic function,

$$y = \wp(z - z_0; g_2, g_3) \tag{7.5}$$

with the constants g_2, g_3 being the invariants. The solution (7.5) has infinitely many movable double poles, at $z = z_0$ and at all congruent points $z = z_0 + 2m\omega_1 + 2n\omega_2 \in \mathbb{C}$ for $(m,n) \in \mathbb{Z}^2$ on the period lattice defined by the half-periods ω_1, ω_2. (For an introduction to Weierstrass elliptic functions see Chap. 20 in [112] or Chap. VI in [98].) We shall return to Eq. (7.3) in the next section.

The first of the *Painlevé equations* is PI, which is a non-autonomous version of (7.3) given by

$$y'' = 6y^2 + z. \tag{7.6}$$

Its general solution is a meromorphic function of z, and the solution of PI (or sometimes the equation itself) may be referred to as a Painlevé transcendent, since it essentially defines a new transcendental function. The other equations PII–PVI also contain parameters; for example the second Painlevé equation (PII) is

$$y'' = 2y^3 + zy + \alpha, \tag{7.7}$$

where α is a constant parameter. The general solution of each of the Painlevé equations cannot be expressed in terms of elliptic functions or other classical special functions [61], although for special parameter values they can be solved in this way; e.g. when α is an integer, Eq. (7.7) has particular solutions given by rational functions of z, and it has solutions in terms of Airy functions for half-integer values of α.

An important early result was the connection of PVI with the isomonodromic deformation of an associated linear system [32, 33]. After the work of Painlevé and his colleagues around the turn of the last century, the Painlevé equations were probably only of interest to experts on differential equations. However, in the latter half of the 20th century the Painlevé transcendents enjoyed something of a renaissance when it was discovered that they gave exact formulae for correlation functions in solvable models of statistical mechanics [113], quantum field theory [59, 60] and random matrix models [28, 63], and also arose as symmetry reductions of partial differential equations (PDEs) integrable by the inverse scattering transform (see [4] and Sect. 3 below). The link with integrable PDEs and linear Lax pairs established the exact solution of the Painlevé equations by the isomonodromy method [31]. More recently a weakened version of the Painlevé property has been used to find exact metrics for relativistic fluids [41]. With this wide variety of physical applications, the Painlevé transcendents have acquired the status of nonlinear special functions (see the review and references in Chap. 7 of [5]).

The continuation of Painlevé's classification programme to higher order equations becomes increasingly difficult as the order increases. Even at third order a new phenomenon can be encountered, in the form of a movable natural barrier or boundary beyond which the solution cannot be analytically continued; this occurs in Chazy's equation

$$y''' = 2yy'' - 3\left(y'\right)^2. \tag{7.8}$$

A variety of results for third or higher order equations have been obtained by Chazy [14], Gambier, Bureau and most recently by Cosgrove; see [24] and references therein. Chazy's equation (7.8) has some higher order relatives known as Darboux–Halphen systems, which have a very complicated singularity structure, and occur as reductions of the integrable self-dual Yang–Mills equations (see the contribution of Ablowitz et al. in [23]).

It should be clear from the above that the Painlevé property has a very deep connection with the concept of integrability. This connection is by no means straighforward and continues to be the subject of active research [23]. In the rest of this brief review article, we will introduce the basic techniques for testing the singularity structure of differential equations (both ODEs and PDEs), which are often referred to collectively as *Painlevé analysis*. The basic method for testing ODEs by expansions in Laurent series is treated in Sect. 2. This method should probably be referred to as the *Kowalewski–Painlevé test* to honour both pioneers of the subject, but most commonly only Painlevé is mentioned in this context. Section 3 describes the conjecture of Ablowitz, Ramani and Segur [4] on the connection between integrable PDEs and Painlevé-type ODEs, and how this can be used as an integrability test for PDEs. In the fourth section we explain how the preceding analysis can be bypassed by a direct consideration of the singularity structure of a PDE, by using the method of Weiss, Tabor and Carnevale [89]. This is followed in Sect. 5 by associated truncation techniques related to Bäcklund transformations, Lax pairs and Hirota bilinear equations for integrable systems of PDEs. In Sect. 6 we highlight the limitations of the Painlevé property as a criterion for integrability, in the context of integrable systems with movable algebraic branching and the weak Painlevé property [93]. In the final section we give our outlook on methods of singularity analysis for differential equations and mention how some of these methods and concepts have been extended to the discrete context of maps or difference equations.

7.2 Painlevé Analysis for ODEs

Consider an ODE for a dependent variable $y(z)$, which may be a single scalar or a vector quantity. If the ODE has the Painlevé property then its solutions must have local Laurent expansions around movable singularities at $z = z_0$, where z_0 is arbitrary. However, if branching occurs then this can be detected by local singularity analysis. The basic Painlevé test for ODEs consists of the following steps:

- **Step 1:** Identify all possible *dominant balances*, i.e. all singularities of form $y \sim c_0 (z - z_0)^\mu$.
- **Step 2:** If all exponents μ are integers, find the *resonances* where arbitrary constants can appear.
- **Step 3:** If all resonances are integers, check the *resonance conditions* in each Laurent expansion.
- **Conclusion:** If no obstruction is found in steps 1–3 for every dominant balance then the Painlevé test is satisfied.

Note that the exponents μ and leading coefficients c_0 must have as many components as the vector y, and if the ODE is polynomial then at least one of the exponents must be a *negative* integer for a leading order pole-type singularity. Rather than give formal definitions of the terms introduced in steps 1–3 above (which can be found in [13] and elsewhere in the references), we would like to illustrate them with a couple of examples.

First of all we describe the Painlevé test applied to Eq. (7.3), in which case y is just a scalar. Applying step 1 we look for leading order behaviour which produces a singularity in the ODE, so we require $y \sim c_0 (z - z_0)^\mu$ and μ to be a negative integer for a movable pole with no branching. This gives immediately

$$y \sim \frac{1}{(z - z_0)^2} \tag{7.9}$$

as the only possible dominant balance. Note that we could also have obtained this balance by assuming that y blows up as $z \to z_0$, and then (since g_2 is constant) $y^2 \gg g_2$ on the right-hand side of the ODE, so the y^2 term must balance with the left-hand side of (7.3), giving

$$y'' \sim 6y^2, \qquad \text{as} \quad z \to z_0. \qquad (7.10)$$

We can multiply by y' on both sides of (7.10) and integrate to find

$$\frac{1}{2}(y')^2 \sim 2y^3, \qquad \text{as} \quad z \to z_0 \qquad (7.11)$$

(throwing away the integration constant, which is strictly dominated by the other terms), and after taking a square root in (7.11) and integrating we find (7.9).

We now seek a solution of (7.3) given locally by a Laurent expansion around a double pole at $z = z_0$, in the form

$$y = \sum_{j=0}^{\infty} c_j (z - z_0)^{j-2}, \qquad c_0 = 1, \qquad (7.12)$$

where the value of c_0 has been fixed as in (7.9). We wish to determine the *resonances*, which are the positions in the Laurent series (7.12) where arbitrary coefficients c_j can appear. Since the ODE (7.3) is of second order, there must be two arbitrary constants in a local representation of the general solution: z_0, the arbitrary position of the movable pole, and one other. To apply step 2 of the Painlevé test we take a perturbation of the leading order with small parameter ϵ, in the form

$$y \sim (z - z_0)^{-2} \left(1 + \epsilon t (z - z_0)^r \right). \qquad (7.13)$$

To first order in ϵ we have

$$y^2 \sim (z - z_0)^{-4} \left(1 + 2\epsilon (z - z_0)^r \right), \quad y'' \sim (z - z_0)^{-4} \left(6 + \epsilon (r - 2)(r - 3)(z - z_0)^r \right).$$

Thus when we substitute the perturbation (7.13) into the dominant terms (7.10) and retain only first-order terms in ϵ we find

$$y'' - 6y^2 \sim \epsilon \left((r - 2)(r - 3) - 12 \right)(z - z_0)^{r-4} = 0.$$

Since the perturbation ϵ is arbitrary, corresponding to the first appearance of a new arbitrary constant in the Laurent expansion (7.12), the expression in large brackets must vanish, giving the resonance polynomial

$$r^2 - 5r - 6 = 0, \qquad \text{whence} \quad r = -1 \quad \text{or} \quad r = 6.$$

The first resonance at $r = -1$ must *always* be present in any expansion around a movable singularity, since it corresponds to the arbitrariness of z_0. The second resonance value $r = 6$ indicates that the coefficient c_6 should be arbitrary.

In order to complete the Painlevé test, we must now substitute in the full Laurent expansion and check that it is consistent up to the coefficient c_6. In this case we find that the expansion is precisely

$$y = \frac{1}{(z-z_0)^2} + \frac{1}{20}g_2 (z-z_0)^2 + \frac{1}{28}g_3 (z-z_0)^4 + \dots, \qquad (7.14)$$

so that $c_6 = g_3/28$ is the arbitrary constant that appears after integrating (7.3) to obtain (7.4). In fact only even powers of $(z-z_0)$ occur in this expansion, since the Weierstrass function (7.5) is an even function of its argument. The higher coefficients in (7.14) can be found recursively in terms of the invariants g_2, g_3. (Up to overall multiples these coefficients are the Eisenstein series associated with the corresponding elliptic curve [98].) The pole position z_0 does not appear in the coefficients because the ODE (7.3) is autonomous.

Here we should point out that passing the basic Painlevé test is only a *necessary* condition for an ODE to have the Painlevé property. Proving the Painlevé property requires showing that the local Laurent expansions can be analytically continued globally to a single-valued function (or one with only fixed branched points), in the absence of movable essential singularities. For the ODE (7.3) this follows from the fact that the general solution (7.5) is given by a Weierstrass elliptic function, which is meromorphic (for a proof see e.g. [98, 112]). Painlevé's proof that the first Painlevé transcendent (7.6) is free from movable essential singularities is outlined by Ince in Chap. 14 of [58], but the proof is unclear and this has prompted recent efforts to find a more straightforward approach [46, 65, 100].

Having seen an example where the Painlevé test is passed, we now move on to an example for which it fails, by considering the following coupled second-order system:

$$y_1'' = 2y_1^2 - 12y_2, \qquad y_2'' = 2y_1 y_2. \qquad (7.15)$$

In [37] this system is associated with an interaction of four particles moving in a plane, subject to velocity-dependent forces, and in that context it is essential that both $y_1(z)$, $y_2(z)$ (denoted $c_2(\tau)$, $c_4(\tau)$ in the original reference) and the independent variable z should be *complex*. To find the dominant balances, we look for leading order singular behaviour of the form

$$y_1 \sim aZ^\mu, \qquad y_2 \sim bZ^\nu, \qquad (7.16)$$

corresponding to a singularity in the solution at $Z = z - z_0 = 0$ for at least one of μ, ν negative. Because the system (7.15) is *autonomous*, we can expand in the variable Z, since the position z_0 of the movable singularity will not appear in the coefficients of local expansions around $z = z_0$.

There are three possible dominant balances for the system (7.15), namely

(i) $y_1 \sim 3Z^{-2}$, $y_2 \sim bZ^{-2}$, b arbitrary;
(ii) $y_1 \sim 3Z^{-2}$, $y_2 \sim bZ^3$, b arbitrary;
(iii) $y_1 \sim 10Z^{-2}$, $y_2 \sim \frac{35}{3}Z^{-4}$.

Other possible power law behaviour around $Z = 0$ corresponds to μ, ν both non-negative integers and leads to Taylor series expansions, which are not relevant to our analysis of singular points.

The second step in applying the Painlevé test is to find the resonances. For the system (7.15) to possess the Painlevé property we require that all resonances for all dominant balances be integers, and at least one balance must have one resonance value of -1 with the rest being non-negative integers, in which case this is a *principal balance* for which the Laurent expansion should provide a local representation of the general solution. To find the resonance numbers r we substitute

$$y_1 \sim aZ^\mu (1 + \delta Z^r), \qquad y_2 \sim bZ^\nu (1 + \epsilon Z^r)$$

into the dominant terms of the system (7.15) for each of the balances (i)–(iii) and take only the terms linear in δ and ϵ. This yields a pair of homogeneous linear equations for δ, ϵ (which correspond to the arbitrary coefficients appearing at the resonances). The determinant of this 2×2 system must vanish, which gives in each case a fourth-order polynomial in r.

Principal balance (i): It turns out that the balance (i) is the only principal balance, with resonances

$$(i) \qquad r = -1, 0, 5, 6.$$

As mentioned before, the resonance -1 is always present, since it corresponds to the arbitrary position z_0 of the pole, while $r = 0$ comes from the arbitrary constant b in the leading order term of the expansion for y_2; the other two values arise from arbitrary coefficients higher up in the series for y_1, y_2, so that altogether there should be four arbitrary constants appearing in these Laurent series. However, for step 3 of the test we also require that all resonance conditions hold: so far we have only found the orders in the series where arbitrary constants may appear, but it is necessary to check that all other terms vanish at this order when the series are substituted into the equations. Taking

$$y_1 \sim L_1(Z) := \sum_{j=-2}^{\infty} k_{1,j} Z^j, \qquad y_2 \sim L_2(Z) := \sum_{j=-2}^{\infty} k_{2,j} Z^j \qquad (7.17)$$

in each of Eqs. (7.15) we know already that the leading order terms require

$$k_{1,-2} = 3, \qquad k_{2,-2} = b \text{ (arbitrary)},$$

giving the resonant term at $r = 0$ in the expansion for y_2, while at subsequent orders we find

$$k_{1,-1} = 0 = k_{2,-1}; \quad k_{1,0} = b, k_{2,0} = -b^2/3; \quad k_{1,1} = 0 = k_{2,1}.$$

At the next orders we further obtain

$$k_{1,2} = -3b^2/5, k_{2,2} = 7b^3/15; \quad k_{1,3} = 0, k_{2,3} \quad \text{arbitrary},$$

so that the resonance condition at $r = 5$ corresponding to $k_{2,3}$ is satisfied. However, at the next order in the first equation of the system (7.15), at the first appearance of the resonance coefficient $k_{1,4}$, we find the additional relation

$$k_{2,2} = -b^3/5,$$

which means that the resonance condition is not satisfied unless $b = 0$, contradicting the fact that b should be arbitrary. Thus the Painlevé test is failed by this principal balance.

The only way to rectify the failure of the resonance condition and leave b as a free parameter is to modify (7.17) by adding logarithm terms. More precisely taking

$$y_1 \sim L_1(Z) + \Delta_1(Z), \qquad y_2 \sim L_2(Z) + \Delta_2(Z), \qquad (7.18)$$

the resonance condition is resolved by taking

$$\Delta_1 \sim -\frac{8}{7}b^3 Z^4 \log Z, \qquad \Delta_2 \sim -\frac{8}{21}b^4 Z^4 \log Z. \qquad (7.19)$$

However, the additional terms Δ_1, Δ_2 in (7.18) must then consist of a doubly infinite series in powers of Z and $\log Z$, with the leading order behaviours given by (7.19). Only in this way is it possible to represent the general solution of the system (7.15) as an expansion in the neighbourhood of a singular point containing four arbitrary parameters. Such infinite logarithmic branching is a strong indicator of non-integrability [79, 94].

Non-principal balance (ii): The second balance denoted (*ii*) above has resonances

$$r = -5, -1, 0, 6.$$

The presence of the negative integer value $r = -5$ means that this is a non-principal balance. (For an extensive discussion of negative resonances see [21].) This gives Laurent expansions

$$y_1 \sim 3Z^{-2} + kZ^4 - \frac{3}{2}bZ^5 + O(Z^7), \qquad y_2 \sim bZ^3 + O(Z^5). \qquad (7.20)$$

In this case all resonance conditions are satisfied and all higher coefficients in (7.20) are determined uniquely in terms of k and b. However, because it only contains three arbitrary constants (namely b, k and the position z_0 of the pole), it cannot represent the general solution, but can correspond to a particular solution which is meromorphic.

Non-principal balance (iii): For the balance (*iii*) the resonances are given by $r = -1$ and the roots of the cubic equation

$$r^3 - 15r^2 + 26r + 280 = 0,$$

which turn out to be a real irrational number and a complex conjugate pair, approximately

$$r = -3.2676, \quad 9.1338 \pm 1.5048i.$$

While non-integer rational resonances are allowed within the weak extension of the Painlevé test (see [93] and Sect. 6), irrational or complex resonances lead to infinite branching, and (as already evidenced by the principal balance (i)) the system (7.15) cannot possess the Painlevé property. This non-principal balance may be interpreted as a particular solution corresponding to a degenerate limit of the general solution, and perturbation of this particular solution (within the framework of the Conte–Fordy–Pickering perturbative Painlevé test [21]) will pick up the logarithmic branching present in the general solution. Clearly it would have been sufficient to stop the test after the failure of the resonance condition in the principal balance (i), but we wanted to present the details of the other balances to show the different possibilities that can arise in the singularity analysis of ODEs.

7.3 The Ablowitz–Ramani–Segur Conjecture

Having considered how to test for the Painlevé property in ODEs, we now turn to the connection with integrable PDEs. In the 1970s it was discovered that ODEs of Painlevé type, and in particular some of the Painlevé transcendents, appeared as symmetry reductions of PDEs solvable by the inverse scattering technique. This led Ablowitz, Ramani and Segur [4] to formulate the following:

Ablowitz–Ramani–Segur conjecture: *Every exact reduction of a PDE which is integrable (in the sense of being solvable by the inverse scattering transform) yields an ODE with the Painlevé property, possibly after a change of variables.*

To obtain ODE reductions of PDEs one can use the classical Lie symmetry method or its non-classical variants (see [83] for details), or the direct method of Clarkson and Kruskal [16]. The idea is that having found the symmetry reductions of the PDE, one can either solve the ODEs that are obtained or apply the Painlevé test to them, to see if branching occurs. If all the ODE reductions are of Painlevé type, then this suggests that the original PDE may be integrable. However, the need to allow for a possible change of variables will become apparent in Sect. 6. Indeed, the most difficult aspect of this conjecture, if one would like to provide a proof of it, is in defining exactly what class of variable transformations should be allowed.

As an example, consider the Korteweg–de Vries (KdV) equation for long waves on shallow water, which we write in the form

$$u_t = u_{xxx} + 6uu_x. \tag{7.21}$$

This has three essentially different reductions to ODEs; details of their derivation are given in Chap. 3 of [83]. The first is the travelling wave solution

$$u(x,t) = w(z), \qquad z = x - ct, \tag{7.22}$$

where c is the (arbitrary) wave speed and $w(z)$ satisfies

$$w''' + 6ww' + cw' = 0. \tag{7.23}$$

After an integration and a shift in w this is equivalent to (7.3), and the solution of (7.23) is given by

$$w = -2\wp(z - z_0) - c/6, \tag{7.24}$$

where z_0 and the invariants g_2 and g_3 of the $\wp$-function are arbitrary constants. In the special case $g_2 = 4k^4/3$, $g_3 = -8k^6/27$ the elliptic function degenerates to a hyperbolic function, and for $c = -4k^2$ the reduction (7.22) yields the one-soliton solution

$$u(x,t) = 2k^2 \operatorname{sech}^2(kx + 4k^3 t). \tag{7.25}$$

(Of course there is the additional freedom to shift the position of the soliton (7.25) by the transformation $x \to x - x_0$.)

The second reduction of KdV is the Galilean-invariant solution

$$u(x,t) = -2(w(z) + t), \qquad z = x - 6t^2, \tag{7.26}$$

where $w(z)$ satisfies

$$w''' - 12ww' - 1 = 0. \tag{7.27}$$

Upon integration, and making a shift in z to remove the constant of integration, the ODE (7.27) becomes the first Painlevé equation (7.6).

The third reduction of (7.21) is the scaling similarity solution

$$u(x,t) = (-3t)^{-\frac{2}{3}} w(z), \qquad z = (-3t)^{-\frac{1}{3}} x. \tag{7.28}$$

This solution arises from the invariance of the PDE (7.21) under the group of scaling symmetries

$$(x,t,u) \longrightarrow (\lambda x, \lambda^3 t, \lambda^{-2} u).$$

After substituting the similarity form (7.28) into KdV and integrating once we find the ODE for w:

$$w'' + 2w^2 - zw + \frac{\ell^2 - 1/4 + w' - (w')^2}{2w - z} = 0. \tag{7.29}$$

The parameter ℓ^2 is the constant of integration, and (7.29) turns out to be equivalent to the equation P34, so called because it is labelled XXXIV in the Painlevé classification of second-order ODEs as detailed by Ince [58]. The equation P34 can be solved in terms of the second Painlevé equation (7.7), according to the relation

$$w = -y' - y^2, \qquad \text{with} \quad \ell = \alpha + 1/2. \tag{7.30}$$

The above formula defines a Bäcklund transformation between the two equations (7.29) and (7.7), and in fact there is a one–one correspondence between their solutions. With the parameters of the two ODEs related as in (7.30), the inverse of this transformation (defined for $w \neq z/2$) is given by

$$y = \frac{w' + \alpha}{2w - z}.$$

For more details, and higher order analogues, see [51] and references.

Thus we have seen that the ODE reductions of the KdV equation (7.21) are solved either in elliptic functions or in terms of Painlevé transcendents, and hence these reductions certainly have the Painlevé property. So the KdV equation clearly fulfils the necessary condition for integrability required by the Ablowitz–Ramani–Segur conjecture, as it should do because it is integrable by means of the inverse scattering transform. In contrast to KdV, we consider another equation that models long waves in shallow water, namely the Benjamin–Bona–Mahoney (often referred to as BBM) equation [9], which takes the form

$$u_t + u_x + uu_x - u_{xxt} = 0. \tag{7.31}$$

The Benjamin–Bona–Mahoney equation is also known as the regularized long-wave equation and was apparently first proposed by Peregrine [86]. The travelling wave reduction of the Benjamin–Bona–Mahoney equation is very similar to that for KdV: the PDE (7.31) has the solution

$$u(x,t) = -12c\wp(z - z_0) + c - 1, \qquad z = x - ct, \tag{7.32}$$

given in terms of the Weierstrass $\wp$-function (with arbitrary values of the invariants g_2, g_3 and the constant z_0). In the hyperbolic limit with $c = (1 - 4k^2)^{-1}$ for $k \neq \pm 1/2$ this gives the solitary wave solution

$$u(x,t) = \frac{12k^2}{1 - 4k^2} \operatorname{sech}^2(kx - k(1 - 4k^2)^{-1}t),$$

but in contrast to (7.25) this is not a soliton because the Benjamin–Bona–Mahoney equation is not integrable and collisions between such waves are inelastic: see the discussion and references in Chap. 10 of [27].

Evidence for the non-integrable nature of the Benjamin–Bona–Mahoney equation is provided by another symmetry reduction, namely

$$u(x,t) = \frac{1}{t}w(z) - 1, \qquad z = x + \kappa \log t, \tag{7.33}$$

where κ is a constant. Upon substitution of (7.33) into (7.31), w is found to satisfy the ODE

$$\kappa w''' - w'' - ww' - \kappa w' + w = 0. \tag{7.34}$$

For all values of the parameter κ, Eq. (7.34) does not have the Painlevé property, which means that (at least in these variables) the Benjamin–Bona–Mahoney equation fails the necessary condition required by the Ablowitz–Ramani–Segur conjecture. In the case $\kappa = 0$, (7.34) just becomes second order, so it is possible to compare with the list in Chap. 14 of Ince's book [58] to see that $w'' + ww' - w = 0$ is not an ODE of Painlevé type. A direct method, which works for any κ, is to apply Painlevé analysis directly to the equation and show that a resonance condition is failed. In fact the analysis can be greatly simplified by integrating in (7.34) to obtain

$$\kappa w'' - w' - \frac{w^2}{2} - \kappa w = - \int_{z_1}^{z} w(s)\, ds. \tag{7.35}$$

(The lower endpoint of integration z_1 is an arbitrary constant.)

Now we can perform a Painlevé test on the integro-differential equation (7.35). For $\kappa \neq 0$, in the neighbourhood of a movable singularity at $z = z_0$ the dominant balance is between the w'' and w^2 terms, giving

$$w(z) \sim 12\kappa(z - z_0)^{-2}, \qquad z \to z_0.$$

If we suppose that this is the leading order in a Laurent expansion around $z = z_0$, i.e.

$$w(z) = \sum_{j=0}^{\infty} w_j (z - z_0)^{j-2}, \qquad w_0 = 12\kappa, \tag{7.36}$$

then at the next order we see that the coefficient of $(z - z_0)^{-1}$ is

$$w_1 = \frac{12\kappa}{6\kappa - 1}, \qquad \kappa \neq 1/6.$$

(For $\kappa = 1/6$ the Laurent expansion immediately breaks down.) However, substituting the expansion (7.36) into the left-hand side of (7.35) gives a Laurent series, while on the right-hand side there is a term $\log(z - z_0)$ arising from the non-zero residue $w_1 \neq 0$. Hence the expansion (7.36) cannot satisfy Eq. (7.35), or equivalently (7.34), and the Painlevé test is failed.

Thus we have seen that all of the ODE reductions of the KdV equation possess the Painlevé property, but not all the reductions of the non-integrable Benjamin–Bona–Mahoney equation (7.31) are of Painlevé type. We leave it as an exercise for the reader to check whether the Benjamin–Bona–Mahoney equation has other reductions apart from (7.32) and (7.33) (for hints see exercise 3.2 in [83]). However, it should be clear from the above that a fair amount of work is required when analysing a PDE in the light of the Ablowitz–Ramani–Segur conjecture, since one must first find all possible reductions to ODEs and then perform Painlevé analysis on each of them separately. Finding the symmetry reductions can be a difficult enterprise in itself (see [17] for example), but in the next section we shall see how this complication can be avoided by using the direct method due to Weiss, Tabor and Carnevale [107].

7.4 The Weiss–Tabor–Carnevale Painlevé Test

While the symmetry reductions of PDEs are clearly indicative of their integrability or otherwise, it is more convenient to analyse the singularity structure of PDEs directly. This approach was pioneered by Weiss, Tabor and Carnevale [107] (hence it is usually referred to as the WTC Painlevé test). However, in the context of PDEs with d independent (complex) variables $z_1, \ldots, z_d$ the singularities of the solution no longer occur at isolated points but rather on an analytic hypersurface $\mathscr{S}$ of codimension one, defined by an equation

$$\phi(\mathbf{z}) = 0, \qquad \mathbf{z} = (z_1, \ldots, z_d) \in \mathbb{C}^d, \tag{7.37}$$

where ϕ is analytic in the neighbourhood of $\mathscr{S}$. The hypersurface where the singularities lie is known as the *singular manifold*, and it can be used to define a natural extension of the Painlevé property for PDEs, which we state here in the form given by Ward [105]:

Definition 2. The Painlevé property for PDEs: If $\mathscr{S}$ is an analytic non-characteristic complex hypersurface in $\mathbb{C}^d$, then every solution of the PDE which is analytic on $\mathbb{C}^d \backslash \mathscr{S}$ is meromorphic on $\mathbb{C}^d$.

With the above definition in mind, it is natural to look for the solutions of the PDE in the form of a Laurent-type expansion near $\phi(\mathbf{z}) = 0$:

$$u(\mathbf{z}) = \frac{1}{\phi(\mathbf{z})^\mu} \sum_{j=0}^{\infty} \alpha_j(\mathbf{z}) \phi(\mathbf{z})^j. \tag{7.38}$$

If the PDE has the Painlevé property, then the leading order exponent μ appearing in the denominator of (7.38) should be a positive integer, with the expansion coefficients α_j being analytic near the singular manifold $\phi = 0$, and sufficiently many of these must be arbitrary functions together with the arbitrary non-characteristic function ϕ. As mentioned in [64] in the context of the self-dual Yang–Mills equations, and further explained in [105], it is important to state that ϕ should be non-characteristic because (even for linear equations) the solutions of PDEs can have arbitrary singularities along characteristics.

The application of the Weiss–Tabor–Carnevale test using series of the form (7.38) proceeds as for the usual Painlevé test for ODEs: when the series is substituted into the PDE, equations arise at each order in ϕ which determine the coefficients α_j successively, except at resonant values $j = r$, where the corresponding α_r are required to be arbitrary (subject to compatibility conditions being satisfied). The Weiss–Tabor–Carnevale test is only passed if all resonance conditions are fulfilled for every possible balance in the PDE (i.e. all consistent choices of μ). Note that, just as for ODEs, passing the test merely constitutes a *necessary* condition for the Painlevé property: a complete proof is much harder in general, although in the particular case of the self-dual Yang–Mills equations Ward [105] was able to use twistor methods to prove that they satisfy the requirements of Definition 2.

To see how the Weiss–Tabor–Carnevale test works, we will indicate the first steps of the analysis for the example of the KdV equation (7.21). In that case, there are just two independent variables x and t, so $d = 2$, and there is only one dominant balance where the degree of the singularity for the linear term u_{xxx} matches that for the nonlinear term uu_x. Substituting an expansion of the form (7.38) into (7.21), with

$\mathbf{z} = (x,t) \in \mathbb{C}^2$, it is clear that this gives $\mu = 2$ as the only possibility, and for the leading order and next-to-leading order the first two coefficients are determined as

$$\alpha_0 = -2\phi_x^2, \qquad \alpha_1 = 2\phi_{xx}. \tag{7.39}$$

This means that the expansion around the singular manifold for KdV can be written concisely as

$$u(x,t) = 2(\log\phi)_{xx} + \sum_{k=0}^{\infty} \alpha_{k+2}(x,t)\,\phi(x,t)^k, \tag{7.40}$$

where it is necessary to assume $\phi_x \not\equiv 0$ so that ϕ is non-characteristic.

In general, at each order j there is a determining equation for the coefficients of the series given by

$$(j+1)(j-4)(j-6)\alpha_j = F_j[\phi_x, \phi_t, \phi_{xt} \dots, \alpha_k; k < j], \tag{7.41}$$

where the functions F_j depend only on the previous coefficients α_k for $k < j$ and their derivatives, as well as the various x and t derivatives of ϕ. It is clear from (7.41) that the resonance values are $r = -1, 4, 6$, meaning that we require ϕ, α_4 and α_6 to be arbitrary functions of x and t. For the KdV equation, apart from the standard resonance at -1 corresponding to the arbitrariness of ϕ, the other necessary conditions for $r = 4, 6$, namely $F_4 \equiv 0$, $F_6 \equiv 0$ are satisfied identically, and so in accordance with the Cauchy–Kowalewski theorem these three arbitrary functions are the correct number to provide a local representation (7.40) for the general solution of the third-order PDE (7.21). We leave it to the reader to calculate the expressions for the higher F_j in (7.41) and verify the resonance conditions for F_4 and F_6; this is a standard calculation, so we omit further details which can be found in several sources, e.g. [79, 89]. For completeness we note that the issue of convergence of the expansion (7.40) for KdV has also been completely resolved [66].

We shall return briefly to the KdV equation in the next section, where we discuss how series such as (7.38) can be truncated within the *singular manifold method*, leading to Bäcklund transformations and Lax pairs for integrable PDEs, and by further truncation to Hirota bilinear equations for the associated tau-functions. Before doing so, we would like to illustrate ways in which the basic Weiss–Tabor–Carnevale test may be further simplified, taking the non-integrable Benjamin–Bona–Mahoney equation (7.31) as our example. Applying the test as outlined above directly to Eq. (7.31) leads to an expansion (7.38) very similar to that for KdV: it also has a single dominant balance with $\mu = 2$ for a non-characteristic singular manifold (where $\phi_x \not\equiv 0 \not\equiv \phi_t$), and the same resonances $r = -1, 4, 6$, but for the Benjamin–Bona–Mahoney equation not all resonance conditions are satisfied and the test is failed. It is a good exercise to perform this calculation and compare it with the corresponding results for KdV. Rather than presenting such a comparison here, we wish to give two shortcuts to the conclusion that Eq. (7.31) does not possess the Painlevé property for PDEs. First of all, observe that if $\phi_x \not\equiv 0$ then locally we can apply the implicit function theorem and solve Eq. (7.37) for x. Thus we set

$$\phi(x,t) = x - f(t) \tag{7.42}$$

with $\dot{f}(t) := df/dt \not\equiv 0$, and then we can take the coefficients in the expansion (7.38) to be functions of t only; this is referred to as the 'reduced ansatz' of Kruskal, first suggested in [64]. With this ansatz, the Weiss–Tabor–Carnevale analysis for PDEs becomes only slightly more involved than applying the Painlevé test for ODEs, and so constitutes a very effective way to decide if a PDE is likely to be integrable.

For the Benjamin–Bona–Mahoney equation there is a second shortcut that can be made, which is to take the potential form of the equation by making use of the fact that it has a conservation law. This approach is widely applicable, since nearly all physically meaningful PDEs admit one or more conservation laws. For Eq. (7.31) it is immediately apparent that it can be put in conservation form as

$$\frac{\partial u}{\partial t} = \frac{\partial}{\partial x}\left(u_{xt} - \frac{1}{2}u^2 - u\right),$$

which implies that

$$C = \int_{-\infty}^{\infty} u\,dx$$

is a conserved quantity for the Benjamin–Bona–Mahoney equation, i.e. $dC/dt = 0$ for $u(x,t)$ defined on the whole real x-axis with vanishing boundary conditions at $x = \pm\infty$. It follows that upon introducing the potential v as the new dependent variable, with

$$v = \int_{-\infty}^{x} u\,dx \longrightarrow C \quad \text{as} \quad x \to \infty,$$

we can replace u by v and its derivatives in (7.31) to obtain the potential form of the PDE, namely

$$v_t - v_{xxt} + v_x + \frac{1}{2}v_x^2 = 0 \tag{7.43}$$

(where we have integrated once and applied the boundary conditions to eliminate the arbitrary function of t). If we now apply the Weiss–Tabor–Carnevale test to (7.43), at the same time using the 'reduced ansatz' (7.42), then we see that the only possible leading exponent in a Laurent-type expansion for v is $\mu = 1$, giving

$$v(x,t) = \sum_{j=0}^{\infty} \beta_j(t)\,(x - f(t))^{j-1}. \tag{7.44}$$

The equations for the coefficients $\beta_j(t)$ at each order take the form

$$(j-1)(j+1)(j-6)\beta_j = F_j[\dot{f},\ddot{f},\ldots,\beta_k; k < j],$$

so the resonances are $r = -1, 1, 6$ which compares with $r = -1, 4, 6$ for the original equation (7.31): clearly one of the resonances has shifted to a lower value by taking the equation in potential form (7.43). Upon substituting the series (7.44) into the potential Benjamin–Bona–Mahoney equation, the leading order term is at order ϕ^{-4}, giving the equation

$$-6\beta_0 \dot{f} + \frac{1}{2}\beta_0^2 = 0.$$

Since $\beta_0 \neq 0$, this determines the first coefficient as

$$\beta_0 = 12\dot{f}.$$

However, at the next order ϕ^{-3} in Eq. (7.43), we have the resonance $r = 1$ with the condition

$$-2\dot{\beta}_0 = 0, \qquad \text{whence} \qquad \ddot{f} = 0. \tag{7.45}$$

Since f is supposed to be an arbitrary non-constant function of t, we see that the resonance condition (7.45) is not satisfied, so Eq. (7.43) fails the Weiss–Tabor–Carnevale Painlevé test, indicating the non-integrability of the Benjamin–Bona–Mahoney equation. However, observe what happens if f is a linear function of t: then (7.45) is satisfied, corresponding to the travelling wave reduction (7.32), which does have the Painlevé property.

The only way to remove the restriction (7.45) on the function f would be to add a term $-(\dot{\beta}/\dot{f})\log(x - f(t))$ to the expansion (7.44). It has been observed [90] that the inclusion of terms linear in $\log\phi$ for PDEs in potential form is not incompatible to integrability. However, in this case terms of all powers of $\log(x - f(t))$ are required to ensure a consistent expansion in the potential Benjamin–Bona–Mahoney equation (7.43) with three arbitrary functions f, β_1 and β_6 corresponding to the three resonances.

For the reader who is interested in applying either the Painlevé test for ODEs, as described in Sect. 2, or the Weiss–Tabor–Carnevale Painlevé test for PDEs, it is worth remarking that software implementations of these tests are now freely available. The web page www.mines.edu/fs_home/whereman has algorithms written by D. Baldwin and W. Hereman, for instance.

7.5 Truncation Techniques

Aside from the obvious application of the various Painlevé tests in isolating potentially integrable equations (for example, in the classification of integrable coupled KdV equations [67]), their usefulness can be extended by means of truncation techniques. The first of these is known as the *singular manifold method*, which was primarily developed in a series of papers by Weiss [108–111]. The idea behind the method is that by truncating an expansion such as (7.38), usually at the zero order (ϕ^0) term, it is possible to obtain a Bäcklund transformation for the PDE. For such truncated expansions the singular manifold function ϕ is no longer arbitrary, but satisfies constraints. In the case of integrable equations that are solvable by the inverse scattering transform, the singular manifold method can be used to derive the associated Lax pair; for directly linearizable equations, such as Burger's equation or its hierarchy [87], the method instead leads to the correct linearization. Even for non-integrable PDEs, where the constraints on ϕ are much stronger, the singular manifold method can still be used to obtain exact solutions. Furthermore, for integrable PDEs the truncation approach can be carried further by cutting off the series *before* the zero-order term, to yield tau-functions satisfying bilinear equations [35].

We will outline the basic truncation results for the KdV equation (7.21), before presenting more detailed calculations for the nonlinear Schrödinger (NLS) equation. For KdV, the Laurent-type expansion (7.40) can be consistently truncated at the zero-order term to yield

$$u = 2(\log \phi)_{xx} + \tilde{u}, \qquad \tilde{u} \equiv \alpha_2. \tag{7.46}$$

While substituting the full expansion (7.40) into KdV gives an infinite set of Eqs. (7.41) for ϕ and the α_j, the truncated expansion gives only a finite number. The last of these equations does not involve ϕ, and just says that $\tilde{u}$ is also a solution of KdV, i.e.

$$\tilde{u}_t = \tilde{u}_{xxx} + 6\tilde{u}\tilde{u}_x.$$

The other equations (after some manipulation and integration) boil down to just two independent equations for ϕ and $\tilde{u}$, as follows:

$$\tilde{u} = k^2 - \frac{(\sqrt{\phi_x})_{xx}}{\sqrt{\phi_x}}; \tag{7.47}$$

$$\frac{\phi_t}{\phi_x} = 6k^2 + \left(\frac{\phi_{xxx}}{\phi_x} - \frac{3\phi_{xx}^2}{2\phi_x^2} \right). \tag{7.48}$$

In the above, k is a constant parameter. The important feature to note is that since u and $\tilde{u}$ are both solutions of (7.21), Eq. (7.46) constitutes a Bäcklund transformation for KdV, provided that ϕ satisfies (7.47) and (7.48). For example, starting from the seed solution $\tilde{u} = 0$, the Bäcklund transformation defined by (7.46), (7.47) and (7.48) can be used to generate the one-soliton solution (7.25) or even a mixed rational-solitonic solution by taking $\phi = (x - 12k^2t) + (2k)^{-1} \sinh(2kx + 8k^3t)$.

Maybe it is not immediately obvious that the system comprised of the two equations (7.47) and (7.48) is equivalent to the standard Lax pair for KdV. This can be seen by making the squared eigenfunction substitution $\phi_x = \psi^2$, so that (7.47) becomes a linear (time-independent) Schrödinger equation. In the context of quantum mechanics in one dimension, ψ is the wave function with potential $-\tilde{u}$ and energy $-k^2$, i.e. (7.47) is equivalent to

$$\psi_{xx} + \tilde{u}\psi = k^2\psi.$$

The second equation (7.48) is known as the Schwarzian KdV equation [80], and in its own right it constitutes a nonlinear integrable PDE for the dependent variable ϕ; with the squared eigenfunction substitution it leads to the linear equation for the time evolution ψ_t. All these results for KdV are well known and have been extended to the whole KdV hierarchy; the interested reader who wishes to check these calculations is referred to [79] for more details.

Perhaps less well understood, however, is the interesting connection [35] between the singularity structure of PDEs and the tau-function approach to soliton equations pioneered by Hirota [48, 81], which culminated in the Sato theory relating integrable systems to representations of affine Lie algebras [77, 82]. The link with the singular manifold method is made by truncating the expansion (7.40) at the last singular term in ϕ, and setting $\phi = \tau$, to give

$$u = 2(\log \tau)_{xx}, \tag{7.49}$$

which is the standard substitution for the KdV variable u in terms of its tau-function. From (7.21), after substituting (7.49) and performing an integration (subject to suitable boundary conditions), a bilinear equation is obtained for the new dependent variable τ. This bilinear equation may be written concisely as

$$(D_x D_t - D_x^4)\tau \cdot \tau = 0, \tag{7.50}$$

by making use of the Hirota derivatives:

$$D_x^j D_t^k g \cdot f := \left(\frac{\partial}{\partial x} - \frac{\partial}{\partial x'}\right)^j \left(\frac{\partial}{\partial t} - \frac{\partial}{\partial t'}\right)^k g(x,t)f(x',t')|_{x'=x,t'=t}.$$

The bilinear form is particularly convenient for calculating multisoliton solutions [48], and leads to the connection with vertex operators [77, 81, 82]. For solitons the tau-function is just a polynomial in exponentials. In general τ is holomorphic, so from (7.49) it is clear that the places where τ vanishes correspond to the singularities of u.

We now present details on the application of the singular manifold method to the nonlinear Schrödinger equation

$$i\psi_t + \psi_{xx} - 2|\psi|^2\psi = 0. \tag{7.51}$$

This PDE (commonly referred to as NLS) describes the evolution of a complex wave amplitude ψ, and due to the minus sign in front of the cubic nonlinear term this is the non-focusing case of the nonlinear Schrödinger equation; the focusing case has $+2|\psi|^2\psi$ instead, and describes a different physical context. The following results on the singular manifold method for the nonlinear Schrödinger equation appeared in [49]. Seeking an expansion of the form (7.38) for (7.51), at leading order we find the behaviour

$$\psi \sim \frac{\alpha_0}{\phi}, \qquad |\alpha_0|^2 = \phi_x^2.$$

Thus, truncating the expansion at the zero order (ϕ^0) level, we find

$$\psi = \frac{\alpha_0}{\phi} + \hat{\psi}, \qquad \hat{\psi} \equiv \alpha_1. \tag{7.52}$$

To proceed with the singular manifold method we substitute the truncated expansion (7.52) into (7.51), and set the terms at each order in ϕ to zero. This yields the following four equations (the singular manifold equations):

$$\phi^{-3}: \qquad\qquad\qquad\qquad |\alpha_0|^2 - \phi_x^2 = 0;$$

$$\phi^{-2}: i\phi_t + 2\phi_x(\log\alpha_0)_x + \phi_{xx} + 2\alpha_0\overline{\hat{\psi}} + 4\overline{\alpha}_0\hat{\psi} = 0;$$

$$\phi^{-1}: \qquad i\alpha_{0,t} + \alpha_{0,xx} - 4\alpha_0|\hat{\psi}|^2 - 2\overline{\alpha}_0\hat{\psi}^2 = 0;$$

$$\phi^0: \qquad\qquad\qquad i\hat{\psi}_t + \hat{\psi}_{xx} - 2|\hat{\psi}|^2\hat{\psi} = 0.$$

(7.53)

Clearly the coefficient of ϕ^{-3} just gives the leading order behaviour, while the ϕ^0 equation in (7.53) means that the truncated expansion (7.52) constitutes an auto-Bäcklund transformation for the nonlinear Schrödinger equation, since $\hat{\psi}$ is another solution of (7.51). Observe that for x and t real, the singular manifold function ϕ is seen to be real valued from the leading order behaviour. Since the Painlevé analysis is really concerned with singularities in the space of complex x, t variables, it is more consistent to write the nonlinear Schrödinger equation, together with its complex conjugate, as the system

$$i\psi_t + \psi_{xx} - 2\psi^2\overline{\psi} = 0,$$
$$-i\overline{\psi}_t + \overline{\psi}_{xx} - 2\overline{\psi}^2\psi = 0,$$

(7.54)

and then treat ψ and $\overline{\psi}$ as independent quantities. The system (7.54) is the first non-trivial flow in the Ablowitz–Kaup–Newell–Segur (AKNS) hierarchy [2]. For this full system the singular manifold equations (7.53) should be augmented with the corresponding 'conjugate' equations: formally these are obtained by taking the complex conjugate with ϕ real (as for real x and t), and α_0, ψ and $\hat{\psi}$ complex. By formally taking the real and imaginary parts of the second equation in (7.53), which are equivalent to linear combinations of that equation together with its conjugate, the following consequences arise:

$$\phi_{xx} + \overline{\alpha}_0\hat{\psi} + \alpha_0\overline{\hat{\psi}} = 0;$$
$$i\phi_t + \phi_x(\log[\alpha_0/\overline{\alpha}_0])_x + \overline{\alpha}_0\hat{\psi} - \alpha_0\overline{\hat{\psi}} = 0.$$

(7.55)

Further manipulation of the singular manifold equations (7.53) and their conjugates, together with (7.55), leads to the two equations

$$\alpha_{0,x} = -2i\lambda\alpha_0 - 2\hat{\psi}\phi_x, \qquad\qquad\qquad (7.56)$$

$$i\alpha_{0,t} = (4\lambda^2 + 2\hat{\psi}\overline{\hat{\psi}})\alpha_0 + (-4i\lambda\hat{\psi} + 2\hat{\psi}_x)\phi_x \qquad (7.57)$$

and their corresponding conjugates, where λ is a constant. Upon substitution of the rearrangement

$$\alpha_0 = (\psi - \hat{\psi})\phi$$

of (7.52) into (7.56), we find

$$(\psi - \hat{\psi})_x = -2i\lambda(\psi - \hat{\psi}) - (\psi + \hat{\psi})|\psi - \hat{\psi}|, \tag{7.58}$$

where we have used the first equation (7.53) to substitute $\phi_x = |\alpha_0| = |\psi - \hat{\psi}|$ in the reduction to real x and t. A similar equation for $(\psi - \hat{\psi})_t$ is obtained by eliminating α_0 and ϕ from (7.57), and the resulting relations between ψ and $\hat{\psi}$ together with (7.58) constitute a Bäcklund transformation for the nonlinear Schrödinger equation in the form studied by Boiti and Pempinelli, taking the special case $\sigma = 0$ in the formulae of [10]. Starting from the vacuum solution $\hat{\psi} = 0$, and with zero Bäcklund parameter $\lambda = 0$, this BT can be applied repeatedly to obtain a sequence of singular rational solutions of the nonlinear Schrödinger equation, which are described in [50].

The simplest singular rational solution has a single pole, which can be fixed at $x = 0$. If we denote the sequence of these rational solutions $\{\psi_n\}_{n\geq 0}$, then applying the BT (7.58) with $\lambda = 0$ starting from the vacuum solution the first three are

$$\psi_0 = 0, \qquad \psi_1 = \frac{1}{x}, \qquad \psi_2 = \frac{-2x^3 + 12itx + \tau_3}{x^4 + \tau_3 - 12t^2}, \tag{7.59}$$

with τ_3 being an arbitrary constant parameter which is real for real x and t. In general these rational functions can be written as a ratio of polynomial tau-functions $\psi_n = G_n/F_n$ satisfying bilinear equations (see below). The zeros and poles of each ψ_n, which are the roots of the polynomials G_n and F_n, respectively, evolve in t according to the equations of Calogero–Moser dynamical systems [50].

As well as leading to the Bäcklund transformation (7.58) for the nonlinear Schrödinger equation, the singular manifold equations also yield the Lax pair, upon making the squared eigenfunction substitution

$$\alpha_0 = -\chi_1^2, \qquad \overline{\alpha}_0 = -\chi_2^2. \tag{7.60}$$

Fixing a sign we find immediately from the first Eq. (7.53) that

$$\phi_x = \chi_1\chi_2,$$

and then putting (7.60) into (7.56), (7.57) and their conjugates gives a matrix system for the vector $\chi = (\chi_1, \chi_2)^T$, that is

$$\chi_x = \mathbf{U}\chi,$$
$$\chi_t = \mathbf{V}\chi, \tag{7.61}$$

with the matrices

$$\mathbf{U} = \begin{pmatrix} -i\lambda & \psi \\ \overline{\psi} & i\lambda \end{pmatrix}, \qquad \mathbf{V} = \begin{pmatrix} -2i\lambda^2 - i|\psi|^2 & 2\lambda\psi + i\psi_x \\ 2\lambda\overline{\psi} - i\overline{\psi}_x & 2i\lambda^2 + i|\psi|^2 \end{pmatrix}$$

(where we have replaced $\hat{\psi}$ by ψ in U, V). The system (7.61) is the non-focusing analogue of the Lax pair for the nonlinear Schrödinger equation found by Zakharov and Shabat [114], and for U, V as above the PDE (7.51) follows from the compatibility condition for the matrix system, which is the zero curvature equation

$$\mathbf{U}_t - \mathbf{V}_x + [\mathbf{U}, \mathbf{V}] = 0.$$

For real λ, these matrices are elements of the Lie algebra $su(1,1)$, as opposed to $su(2)$ for the case of the focusing nonlinear Schrödinger equation.

To obtain the Hirota bilinear form of the nonlinear Schrödinger equation we can make a further truncation in (7.52), setting $\hat{\psi} = 0$, $\alpha_0 = G$, $\phi = F$, so that (7.51) becomes

$$\frac{1}{F^2}\left((iD_t + D_x^2)G \cdot F\right) - \frac{G}{F^3}\left(D_x^2 F \cdot F + 2|G|^2\right) = 0.$$

The two equations in brackets can be consistently decoupled to give the bilinear system for the two tau-functions F, G:

$$\begin{aligned}(iD_t + D_x^2)\,G \cdot F &= 0; \\ D_x^2 F \cdot F + 2|G|^2 &= 0.\end{aligned} \tag{7.62}$$

It is easy to check that the numerators and denominators in the rational functions (7.59) are particular solutions of the system (7.62). The bilinear form of the nonlinear Schrödinger equation was used by Hirota to derive compact expressions for the multisoliton solutions [47]. A further consequence of (7.62) is the bilinear equation

$$iD_x D_t F \cdot F - 2D_x G \cdot \overline{G} = i\gamma F^2, \tag{7.63}$$

with a constant γ. This constant can be removed by a gauge transformation of the tau-functions, rescaling both F and G by $\exp[\gamma xt/2]$. Eliminating G between (7.63) and (7.62), the nonlinear Schrödinger equation is then rewritten as a single trilinear equation, expressed as a sum of two determinants, namely

$$\begin{vmatrix} F & F_x & F_t \\ F_x & F_{xx} & F_{xt} \\ F_t & F_{xt} & F_{tt} \end{vmatrix} + \begin{vmatrix} F & F_x & F_{xx} \\ F_x & F_{xx} & F_{3x} \\ F_{xx} & F_{3x} & F_{4x} \end{vmatrix} = 0. \tag{7.64}$$

The tau-function solution of the trilinear equation (7.64) is sufficient to determine both the modulus and the argument of the complex amplitude ψ (see [50] and references).

From the preceding results for the KdV and nonlinear Schrödinger equations it should be clear that truncation methods can be extremely powerful in extracting information about integrable PDEs. There are several refinements of the singular manifold method, in particular those involving truncations using Möbius-invariant combinations of ϕ and its derivatives [20, 78], and the use of two singular manifolds for PDEs with two different leading order behaviours [22]. Probably the most elegant and general synthesis of these extended methods is the approach formulated by Pickering [88], who uses expansions in a modified variable satisfying a system of Riccati equations. Truncation methods have even been used to derive Bäcklund transformations for ODEs, in particular Painlevé equations [18, 19]. However, it is uncertain whether such methods can really be made sufficiently general in order to constitute an algorithmic procedure for deriving Lax pairs for integrable systems. In particular, truncation methods are not directly applicable to integrable PDEs which exhibit movable algebraic branching in their solutions, which are the subject of the next section.

7.6 Weak Painlevé Tests

There are numerous examples of integrable systems which do not have the strong Painlevé property, but which satisfy the weaker criterion that their general solution has at worst movable algebraic branching. Perhaps the simplest example is to consider an Hamiltonian system with one degree of freedom defined by the Hamiltonian (total energy)

$$H = \frac{1}{2}p^2 + V(q),$$

where the potential energy V is a polynomial in q of degree $d \geq 5$. The equations of motion (Hamilton's equations) are

$$\frac{dq}{dt} = p, \qquad \frac{dp}{dt} = -V'(q),$$

which are trivially integrable by a quadrature:

$$t = t_0 + \int^q \frac{dQ}{\sqrt{2(H - V(Q))}}. \tag{7.65}$$

If the potential energy is normalized so that the leading term of the polynomial is $-2q^d/(d-2)^2$, then with $q(t)$ having a singularity at $t = t_0$ the integral in (7.65) gives

$$t - t_0 \sim \pm \int^q \frac{(2-d)\,dQ}{2Q^{d/2}} = \pm q^{1-d/2}, \qquad \text{as} \quad q \to \infty$$

(for a suitable choice of branch in the square root). Thus at leading order we have

$$q \sim \pm(t - t_0)^{2/(2-d)}. \tag{7.66}$$

For both $d = 2g+1$ (odd) and $d = 2g+2$ (even) q is determined by the hyperelliptic integral (7.65) corresponding to an algebraic curve of genus g. When $g = 1$ the solution is given in terms of Weierstrass or Jacobi elliptic functions, and both q and p are meromorphic functions of t. However, for a potential of degree 5 or more we have $g \geq 2$, and it is clear from (7.66) that q has an algebraic branch point at $t = t_0$, since in that case $2/(2-d)$ is a non-integer, negative rational number. In fact it is easy to verify that (7.66) is the leading order term of an expansion in powers of $(t - t_0)^{2/(d-2)}$. Rather than being meromorphic as in the elliptic case, for $d \geq 5$ the function $q(t)$ is generically single valued only on a covering of the complex t-plane with an *infinite* number of sheets and has an infinite number of algebraic branch points (see [2]).

Clearly for potentials of degree 5 or more, this simple Hamiltonian system fails the basic Painlevé test, and yet it is certainly integrable according to any reasonable definition. (Indeed, any Hamiltonian system with one degree of freedom is integrable in the sense that Liouville's theorem holds.) In order to avoid excluding such basic integrable systems from singularity classification, Ramani et al. [93] proposed an extension of the Painlevé property.

Definition 3. The weak Painlevé property: An ODE has the weak Painlevé property if all movable singularities of the general solution have only a finite number of branches.

There are many examples of finite-dimensional many-body Hamiltonian systems which are Liouville integrable and yet have algebraic branching in their solutions [1, 2]. Among these examples [2] is the geodesic flow on an ellipsoid, which was solved classically by Jacobi [62]. Many other examples, such as those considered by Abenda and Fedorov in [1], arise naturally as stationary or travelling wave reductions of PDEs derived from Lax pairs, in particular those obtained from energy-dependent Schrödinger operators [52]. Thus the corresponding Lax-integrable PDEs have algebraic branching in their solutions and fail the Weiss–Tabor–Carnevale test described in Sect. 4. It is natural to extend the notion of the weak Painlevé property to PDEs as well and perform Painlevé analysis on ODEs and PDEs with this property by allowing algebraic branching and rational (not necessarily integer) values for the resonances. We illustrate this procedure with the example of the Camassa–Holm equation and a related family of PDEs [53] which have peaked solitons (peakons).

The Camassa–Holm equation was derived in [12] by asymptotic methods as an approximation to Euler's equation for shallow water waves and was shown to be an integrable equation with an associated Lax pair. In the special case when the linear dispersion terms are removed the equation takes the form

$$u_t - u_{xxt} + 3uu_x = 2u_x u_{xx} + uu_{xxx}, \tag{7.67}$$

and in this dispersionless limit it admits a weak solution known as a peakon, which has the form

$$u(x,t) = ce^{-|x-ct|}. \tag{7.68}$$

Note that the notion of a 'weak solution' (as defined in [39], for instance) is completely unrelated to the 'weak' Painlevé property. The peakon solution has a discontinuous derivative at the position of the peak, and the dispersionless Camassa–Holm equation (7.67) has exact solutions given by a superposition of an arbitrary number of peakons which interact and scatter elastically, just as for ordinary solitons. A detailed analysis of weak solutions of (7.67) has been performed by Li and Olver [74].

However, the Camassa–Holm equation is an example of an integrable equation which does not satisfy the requirements of Definition 2, but instead passes the *weak* Painlevé test. In the neighbourhood of an arbitrary non-characteristic hypersurface $\phi(x,t) = 0$ where the derivatives of u blow up, it admits an expansion with algebraic branching:

$$u(x,t) = -\phi_t/\phi_x + \sum_{j=0}^{\infty} \alpha_j(x,t)\, \phi^{2/3+j/3}. \qquad (7.69)$$

If we regard the branching part $\phi^{2/3}$ as the leading term (since it produces the singularity in the derivatives u_x, u_t on $\phi = 0$), then the resonances are $r = -1, 0, 2/3$ which correspond to the functions ϕ, α_0, α_2 being arbitrary. The Camassa–Holm equation thus satisfies the weak extension of the Weiss–Tabor–Carnevale test, since the expansion (7.69) is consistent, with the resonance conditions at $r = 0$ and $r = 2/3$ being satisfied. Of course the test is only local, whereas the weak Painlevé property is a global phenomenon, and to prove it rigorously for this PDE would require considerable further analysis. The weak extension of the Painlevé test is still a useful tool, in the sense that if an equation has irrational or complex branching (either at leading order or in its resonances), or if a failed resonance condition introduces logarithmic branching into the general solution, then this is a good indication of non-integrability. Nevertheless, even for ODEs the weak Painlevé property should be applied cautiously as an integrability criterion. For an excellent discussion see [94].

We would now like to apply the weak Painlevé test to a one-parameter family of PDEs that includes (7.67), before showing the effect that changes of variables can have on singularity structure. We shall consider the family of PDEs

$$u_t - u_{xxt} + (b+1)uu_x = bu_xu_{xx} + uu_{xxx}, \qquad (7.70)$$

where the parameter b is constant. These are non-evolutionary PDEs: due to the presence of the u_{xxt} term, (7.70) is not an evolution equation for u. The (dispersionless) Camassa–Holm equation is the particular member of this family corresponding to $b = 2$. The original reason for interest in this family is that Degasperis and Procesi applied the method of asymptotic integrability [25] and isolated a new equation as satisfying the necessary conditions for integrability up to some order in a multiple-scales expansion. After removing the dispersion terms by combining a Galilean transformation with a shift in u and rescaling, the Degasperis–Procesi equation can be written as

$$u_t - u_{xxt} + 4uu_x = 3u_xu_{xx} + uu_{xxx}, \qquad (7.71)$$

which is the $b = 3$ case of (7.70), and it was proved in [26] by construction of the Lax pair that this new equation is integrable. A powerful perturbative extension of the symmetry approach was also applied to the non-evolutionary PDEs (7.70) in [76], and it was confirmed that only the special cases $b = 2$ (Camassa–Holm) and $b = 3$ (Degasperis–Procesi) fulfil the necessary conditions to be integrable. Hamiltonian structures and the Wahlquist–Estabrook prolongation algebra method for these PDEs have also been treated in detail [53]. Subsequently it has been shown that (after including dispersion) every member of the family (7.70) arises as a shallow water wave equation [29], except for the special case $b = -1$.

For Painlevé analysis it is convenient to rewrite (7.70) in the form

$$m_t + u m_x + b u_x m = 0, \qquad m = u - u_{xx}. \tag{7.72}$$

To apply the weak Painlevé test, we look for algebraic branching similar to the leading order in (7.69), with the derivatives of u blowing up on a singular manifold $\phi(x,t) = 0$. Thus we seek the following leading behaviour:

$$u \sim u_0 + \alpha \phi^\mu, \qquad \mu \in \mathbb{Z}, \qquad 0 < \mu < 1. \tag{7.73}$$

Then for the derivatives of u and m as defined in (7.72) the most singular terms are as follows:

$$u_x \sim \alpha \phi_x \mu \phi^{\mu-1}, \qquad m \sim -\alpha \phi_x^2 \mu (\mu - 1) \phi^{\mu-2},$$
$$m_x \sim -\alpha \phi_x^3 \mu (\mu - 1)(\mu - 2) \phi^{\mu-3}, \qquad m_t \sim -\alpha \phi_x^2 \phi_t \mu (\mu - 1)(\mu - 2) \phi^{\mu-3}.$$

Substituting these leading orders into (7.72) we find a balance at order $\phi^{\mu-3}$ between the m_t and $u m_x$ terms provided that

$$u_0 = -\phi_t / \phi_x.$$

The next most singular term in the PDE is then at order $\phi^{2\mu-3}$, corresponding to a balance between the $u m_x$ and $u_x m$ terms in (7.72), with coefficient

$$-\alpha^2 \phi_x^3 \mu (\mu - 1)(\mu - 2 + b\mu),$$

and this is required to vanish giving

$$\mu = \frac{2}{1+b}. \tag{7.74}$$

Thus we see that for a weak Painlevé expansion with the leading exponent μ being a rational number between zero and one, the most singular terms require that the parameter b should also be rational with

$$b = \frac{2}{\mu} - 1 > 1.$$

To find and test the resonances in an expansion with this leading order, it is sufficient to take the reduced ansatz (7.42) for ϕ, and then make a perturbation of the leading order terms with parameter ϵ:

$$u \sim \dot{f}(t) + \alpha(t) \phi^\mu (1 + \epsilon \phi^r), \qquad \phi = x - f(t). \tag{7.75}$$

Substituting the perturbed expression into (7.72) and keeping only terms linear in ϵ, we see that terms possibly appearing at order $\phi^{\mu+r-3}$ cancel out automatically (due to the form of u_0), leaving the resonance equation coming from the coefficient of $\phi^{2\mu+r-3}$, which is

$$-\epsilon\alpha^2(r^3 + (2\mu - 1)r^2 + 2(\mu - 1)r) = 0.$$

Hence the resonances are

$$r = -1, 0, 2(1 - \mu),$$

with μ given in terms of the parameter b by (7.74).

Having applied the first part of the weak Painlevé test and found a dominant balance and the corresponding values for the resonances, it becomes apparent that the test is completely ineffective as a means to isolate the two integrable cases $b = 2$ and $b = 3$ of (7.72). Although the leading order resonance $r = 0$ (corresponding to α being arbitrary) is automatically satisfied, the second resonance condition at $r = 2(1 - \mu)$ must be checked for every rational value of μ with $0 < \mu < 1$ (or equivalently every rational value of the parameter $b > 1$). If we write μ in its lowest terms as a ratio of positive integers, $\mu = N_1/N_2$, then (7.73) is the leading part of an expansion for u in all powers of ϕ^{1/N_2}, and as the difference $N_2 - N_1$ increases there is an increasingly large number of terms to compute before the final resonance is reached. Checking this resonance for the whole countable infinity of rational numbers $b > 1$ seems to be a totally intractable task. Gilson and Pickering showed that all the PDEs within a class including (7.72) failed every one of a combination of *strong* Painlevé tests [36]. Nevertheless, it is simple to verify that the weak Painlevé test is satisfied for the two particular cases $b = 2, 3$ which are known to be integrable.

However, after a judicious change of variables, involving a transformation of hodograph type, it is still possible to use Painlevé analysis to isolate the two integrable peakon equations. Such transformations have been applied to integrable PDEs with algebraic branching (see [15]) in order to obtain equivalent systems with the strong Painlevé property. That this should be possible is in accordance with the Ablowitz–Ramani–Segur conjecture, but the difficulty lies in finding the correct change of variables. In fact, for a general class of systems that display weak Painlevé behaviour (related to energy-dependent Schrödinger operators) we presented a particular transformation in [52] and, from an examination of a principal balance, we asserted (without proof) that this transformation produced equivalent systems with the strong Painlevé property. However, from a more careful calculation of other balances we have recently observed that this earlier assertion was incorrect [56]. In the case of the Camassa–Holm equation (7.67), a link to the first negative flow in the KdV hierarchy was found by Fuchssteiner [34], and in [53] it was shown that the appropriate transformation can be extended to (almost) every member of the family of non-evolutionary PDEs (7.72).

The key to a suitable change of variables for (7.72) is the fact that for any $b \neq 0$, $\int m^{1/b} dx$ is a conserved quantity, with the conservation law

$$p_t = -(pu)_x, \qquad m = -p^b. \tag{7.76}$$

This allows a reciprocal transformation, defining new independent variables X, T via

$$dX = p\,dx - pu\,dt, \qquad dT = dt. \tag{7.77}$$

Observe that the closure condition $d^2X = 0$ for the exact one-form dX is precisely (7.76), and transforming the derivatives yields the new conservation law

$$(p^{-1})_T = u_X. \tag{7.78}$$

In the old variables, p is related to u by

$$p^b = (\partial_x^2 - 1)u, \tag{7.79}$$

Replacing ∂_x by $p\partial_X$ and using (7.78), this means that (7.79) can be solved for u to give the identity

$$u = -p(\log p)_{XT} - p^b. \tag{7.80}$$

Finally the conservation law (7.78) can be written as an equation for p alone, by substituting back for u as in (7.80) to obtain

$$\frac{\partial}{\partial T}\left(\frac{1}{p}\right) + \frac{\partial}{\partial X}\left(p(\log p)_{XT} + p^b\right) = 0. \tag{7.81}$$

Thus we have seen that for each $b \neq 0$, the Eq. (7.72) is reciprocally transformed to (7.81), with the new dependent variable p and new independent variables X, T as in (7.77). (For more background on reciprocal transformations, see [68, 69].) By making the substitution $p = \exp(i\eta)$, (7.81) becomes a generalized equation of sine-Gordon type [53]. The point of making the reciprocal transformation is that we may now apply the *strong* Weiss–Tabor–Carnevale Painlevé test to the equation in these new variables. At leading order near a hypersurface $\phi(X, T) = 0$ there are two types of singularity that can occur in Eq. (7.81), corresponding to p either vanishing or blowing up there:

- $p \sim \alpha\phi$, for $b \geq -1$, with $\alpha = \pm\phi_X^{-1}$ for $b \neq -1$;
- $p \sim \beta\phi^\mu$, for $\mu = 2/(1-b) < 1$.

In the first balance, the resonances are $r = -1, 1, 2$. However, if we require the strong Painlevé test to hold we see that we must have $b \in \mathbb{Z}$, since otherwise the p^b term will introduce branching into the expansion in powers of ϕ. The second balance can only hold for $|b| > 1$, but if $b < -1$ then $\mu \notin \mathbb{Z}$, while if $b > 1$ then requiring $\mu = 1 - M$ to be a (negative) integer gives

$$b = \frac{M+1}{M-1}, \qquad M = 2, 3, 4, \ldots \tag{7.82}$$

From the first balance we require b to be an integer, and the only integer values in the sequence (7.82) are $b = 2, 3$ (corresponding to $M = 3, 2$, respectively). Interestingly, when the Wahlquist–Estabrook method is applied to (7.72), this same sequence crops up from purely algebraic considerations [53].

The above analysis shows that the two integrable cases $b = 2, 3$ are isolated immediately just by looking at the leading order behaviour. It is then straightforward to show that for both types of singularity in Eq. (7.81), these two cases fulfil the resonance conditions and thus satisfy the strong Painlevé test. However, the observant reader will notice that further analysis is required to exclude the two special integer values $b = \pm 1$, for which only the first type of singularity arises; this is left as a challenge to the reader.

7.7 Outlook

It should be apparent from our discussion that the various Painlevé tests are excellent heuristic tools for identifying whether a given system of differential equations is likely to be integrable or not. However, the strong Painlevé property is clearly too stringent a requirement, since it is not satisfied by a large class of integrable systems which have movable algebraic branch points in their solutions. On the other hand, checking all possible resonances in the weak Painlevé test can be impractical as a means to isolate integrable systems, and if there are negative resonances then more detailed analysis may be necessary to pick up logarithmic branching [89]. In this short review we have concentrated on methods for detecting movable poles and branch points. However, for equations like (7.67), the existence of the peakon solution (7.68) has led to the promising suggestion that Dirichlet series (sums of exponentials) may be a useful means of testing PDEs [91]. Also, although we have only considered singularities of ODEs in the finite complex plane, there are extensive techniques for analysing asymptotic behaviour at infinity [97, 103, 106].

Before closing, we should like to give a brief mention to the fruitful connection between the singularity structure and integrability of discrete systems, in the context of birational maps or difference equations. In the last 20 years, there has been increased interest in discrete integrable systems. Liouville's theorem on integrable Hamiltonian systems extends naturally to the setting of symplectic maps or more generally to Poisson maps or correspondences [11, 104], and many new examples of integrable maps have been found [101]. Grammaticos, Ramani and Papageorgiou introduced a notion of singularity confinement for maps or difference equations [40], which they used very successfully as a criterion to identify discrete analogues of the Painlevé equations, and they proposed that it should be regarded as a discrete version of the Painlevé property.

In order to illustrate singularity confinement, we shall consider the second-order discrete equation

$$u_{n+1}(u_n)^2 u_{n-1} = \alpha q^n u_n + \beta, \tag{7.83}$$

which is a non-autonomous version of an equation of the Quispel–Roberts–Thompson type [92] and can be explicitly solved in elliptic functions in the autonomous case $q = 1$ [54, 55]. For $q \neq 1$, Eq. (7.83) can be regarded as a discrete analogue of the first Painlevé equation, because if we set $u_n = h^{-2} - y(nh)$,

$\alpha = 4h^{-6}$, $\beta = -3h^{-8}$, $q = 1 - h^5/4$ and take the continuum limit $h \to 0$, with $z = nh$ held fixed, then Eq. (7.6) arises at leading order in h.

The idea of singularity confinement is that if a singularity is reached upon iteration of a discrete equation or map, then it is possible to analytically continue through it. (This is by analogy with the fact that the solution of an ODE with the Painlevé property has a unique analytic continuation around a movable pole.) In the case of (7.83), a singularity will be reached if one of the iterates, say u_N, is zero, because this means that the next iterate u_{N+1} is not defined. By redefining α and shifting the index n if necessary, we can take $N = 1$ without loss of generality, so $u_1 = 0$. The vanishing of u_1 requires that at the previous stage $\alpha u_0 + \beta = 0$ must hold. Setting $u_{-1} = a$ (arbitrary) and

$$\alpha u_0 + \beta = \epsilon$$

gives $u_1 \sim \alpha^2 \beta^{-2} a^{-1} \epsilon \to 0$ as $\epsilon \to 0$, and the singularity appears at

$$u_2 \sim -\beta^4 a^2 \alpha^{-3} \epsilon^{-2}.$$

However, subsequently we have $u_3 \sim -q^2 \alpha^2 \beta^{-2} a^{-1} \epsilon$, $u_4 = O(1)$ and further iterates are regular in the limit $\epsilon \to 0$. In this sense, we say that the singularity is confined.

Although the singularity confinement criterion led to the discovery of many new discrete integrable systems (see [95] and references), it was shown by Hietarinta and Viallet that it is not a sufficient condition for integrability [44]. In fact, they found numerous examples of maps of the plane defined by difference equations of the form

$$u_{n+1} + u_{n-1} = f(u_n),$$

for certain rational functions f, which have confined singularities and yet whose orbit structure displays the characteristics of chaos. Other examples of singularity confinement in non-integrable maps can be found in [57]. Nevertheless, it seems that singularity confinement should be a necessary condition for integrability of a suitably restricted class of maps. In fact, Lafortune and Goriely have shown that for birational maps in d dimensions, singularity confinement is a necessary condition for the existence of $d - 1$ independent algebraic first integrals [73]. Ablowitz, Halburd and Herbst have made an alternative proposal for extending the Painlevé property to difference equations, by using Nevanlinna theory [7, 43], and this has deep connections with various algebraic or arithmetic measures of complexity in discrete dynamics (see [42, 44, 96, 99] and references).

For the reader who is interested in pursuing the subject of Painlevé analysis and its applications to both integrable and non-integrable equations, a number of excellent review articles are to be recommended [30, 72, 79, 94, 102], as well as the proceedings volume [23].

Acknowledgments Some of this article began as a tutorial on the Weiss–Tabor–Carnevale method during the summer school *What is Integrability?* at the Isaac Newton Institute, Cambridge in 2001. I would like to thank the students for requesting the tutorial and for being such an attentive audience, especially the two mature students who were present (Peter Clarkson and Martin Kruskal). I am also extremely grateful to the many friends and colleagues who have helped me to understand different aspects of all things Painlevé, in particular Harry Braden, Allan Fordy, Nalini Joshi, Frank Nijhoff and Andrew Pickering.

References

1. S. Abenda and Y. Fedorov, Acta Appl. Math. 60, 137–178, 2000.
2. S. Abenda and Y. Fedorov, J. Nonlin. Math. Phys. 8, Supplement 1–4, 2001.
3. M.J. Ablowitz, D.J. Kaup, A.C. Newell, and H. Segur, Stud. Appl. Math. 53, 249–315, 1974.
4. M.J. Ablowitz, A. Ramani, and H. Segur, Lett. Nuovo Cim. 23, 333–338, 1978.
5. M.J. Ablowitz and P.A. Clarkson, Solitons, Nonlinear Evolution Equations and Inverse Scattering, Cambridge University Press, 1991.
6. M.J. Ablowitz and A.S. Fokas, Complex Analysis: Introduction and Applications, Cambridge University Press, 1997.
7. M.J. Ablowitz, R. Halburd, and B. Herbst, Nonlinearity 13, 889–905, 2000.
8. O. Babelon, D. Bernard, and M. Talon, Introduction to Classical Integrable Systems, Cambridge University Press, 2003.
9. T.B. Benjamin, J.L. Bona, and J.J. Mahony, Phil. Trans. Roy. Soc. London 272 47–78, 1972.
10. M. Boiti and F. Pempinelli, Il Nuovo Cimento 59B, 40–58, 1980.
11. M. Bruschi, O. Ragnisco, P.M. Santini, and G.-Z. Tu, Physica D 49, 273–294, 1991.
12. R. Camassa and D.D. Holm, Phys. Rev. Lett. 71, 1661–1664, 1993.
13. Y.F. Chang, J. Tabor, and J. Weiss, J. Math. Phys. 23, 531–538, 1982.
14. J. Chazy, Acta Math. 34, 317–385, 1911.
15. P.A. Clarkson, A.S. Fokas, and M.J. Ablowitz, SIAM J. Appl. Math. 49, 1188–1209, 1989.
16. P.A. Clarkson and M.D. Kruskal, J. Math. Phys. 30, 2201–2213, 1989.
17. P.A. Clarkson and E.L. Mansfield, Nonlinearity 7, 975–1000, 1994.
18. P.A. Clarkson, N. Joshi, and A. Pickering, Inverse Problems 15, 175–187, 1999.
19. P. Gordoa, N. Joshi, and A. Pickering, Glasgow Math. J. 43A, 23–32, 2001.
20. R. Conte, Phys. Lett. A 140, 383–390, 1989.
21. R. Conte, A.P. Fordy, and A. Pickering, Physica D 69, 33–58, 1993.
22. R. Conte, M. Musette, and A. Pickering, J. Phys. A 28, 179–187, 1995.
23. R. Conte (ed.), The Painlev'e Property, One Century Later, CRM Series in Mathematical Physics, Springer, New York, 1999.
24. C.M. Cosgrove, Stud. Appl. Math. 104, 1–65, 2000.
25. A. Degasperis and M. Procesi, Asymptotic Integrability, in Symmetry and Perturbation Theory, A. Degasperis and G. Gaeta (eds.), World Scientific, Singapore, 23–37, 1999.
26. A. Degasperis, D.D. Holm, and A.N.W. Hone, Theor. Math. Phys. 133, 1461–1472, 2002.
27. R.K. Dodd, J.C. Eilbeck, J.D. Gibbon, and H.C. Morris, Solitons and Nonlinear Wave Equations, Academic Press, New York, 1984.
28. M.R. Douglas, Phys. Lett. B 238, 176–180, 1990.
29. H.R. Dullin, G.A. Gottwald, and D.D. Holm, Fluid Dyn. Res. 33, 73–95, 2003.
30. N. Ercolani and E.D. Siggia, Painlev'e Property and Integrability, in What is Integrability? V.E. Zakharov (ed.), Springer Series in Nonlinear Dynamics, Springer-Verlag, 63–72, 1991.
31. H. Flaschka and A.C. Newell, Commun. Math. Phys. 76, 65–116, 1980.
32. R. Fuchs, C.R. Acad. Sci. (Paris) 141, 555, 1905.
33. R. Fuchs, C.R. Math. Ann. 63, 301, 1907.
34. B. Fuchssteiner, Physica D 95, 229–243, 1996.
35. J.D. Gibbon, P. Radmore, M. Tabor, and D. Wood, Stud. Appl. Math. 72, 39–63, 1985.
36. C. Gilson and A. Pickering, J. Phys. A 28, 2871–2888, 1995.
37. D. Gomez-Ullate, A.N.W. Hone, and M. Sommacal, New J. Phys. 6, 24, 2004.
38. A. Goriely, J. Math. Phys. 37, 1871–1893, 1996.

39. P. Grindrod, The Theory and Applications of Reaction-Diffusion Equations, 2nd edition, Oxford University Press, 1996.
40. B. Grammaticos, A. Ramani, and V. Papageorgiou, Phys. Rev. Lett. 67, 1825, 1991.
41. R.G. Halburd, J. Math. Phys. 43, 1966–1979, 2002.
42. R.G. Halburd, J. Phys. A: Math. Gen. 38, L263–L269, 2005.
43. R.G. Halburd and R.J. Korhonen, J. Phys. A 40, R1–R38, 2007.
44. J. Hietarinta and C. Viallet, Phys. Rev. Lett. 81, 325–328, 1998.
45. E. Hille, Ordinary Differential Equations in the Complex Domain, Wiley, New York, 1976.
46. A. Hinkkanen and I. Laine, J. Anal. Math. 79, 345–377, 1999.
47. R. Hirota, J. Math. Phys. 14(7), 805–814, 1973.
48. R. Hirota, Direct methods in soliton theory, in Solitons, R.K. Bullough and P.J. Caudrey (eds.), Springer, Berlin, 1980.
49. A.N.W. Hone, Integrable Systems and their Finite-Dimensional Reductions, PhD thesis, University of Edinburgh, 1997.
50. A.N.W. Hone, J. Phys. A 30, 7473–7483, 1997.
51. A.N.W. Hone, Physica D 118, 1–16, 1998.
52. A.N.W. Hone, Phys. Lett. A 249, 46–54, 1998.
53. A.N.W. Hone and J.P. Wang, Inverse Probl. 19, 129–145, 2003.
54. A.N.W. Hone, Bull. Lond. Math. Soc. 37, 161–171, 2005.
55. A.N.W. Hone, Bull. Lond. Math. Soc. 38, 741–742, 2006.
56. A.N.W. Hone, V. Novikov, and C. Verhoeven, Inverse Probl. 22, 2001–2020, 2006.
57. A.N.W. Hone, Phys. Lett. A 361, 341–345, 2007.
58. E.L. Ince, Ordinary Differential Equations (1926). Reprint: Dover Publications, New York, 1956.
59. A.R. Its, A.G. Izergin, V.E. Korepin, and N.A. Slavnov, Int. J. Mod. Phys. B 4, 1003–1037, 1990.
60. F.H.L. Essler, H. Frahm, A.R. Its, and V.E. Korepin, J. Phys. A 29, 5619–5626, 1996.
61. K. Iwasaki, H. Kimura, S. Shimomura, and M. Yoshida, From Gauss to Painlevè: a modern theory of special functions, Vieweg, Braunschweig, 1991.
62. C.G. Jacobi, Vorlesungen über Dynamik, Königsberg University, 1842, A. Clebsch (ed.), Reimer, Berlin, 1884.
63. M. Jimbo, T. Miwa, Y. Mori, and M. Sato, Physica 1D, 80–158, 1980.
64. M. Jimbo, M.D. Kruskal, and T. Miwa, Phys. Lett. A 92, 59–60, 1982.
65. N. Joshi and M.D. Kruskal, Stud. Appl. Math. 93, 187–207, 1994.
66. N. Joshi and G.K. Srinivasan, Nonlinearity 10, 71–79, 1997.
67. A.K. Karasu, J. Math. Phys. 38, 3616–3622, 1997.
68. J.G. Kingston and C. Rogers, Phys. Lett. A 92, 261–264, 1982.
69. C. Rogers, Reciprocal Transformations and Their Applications, in Nonlinear Evolutions, J. Leon (ed.), World Scientific, Singapore, 109–123, 1988.
70. S. Kowalewski, Acta Math. 12, 177–232, 1889.
71. S. Kowalewski, Acta Math. 14, 81–93, 1889.
72. Kruskal, M.D., Joshi, N., Halburd, R.: In: Grammaticos, B., Tamizhmani, K. (eds.) *Analytic and Asymptotic Methods for Nonlinear Singularity Analysis: A Review and Extensions of Tests for the Painlevé Property, Proceedings of CIMPA Summer School on Nonlinear Systems*. Lect. Notes Phys. **495**, Springer-Verlag, Heidelberg, 171–205 (1997)
73. S. Lafortune and A. Goriely, J. Math. Phys. 45, 1191–1208, 2004.
74. Y.A. Li and P.J. Olver, Discrete Cont. Dyn. Syst. 3, 419–432, 1997.
75. D. Markushevich, J. Phys. A 34, 2125–2135, 2001.
76. A.V. Mikhailov and V.S. Novikov, J. Phys. A 35, 4775–4790, 2002.
77. T. Miwa, Infinite-dimensional Lie algebras of hidden symmetries of soliton equations, in Soliton theory: a survey of results, A.P. Fordy (ed.), Manchester University Press, 338–353, 1990.
78. M. Musette and R. Conte, J. Math. Phys. 32, 1450–1457, 1991.
79. A.C. Newell, M. Tabor, and Y.B. Zeng, Physica D 29, 1–68, 1987.

80. F. Nijhoff, A.N.W. Hone, and N. Joshi, Phys. Lett. A 267, 147–156, 2000.
81. J. Nimmo, Hirota's Method, in Soliton Theory: A survey of Results A.P. Fordy (ed.), Manchester University Press, 75–96, 1990.
82. Y. Ohta, J. Satsuma, D. Takahashi, and T. Tokihiro, Prog. Theor. Phys. Suppl. 94, 210–241, 1988.
83. P.J. Olver, Applications of Lie Groups to Differential Equations, 2nd edition, Springer-Verlag, New York (1993).
84. P. Painlevé, Bull. Soc. Math. France 28, 201–261, 1900.
85. P. Painlevé, Acta Math. 25, 1–85, 1902.
86. D.H. Peregrine, J. Fluid Mech. 25, 321–330, 1966.
87. A. Pickering, J. Math. Phys. 35, 821–833, 1994.
88. A. Pickering, J. Math. Phys. 37, 1894–1927, 1996.
89. A. Pickering, Phys. Lett. A 221, 174–180, 1996.
90. A. Pickering, Inverse Probl. 13, 179–183, 1997.
91. A. Pickering, Prog. Theor. Phys. 108, 603–607, 2002.
92. G.R.W. Quispel, J.A.G. Roberts, and C.J. Thompson, Physica D 34, 183–192, 1989.
93. A. Ramani, B. Dorizzi, and B. Grammaticos, Phys. Rev. Lett. 49, 1538–1541, 1982.
94. A. Ramani, B. Grammaticos, and T. Bountis, Phys. Rep. 180, 159–245, 1989.
95. A. Ramani, B. Grammaticos, and J. Satsuma, J. Phys. A: Math. Gen. 28, 4655–4665, 1995.
96. J.A.G. Roberts and F. Vivaldi, Phys. Rev. Lett. 90, 034102, 2003.
97. P.L. Sachdev, Nonlinear Ordinary Differential Equations and Their Applications, Marcel Dekker, New York, 1991.
98. J.H. Silverman, The Arithmetic of Elliptic Curves, Springer, 1986.
99. J.H. Silverman, The Arithmetic of Dynamical Systems, Springer, 2007.
100. N. Steinmetz, J. d'Analyse Math. 82, 363–377, 2000.
101. Y.B. Suris, The Problem of Integrable Discretization: Hamiltonian Approach, Birkhäuser, 2003.
102. M. Tabor, Painlevé Property for Partial Differential Equations, in Soliton Theory: A Survey of Results, A.P. Fordy (ed.), Manchester University Press, 427–446, 1990.
103. A. Tovbis, J. Diff. Eq. 109, 201–221, 1994.
104. A.P. Veselov, Russ. Math. Surveys 46, 1–51, 1991.
105. R.S. Ward, Phys. Lett. A 102, 279–282, 1984.
106. W. Wasow, Asymptotic Expansions for Ordinary Differential Equations, Wiley, New York, 1965.
107. J. Weiss, M. Tabor, and G. Carnevale, J. Math. Phys. 24, 522, 1983.
108. J. Weiss, J. Math. Phys. 24, 1405–1413, 1983.
109. J. Weiss, J. Math. Phys. 25, 13–24, 1984.
110. J. Weiss, J. Math. Phys. 26, 258–269, 1985.
111. J. Weiss, J. Math. Phys. 26, 2174–2180, 1985.
112. E.T. Whittaker and G.N. Watson, A Course of Modern Analysis, Fourth Edition, Cambridge University Press, 1965.
113. T.T. Wu, B.M. McCoy, C.A. Tracy, and E. Barouch, Phys. Rev. B 13, 316, 1976.
114. V.E. Zakharov and A.B. Shabat, Soviet Phys. JETP 34, 62–69, 1972.

Chapter 8
Hirota's Bilinear Method and Its Connection with Integrability

J. Hietarinta

Abstract We give an introduction to Hirota's bilinear method, which is particularly efficient for constructing multisoliton solutions to integrable nonlinear evolution equations. We discuss in detail how the method works for equations in the Korteweg–de Vries class and then go through some other classes of equations. Finally we discuss how the existence of multisoliton solutions can be used as an integrability condition and therefore as a method of searching for possible new integrable equations.

8.1 Why the Bilinear Form?

In 1971 R. Hirota introduced a new "direct method" for constructing multisoliton solutions to integrable nonlinear evolution equations [1]. The idea was to make a transformation into new variables, so that in these new variables multisoliton solutions appear in a particularly simple form. The equations turned out to be quadratic in the new dependent variables and all derivatives appeared as Hirota's bilinear derivatives, this is called "Hirota bilinear form". The method turned out to be very effective and was quickly shown to give N-soliton solutions to the Korteweg–de Vries (KdV) [1], modified Korteweg–de Vries (mKdV) [2], sine-Gordon (sG) [3] and nonlinear Schrödinger (nlS) [4] equations. Bäcklund transformations also appear naturally in this formalism [5].

Later it was observed that the essential mathematical ingredient that makes this idea work is that the new dependent variables are "τ-functions", which have many good properties. This has become a starting point for further deep mathematical developments (the Sato theory; see, e.g., [6]). Since the bilinear form of the equation is mathematically fundamental, it has appeared in the literature before (but only in passing): For example, in 1902 Painlevé wrote his first three equations in the bilinear form, and in 1907 H.F. Baker obtained such forms in his study of multidimensional

J. Hietarinta (✉)
Department of Physics, University of Turku, FIN-20014 Turku, Finland,
jarmo.hietarinta@utu.fi

Hietarinta, J.: *Hirota's Bilinear Method and Its Connection with Integrability*. Lect. Notes Phys. **767**, 279–314 (2009)
DOI 10.1007/978-3-540-88111-7_8

σ-functions.[1] However, it was R. Hirota who first emphasized the bilinear aspect as a new starting point.

In this lecture our aim is to describe how multisoliton solutions can be constructed using Hirota's method. Multisoliton solutions can, of course, be derived by many other methods, e.g., by the inverse scattering transform (IST). The advantage of Hirota's method over the others is that it is algebraic rather than analytic. The IST method is more powerful in the sense that it can handle general initial conditions, but at the same time it is more complicated and more demanding to the equation. Accordingly, if one just wants to find soliton solutions, Hirota's method is the best for producing results.

We will not touch the converse approach, starting with the unifying mathematical theory (Sato theory) behind the bilinear approach and obtaining integrable equations as specific reductions. Although the Sato theory is the fundamental theory in which equations and solutions have mathematically elegant forms, it is not the first step to take when one has to study a given equation.

8.2 From Nonlinear to Bilinear (KdV)

The (integrable) PDE that appears in some particular (physical) problem is rarely in the best form for further (mathematical) analysis. For constructing soliton solutions the best form is Hirota's bilinear form and soliton solutions appear as polynomials of simple exponentials only in the corresponding new variables. The first problem we face is therefore to find the bilinearizing transformation. This is not algorithmic and can sometimes require the introduction of new dependent and sometimes even independent variables.

8.2.1 Bilinearizing the KdV Equation

We will first discuss in detail the KdV equation

$$u_{xxx} + 6uu_x + u_t = 0. \tag{8.1}$$

One guideline in searching for the transformation is that the leading derivative should go together with the nonlinear term and, in particular, have the same number of derivatives. If we count a derivative with respect to x having degree 1, then to balance the first two terms of (8.1) u should have degree 2. Thus we introduce a new dependent variable w (of degree 0) by

$$u = \partial_x^2 w. \tag{8.2}$$

---

[1] I thank C. Eilbeck for bringing this last reference to my attention.

After this the KdV equation can be written as

$$w_{xxxxx} + 6w_{xx}w_{xxx} + w_{xxt} = 0, \tag{8.3}$$

which can be integrated once with respect to x to give the *potential form* of KdV

$$w_{xxxx} + 3w_{xx}^2 + w_{xt} = 0. \tag{8.4}$$

In principle we should have introduced an integration constant (function of t), but since (8.2) defines w only up to $w \to w + xa(t) + b(t)$, we can use this freedom to absorb it. Note also that Eq. (8.4) is invariant under scaling $x \to \lambda x, t \to \lambda^3 t$. Indeed, scaling invariance is often a good guide in the search for the bilinear form.

Equations in scale-invariant form can usually be bilinearized by introducing a new dependent variable whose natural degree (in the above sense) would be zero, e.g., $\log F$ or g/f. In this case the first one works, so let us define

$$w = \alpha \log F, \tag{8.5}$$

with a free parameter α. When this is substituted into (8.4) we get an equation that is fourth degree in F, with the structure

$$F^2 \times (\text{something quadratic}) + 3\alpha(2 - \alpha)(2FF'' - F'^2)F'^2 = 0. \tag{8.6}$$

Thus we get a quadratic equation if we choose $\alpha = 2$, and the result is

$$F_{xxxx}F - 4F_{xxx}F_x + 3F_{xx}^2 + F_{xt}F - F_xF_t = 0. \tag{8.7}$$

In addition to being quadratic in the dependent variables, an equation in the Hirota bilinear form must also satisfy a condition with respect to the derivatives: they should only appear in combinations that can be expressed using Hirota's D-operator, which is defined as follows:

$$D_x^n f \cdot g = (\partial_{x_1} - \partial_{x_2})^n f(x_1)g(x_2)\big|_{x_2 = x_1 = x}. \tag{8.8}$$

Thus D operates on a product of two functions like the Leibnitz rule, except for a crucial sign difference. For example

$$D_x f \cdot g = f_x g - f g_x,$$
$$D_x D_t f \cdot g = f g_{xt} - f_x g_t - f_t g_x + f_{xt} g.$$

Using the D-operator we can write (8.7) in the following condensed form:

$$(D_x^4 + D_x D_t) F \cdot F = 0. \tag{8.9}$$

To summarize: what we needed in order to obtain the bilinear form (8.9) for (8.1) was a dependent variable transformation

$$u = 2\partial_x^2 \log F, \tag{8.10}$$

and we also had to integrate the equation once.

For a further discussion of bilinearization, see, e.g., [7–10]; unfortunately the process is far from algorithmic.

8.2.2 Gauge Invariance

One important property of equations in Hirota's bilinear form is their gauge invariance. Let us consider a general quadratic expression homogeneous in the derivatives

$$A_N(f,g) := \sum_{i=0}^{N} c_i \left(\partial_x^i f\right)\left(\partial_x^{N-i} g\right)$$

and the gauge transformation

$$f \to e^\theta f,\, g \to e^\theta g,\, \theta = kx.$$

It is now easy to check that A_N is gauge invariant, i.e.,

$$A_N(e^\theta f, e^\theta g) = e^{2\theta} A_N(f,g),$$

if and only if

$$c_i = (-1)^i \binom{N}{i} c_0,$$

which means that we can write

$$A_N(f,g) = c_0 D_x^N f \cdot g.$$

This is a possible point of generalization [11, 12] and we can define multilinear operators also by a gauge condition: For an expression cubic in dependent variables and homogeneous (of degree N) in the x-derivatives we now require

$$\sum_{k+l+m=N} c_{klm} (\partial^k e^\theta f)(\partial^l e^\theta g)(\partial^m e^\theta h) = e^{3\theta} \sum_{k+l+m=N} c_{klm}(\partial^k f)(\partial^l g)(\partial^m h).$$

One then finds that a basis for such gauge-invariant operators is given by $T^n (T^*)^{N-n}$, where

$$T = \partial_1 + j\partial_2 + j^2\partial_3, \quad T^* = \partial_1 + j^2\partial_2 + j\partial_3$$

and $j = e^{2i\pi/3}$. Note that $T^n T^{*m} F \cdot F \cdot F \equiv 0$, unless $n + 2m \equiv 0 \pmod 3$, corresponding to the bilinear property $D^n F \cdot F \equiv 0$, unless $n \equiv 0 \pmod 2$.[2]

The above generalizes to any order of multilinearity and one can introduce the operators

---

[2] For the bilinear equation $P(D) F \cdot F = 0$ the dispersion relation is given by $P(p) = 0$ (see Sect. 8.3.2), while for a trilinear equation $P(T,T^*)F \cdot F \cdot F = 0$ the dispersion relation is $P(p,p) = 0$. Thus some dispersionless equations can be written in trilinear form, e.g., the Monge–Ampere equation $w_{xy}^2 - w_{xx}w_{yy} = 0$ can be written as $(T_x T_x^* T_y T_y^* - T_x^2 T_y^{*2})F \cdot F \cdot F = 0$.

$$M_n^m = \sum_{k=0}^{n-1} e^{2\pi i k m/n} \partial_{k+1}, \text{ for } 0 < m < n.$$

For example, $D = M_2^1$, $T = M_3^1$, $T^* = M_3^2$.

8.2.3 Some Properties of the Bilinear Derivative

Finally in this section we would like to list some properties of the bilinear derivative that are useful for bilinearization [5, 7, 8]:

$$P(D)f \cdot g = P(-D)g \cdot f, \tag{8.11}$$

$$P(D)1 \cdot f = P(-\partial)f, \quad P(D)f \cdot 1 = P(\partial)f, \tag{8.12}$$

$$P(D)e^{px} \cdot e^{qx} = P(p-q)e^{(p+q)x}, \tag{8.13}$$

$$\partial_x \partial_t \log f = \frac{D_x D_t f \cdot f}{2f^2}, \tag{8.14}$$

$$\partial_x^4 \log f = \frac{D_x^4 f \cdot f}{2f^2} - 3 \frac{(D_x^2 f \cdot f)^2}{2f^4}, \tag{8.15}$$

$$\partial_x^6 \log f = \frac{D_x^6 f \cdot f}{2f^2} - 15 \frac{(D_x^4 f \cdot f)(D_x^2 f \cdot f)}{2f^4} + 15 \left(\frac{D_x^2 f \cdot f}{f^2} \right)^3, \tag{8.16}$$

$$\partial_x \log(a/b) = \frac{D_x a \cdot b}{ab}, \tag{8.17}$$

$$\partial_x^2 \log(a/b) = \frac{D_x^2 a \cdot a}{2a^2} - \frac{D_x^2 b \cdot b}{2b^2}, \tag{8.18}$$

$$\partial_x^2 \log(ab) = \frac{D_x^2 a \cdot b}{ab} - \left(\frac{D_x a \cdot b}{ab} \right)^2. \tag{8.19}$$

8.3 Multisoliton Solutions for the KdV Class

8.3.1 The KdV Class

Now that we have the KdV equation in the bilinear form (8.9), let us construct soliton solutions for it. It will turn out that the crucial property of the bilinear derivative is (8.13), which allows easy termination of expansions in exponentials.

Actually, the construction of one- and two-soliton solutions is quite easy for the generic class of bilinear equations of the form

$$P(D_x, D_y, \ldots)F \cdot F = 0, \tag{8.20}$$

where P is a polynomial in the Hirota derivatives D. We may assume that P is even, because the odd terms cancel due to the antisymmetry of the D-operator. (In the following we sometimes use boldface for multi-component objects, e.g., the above could be written as $P(\mathbf{D})F \cdot F = 0$.)

The main known integrable equations of this class are the following:

- The Kadomtsev–Petviashvili equation

$$\partial_x[u_{xxx} + 6uu_x - 4u_t] = \mp 3u_{yy} \tag{8.21}$$

(KPI with $-$ sign, KPII with $+$ sign), which bilinearizes with the substitution (8.10) and two integrations to

$$\left(D_x^4 - 4D_xD_t \pm 3D_y^2\right)F \cdot F = 0. \tag{8.22}$$

As special cases this equation contains the KdV equation (8.1) (with $y \to 0, t \to -t/4$), and the Boussinesq equation

$$\partial_x(u_{xxx} + 6uu_x + u_x) = u_{tt} \tag{8.23}$$

(KPII with $t \to -x/4, y \to t/\sqrt{3}$).

- The Hirota–Satsuma shallow water wave equation

$$u_{xxt} + 3uu_t - 3u_xv_t - u_x = u_t, v_x = -u \tag{8.24}$$

bilinearizes with (8.10) and one integration to

$$\left(D_x^3D_t - D_x^2 - D_tD_x\right)F \cdot F = 0. \tag{8.25}$$

This has an integrable $(2+1)$-dimensional extension (the Hirota–Satsuma–Ito (HSI) equation):

$$\left(D_x^3D_t + aD_x^2 + D_tD_y\right)F \cdot F = 0. \tag{8.26}$$

- The Sawada–Kotera equation (SK)

$$u_{xxxxx} + 15uu_{xxx} + 15u_xu_{xx} + 45u^2u_x + u_t = 0 \tag{8.27}$$

bilinearizes with (8.10) and one integration to

$$\left(D_x^6 + D_xD_t\right)F \cdot F = 0, \tag{8.28}$$

with the integrable $(2+1)$-dimensional extension

$$\left(D_x^6 + 5D_x^3D_t - 5D_t^2 + D_xD_y\right)F \cdot F = 0. \tag{8.29}$$

8.3.2 The Vacuum and the One-Soliton Solution

Let us start with the zero-soliton solution or the vacuum. We know that the KdV equation has a solution $u \equiv 0$ and now we want to find the corresponding F. From (8.10) we see that $F = e^{a(t)x+b(t)}$ yields $u \equiv 0$, and in view of the gauge freedom we can choose $F = 1$ as our vacuum solution. It solves (8.20) provided that

$$P(0,0,\dots) = 0. \tag{8.30}$$

This is then the first condition that we have to impose on the polynomial P in (8.20): it should not have a constant term.

The multisoliton solutions are obtained by finite perturbation expansions around the vacuum $F = 1$:

$$F = 1 + \varepsilon f_1 + \varepsilon^2 f_2 + \varepsilon^3 f_3 + \cdots . \tag{8.31}$$

Here ε is a formal expansion parameter. For the one-soliton solution (1SS) only one term is needed. If we substitute

$$F = 1 + \varepsilon f_1 \tag{8.32}$$

into (8.20) we get

$$P(D_x,\dots)\left\{1\cdot 1 + \varepsilon 1\cdot f_1 + \varepsilon f_1\cdot 1 + \varepsilon^2 f_1\cdot f_1\right\} = 0.$$

The term of order ε^0 vanishes because of (8.30). For the terms of order ε^1 we use property (8.12) so that, since P is even, we get

$$P(\partial_x, \partial_y, \dots) f_1 = 0. \tag{8.33}$$

The soliton solutions correspond to the exponential solutions of (8.33). For a 1SS we take an f_1 with just one exponential

$$f_1 = e^{\eta}, \quad \eta = px + qy + \omega t + \cdots + \text{const}, \tag{8.34}$$

and then (8.33) becomes the *dispersion relation* on the parameters $p, q, \dots$[3]

$$P(p, q, \dots) = 0. \tag{8.35}$$

Finally, the order ε^2 term vanishes automatically because of (8.13,8.30):

$$P(\mathbf{D})e^{\eta}\cdot e^{\eta} = e^{2\eta}P(\mathbf{p} - \mathbf{p}) = 0.$$

In terms of u in (8.10) we get

---

[3] For KdV the dispersion relation (8.35) reads $p(p^3 + \omega) = 0$, but only the branch $\omega = -p^3$ is used in practice.

$$u = 2\partial_x^2 \log(1 + e^\eta) = \frac{2p^2 e^\eta}{(1 + e^\eta)^2} = \frac{p^2/2}{\cosh^2(\eta/2)}. \qquad (8.36)$$

We see in particular that the soliton is located at $\eta = 0$.

8.3.3 The Two-Soliton Solution

The two-soliton solution (2SS) is built from two 1SSs, and the important principle is that

- for integrable systems one must be able to combine *any* pair of 1SSs built on top of the same vacuum.

Thus if we have two 1SSs, $F_1 = 1 + \varepsilon e^{\eta_1}$ and $F_2 = 1 + \varepsilon e^{\eta_2}$, we should be able to combine them into a form $F = 1 + \varepsilon f_1 + \varepsilon^2 f_2$, where $f_1 = e^{\eta_1} + e^{\eta_1}$. Gauge invariance suggests that we should try the combination

$$F = 1 + e^{\eta_1} + e^{\eta_2} + A_{12} e^{\eta_1 + \eta_2}, \qquad (8.37)$$

where there is just one free constant A_{12}. Substituting this into (8.20) yields

$$
\begin{aligned}
P(\mathbf{D})\{\quad 1 \cdot 1 + &\quad 1 \cdot e^{\eta_1} &+&\quad 1 \cdot e^{\eta_2} &+&\quad \underline{A_{12} 1 \cdot e^{\eta_1 + \eta_2}} &+ \\
e^{\eta_1} \cdot 1 + &\quad e^{\eta_1} \cdot e^{\eta_1} &+&\quad e^{\eta_1} \cdot e^{\eta_2} &+&\quad A_{12} e^{\eta_1} \cdot e^{\eta_1 + \eta_2} &+ \\
e^{\eta_2} \cdot 1 + &\quad e^{\eta_2} \cdot e^{\eta_1} &+&\quad e^{\eta_2} \cdot e^{\eta_2} &+&\quad A_{12} e^{\eta_2} \cdot e^{\eta_1 + \eta_2} &+ \\
\underline{A_{12} e^{\eta_1 + \eta_2} \cdot 1} + &\ A_{12} e^{\eta_1 + \eta_2} \cdot e^{\eta_1} &+&\ A_{12} e^{\eta_1 + \eta_2} \cdot e^{\eta_2} &+&\ A_{12}^2 e^{\eta_1 + \eta_2} \cdot e^{\eta_1 + \eta_2} \ \} &= 0.
\end{aligned}
$$

In this equation all non-underlined terms vanish due to (8.30),(8.35). Since P is even, the underlined terms combine as $2A_{12} P(\mathbf{p}_1 + \mathbf{p}_2) + 2P(\mathbf{p}_1 - \mathbf{p}_2) = 0$, from which A_{12} can be solved as

$$A_{12} = -\frac{P(\mathbf{p}_1 - \mathbf{p}_2)}{P(\mathbf{p}_1 + \mathbf{p}_2)}. \qquad (8.38)$$

For the KdV case we obtain (using the dispersion relation $\omega_i = -p_i^3$)

$$
\begin{aligned}
A_{12} &= -\frac{(p_1 - p_2)^4 + (p_1 - p_2)(\omega_1 - \omega_2)}{(p_1 + p_2)^4 + (p_1 + p_2)(\omega_1 + \omega_2)} \\
&= -\frac{(p_1 - p_2)\left[(p_1 - p_2)^3 - (p_1^3 - p_2^3)\right]}{(p_1 + p_2)\left[(p_1 + p_2)^3 - (p_1^3 + p_2^3)\right]} = \frac{(p_1 - p_2)^2}{(p_1 + p_2)^2}.
\end{aligned}
$$

The important thing about the above construction is that we were able to derive a two-soliton solution for a huge class of equations, namely all those whose bilinear form is of type (8.20) for whatever P. In particular this includes many non-integrable systems.

8.3.4 The Soliton Content of Solution (8.37)

Next we will show that (8.37) actually describes the scattering of two solitons.

In terms of u the solution (8.37),(8.34),(8.10) is given as

$$u = \frac{2p_1^2 e^{\eta_1}\left(1+A_{12}e^{2\eta_2}\right)+2p_2^2 e^{\eta_2}\left(1+A_{12}e^{2\eta_1}\right)+2\left[(p_1-p_2)^2+A_{12}(p_1+p_2)^2\right]e^{\eta_1+\eta_2}}{\left(1+e^{\eta_1}+e^{\eta_2}+A_{12}e^{\eta_1+\eta_2}\right)^2}$$

$$= \frac{2\sqrt{A_{12}}\left[p_1^2\cosh(\eta_2+\frac{1}{2}\alpha_{12})+p_2^2\cosh(\eta_1+\frac{1}{2}\alpha_{12})\right]+(p_1-p_2)^2+A_{12}(p_1+p_2)^2}{\left[\cosh(\frac{1}{2}(\eta_1-\eta_2))+\sqrt{A_{12}}\cosh(\frac{1}{2}(\eta_1+\eta_2+\alpha_{12}))\right]^2}, \tag{8.39}$$

where $\alpha_{12} = \log A_{12}$ (thus we have assumed that $A_{12} > 1$). For example, if in the KdV case we take $p_1 = 2$, $p_2 = 4$ (in particular $\eta_1 = 2x - 8t + 2\log 3$, $\eta_1 = 4x - 64t + 2\log 3$) then $A_{12} = 1/9$ and we get the solution

$$u = 12\frac{4\cosh(2x-8t)+\cosh(4x-64t)+3}{[3\cosh(x-28t)+\cosh(3x-36t)]^2}.$$

Let us now discuss the soliton content of the solution (8.39). Since it is just a combination of two solitons, we should see these solitons before and after interaction. This is best seen if we go to a coordinate frame traveling with one of the solitons and observe what happens there as $|t| \to \infty$.

For 1SS recall from (8.36) that the soliton is located at $\eta = 0$. Thus in the frame comoving with soliton 1 the exponent η_1 is finite while $\eta_2 \to \pm\infty$ (later we will analyze in detail how these limits correspond to $t \to \pm\infty$). The $\eta_i \to \pm\infty$ limits are usually easy to determine, in this particular case we obtain from (8.39)

$$\eta_2 \to -\infty : u \to \frac{2p_1^2 e^{\eta_1}}{(1+e^{\eta_1})^2} = \frac{p_1^2/2}{\cosh^2(\frac{1}{2}\eta_1)},$$

$$\eta_2 \to +\infty : u \to \frac{2A_{12}p_1^2 e^{\eta_1}}{(1+A_{12}e^{\eta_1})^2} = \frac{p_1^2/2}{\cosh^2(\frac{1}{2}(\eta_1+\alpha_{12}))},$$

$$\eta_1 \to -\infty : u \to \frac{2p_2^2 e^{\eta_2}}{(1+e^{\eta_2})^2} = \frac{p_2^2/2}{\cosh^2(\frac{1}{2}\eta_2)},$$

$$\eta_1 \to +\infty : u \to \frac{2A_{12}p_2^2 e^{\eta_2}}{(1+A_{12}e^{\eta_2})^2} = \frac{p_2^2/2}{\cosh^2(\frac{1}{2}(\eta_2+\alpha_{12}))}.$$

We observe that the limits differ in the argument of the cosh function. If we write the argument explicitly we get in the KdV case

$$\begin{array}{c|c|c}
 & \eta_{other} \to -\infty & \eta_{other} \to +\infty \\
\hline
\text{soliton1} & \frac{1}{2}[p_1 x - p_1^3 t] & \frac{1}{2}[p_1(x + \alpha_{12}/p_1) - p_1^3 t]. \\
\text{soliton2} & \frac{1}{2}[p_2 x - p_2^3 t] & \frac{1}{2}[p_2(x + \alpha_{12}/p_2) - p_2^3 t]
\end{array} \tag{8.40}$$

The main results here is that the solitons are experiencing a phase shift, i.e., their location changes by $\Delta x_i = -\alpha_{12}/p_i$.

In order to have a dynamical interpretation of the above we must still determine how the various $\eta_i \to \pm\infty$ limits correspond to $t \to \pm\infty$ limits. The detailed analysis goes as follows:

Comoving with soliton 1: We replace x with the new variable $\xi = x + (\omega_1/p_1)t$ and then $\eta_1 = p_1 \xi + \eta_1^0$, $\eta_2 = p_2[\xi + (\omega_2/p_2 - \omega_1/p_1)t] + \eta_2^0$. In this frame η_1 is time independent, but

$$\eta_2 \to \text{sign}\left[p_2\left(\frac{\omega_2}{p_2} - \frac{\omega_1}{p_1}\right)t\right] \cdot \infty, \quad \text{as} \quad |t| \to \infty.$$

Comoving with soliton 2: Now we use $\xi = x + (\omega_2/p_2)t$ and then $\eta_2 = p_2\xi + \eta_2^0$, $\eta_1 = p_1[\xi - (\omega_2/p_2 - \omega_1/p_1)t] + \eta_1^0$. In this frame η_2 is time independent, but

$$\eta_1 \to \text{sign}\left[p_1\left(\frac{\omega_1}{p_1} - \frac{\omega_2}{p_2}\right)t\right] \cdot \infty, \quad \text{as} \quad |t| \to \infty.$$

Example: For KdV we have $\omega_i = -p_i^3$, $\omega_i/p_i = -p_i^2 (= -v_i)$. Let us furthermore assume that $p_i > 0$ and that the solitons are numbered so that $v_1 > v_2$. Then

$$\text{sign}\left[p_2\left(\frac{\omega_2}{p_2} - \frac{\omega_1}{p_1}\right)\right] = \text{sign}\left[p_2\left(-p_2^2 + p_1^2\right)\right] = 1, \ \text{sign}\left[p_1\left(\frac{\omega_1}{p_1} - \frac{\omega_2}{p_2}\right)\right] = -1,$$

With this we can rewrite table (8.40) as

$$\begin{array}{c|c|c}
 & t \to -\infty & t \to +\infty \\
\hline
\text{soliton 1} & \frac{1}{2}[p_1 x - p_1^3 t] & \frac{1}{2}[p_1(x + \alpha_{12}/p_1) - p_1^3 t]. \\
\text{soliton 2} & \frac{1}{2}[p_2(x + \alpha_{12}/p_2) - p_2^3 t] & \frac{1}{2}[p_2 x - p_2^3 t]
\end{array} \tag{8.40'}$$

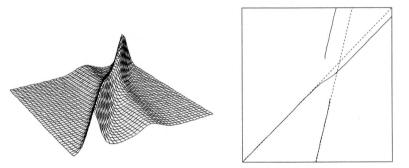

Fig. 8.1 Scattering of Korteweg–de Vries solitons. On the left a profile view, on the right the locations of the maxima, along with the free soliton trajectory as a dotted line. ($p_1 = \frac{1}{2}$, $p_2 = 1$)

Thus, when going from $t = -\infty$ to $t = +\infty$ the position of the soliton 1 is shifted by $-\alpha_{12}/p_1$ with respect to straight line motion, while soliton 2 is shifted by $+\alpha_{12}/p_2$. Since p_1, p_2 are both positive, $A_{12} < 1$ and $\alpha_{12} < 0$, the faster soliton (number 1) will be shifted forward and the slower one (number 2) backward. This is illustrated in Fig. 8.1. On the left we have plotted a profile view of the scattering process directly from the analytical result (8.39). On the right we have plotted the locations of the maxima of the solution. The dotted line shows how a single soliton would have moved. We can see clearly how the faster soliton has advanced and the slower one retarded.

We also note that for some parameter values we may have $A_{12} = 0$ or ∞, it is then said that the solitons *resonate*. In that case the above analysis fails and solitons may appear in other moving frames as well [13–15].

8.3.5 Existence of Multisoliton Solutions

The above shows that for the KdV class (8.20) the existence of 2SS is not strongly related to integrability, but it turns out that the existence of 3SS is actually very restrictive.

A 3SS should start with $f_1 = e^{\eta_1} + e^{\eta_2} + e^{\eta_3}$ and, if the above is any guide, contain terms up to f_3. If we now use the requirement that the solution should reduce to a 2SS when the third soliton goes to infinity (which corresponds to $\eta_k \to \pm\infty$) then one finds that F must have the form

$$F = 1 + e^{\eta_1} + e^{\eta_2} + e^{\eta_3} + A_{12}e^{\eta_1+\eta_2} + A_{13}e^{\eta_1+\eta_3} + A_{23}e^{\eta_2+\eta_3}$$
$$+ A_{12}A_{13}A_{23}e^{\eta_1+\eta_2+\eta_3}. \tag{8.41}$$

Note in particular that this expression contains no additional freedom: The parameters p_i are only required to satisfy the dispersion relation (8.35) and the phase factors A were already determined (8.38). This ansatz extends to NSS [16]:

$$F = \sum_{\substack{\mu_i=0,1 \\ 1\leq i\leq N}} \exp\left(\sum_{1\leq i<j\leq N} \alpha(i,j)\mu_i\mu_j + \sum_{i=1}^{N} \mu_i\eta_i \right), \tag{8.42}$$

with $A_{ij} = e^{\alpha(i,j)}$. Thus the ansatz for a NSS is completely fixed and the requirement that it be a solution of (8.20) implies conditions on the equation itself. Only for integrable equations can we actually combine solitons in this way. More precisely, let us make the

Definition 1. A set of equations written in the Hirota bilinear form is **Hirota integrable**, if one can combine any number N of one-soliton solutions into an NSS without any additional restrictions on the single soliton solutions, and the result is a finite polynomial in the e^{η}s involved.

In all cases studied so far, equations integrable in the above sense have turned out to be integrable in other senses as well.

Since the existence of a 3SS is so restrictive it can be used as a method for searching for new integrable equations. We will return to this question later, here we would just like to note that if one substitutes the 3SS ansatz (8.41) into (8.20) one obtains the condition

$$S_3(P) \equiv \sum_{\sigma_i = \pm 1} P(\sigma_1 \mathbf{p}_1 + \sigma_2 \mathbf{p}_2 + \sigma_3 \mathbf{p}_3) P(\sigma_1 \mathbf{p}_1 - \sigma_2 \mathbf{p}_2)$$
$$\times P(\sigma_2 \mathbf{p}_2 - \sigma_3 \mathbf{p}_3) P(\sigma_3 \mathbf{p}_3 - \sigma_1 \mathbf{p}_1) \doteq 0, \qquad (8.43)$$

where the symbol $\doteq$ means that the equation is required to hold only when the parameters $\mathbf{p}_i$ satisfy the dispersion relation $P(\mathbf{p}_i) = 0$. Since the parameters are not restricted in any other way than by the dispersion relation (which was already demanded for the 1SS and 2SS) Eq. (8.43) should be interpreted as a condition on the polynomial P.

8.4 Soliton Solution for the mKdV and sG Class

8.4.1 The Modified KdV Equation

As mentioned before, Hirota's bilinear method has been applied to many other equations beside KdV. Let us next consider the modified KdV equation (mKdV)

$$u_{xxx} + \epsilon 6u^2 u_x + u_t = 0, \qquad (8.44)$$

where we have explicitly noted the sign $\epsilon = \pm$, because it cannot be scaled away. It is easy to verify that this equation has the traveling wave solutions

$$u = \frac{\pm p}{\cosh(px - p^3 t + c)}, \text{if } \epsilon = 1, \qquad (8.45)$$

$$u = \frac{\pm p}{\sinh(px - p^3 t + c)}, \text{if } \epsilon = -1. \qquad (8.46)$$

Note that the dispersion relation and velocity are as for KdV, but the power of the cosh or sinh term is different. Since $\epsilon = -1$ leads to a singular solution we only consider the $\epsilon = +1$. The sign in $\pm p$ determines whether we have a soliton (+) or antisoliton (−).

For bilinearization we must first introduce the analog of (8.2) to get suitable scale invariance. Indeed, if we let

$$u = \partial_x w \qquad (8.47)$$

we get from (8.44)

$$\partial_x \left[w_{xxx} + 2w_x^3 + w_t \right] = 0, \qquad (8.48)$$

for which the scale transformation $x \to \lambda x, t \to \lambda^3 t, w \to w$ only gives an overall factor. The part in the square bracket is the potential mKdV equation. It now turns out that a good substitution is given by

$$w = 2\arctan(G/F), \text{ i.e., } u = 2\frac{D_x G \cdot F}{F^2 + G^2}, \tag{8.49}$$

and then the potential mKdV gets the form

$$\left(F^2 + G^2\right)\left[(D_x^3 + D_t)G \cdot F\right] + 3\left(D_x F \cdot G\right)\left[D_x^2(F \cdot F + G \cdot G)\right] = 0. \tag{8.50}$$

Note that we now have *two free functions*, G and F, so we can also impose *two equations* on them. The form of (8.50) suggests that we take the two equations as

$$\begin{cases} (D_x^3 + D_t)(G \cdot F) = 0, \\ D_x^2(F \cdot F + G \cdot G) = 0. \end{cases} \tag{8.51}$$

The splitting into a bilinear pair is not unique, we could also have taken

$$\begin{cases} (D_x^3 + D_t + 3\lambda D_x)(G \cdot F) = 0, \\ (D_x^2 + \lambda)(F \cdot F + G \cdot G) = 0, \end{cases}$$

where λ is an arbitrary function of x, t. The 1SS (8.45) corresponds to $F = 1$, $G = \pm e^\eta$, $\eta = px - p^3 t$ and therefore we must take $\lambda = 0$.

8.4.2 Sine-Gordon Equation

The sine-Gordon (sG) equation

$$u_{xx} - u_{tt} = \sin u \tag{8.52}$$

is a nonlinear version of the well-known Klein–Gordon equation $u_{xx} - u_{tt} = 0$.

As before, some information about a possible bilinearizing transformation can be obtained from the 1SS. Let us therefore try the traveling wave ansatz $u = f(z)$, $z = x - vt$; when it is substituted into (8.52) we get $f''(1 - v^2) = \sin f$, which can be integrated once to $\frac{1}{2}(f')^2(1 - v^2) = -\cos f + C$. For solitons we demand $f' \to 0$ and $f'' \to 0$, as $|z| \to \infty$, and therefore we must choose boundary conditions $f \to 2\pi n_\pm$, as $|z| \to \infty$ and correspondingly $C = 1$ and $v < 1$. Then the equation can be integrated yielding a kink-type solution

$$u = 4\overline{\arctan}\left[\exp\left(\pm\frac{x - vt + \delta}{(1 - v^2)^{1/2}}\right)\right]. \tag{8.53}$$

(This shows clearly the relativistic nature of the solution.)

On the basis of this soliton solution we try bilinearization with the ansatz

$$u = 4\arctan(G/F), \tag{8.54}$$

and when this is used in Eq. (8.52) we get

$$[(D_x^2 - D_t^2 - 1)G \cdot F] (F^2 - G^2) - FG [(D_x^2 - D_t^2)(F \cdot F - G \cdot G)] = 0.$$

There is again some ambiguity in splitting this quartic equation into bilinear ones, because the term $\lambda FG(F^2 - G^2)$ could be added to one part and subtracted from the other. For solitons $\lambda = 0$ and we get the bilinearization

$$\begin{cases} (D_x^2 - D_t^2 - 1)G \cdot F = 0, \\ (D_x^2 - D_t^2)(F \cdot F - G \cdot G) = 0. \end{cases} \tag{8.55}$$

8.4.3 Multisoliton Solutions for the mKdV/sG Class

The mKdV and sG equations have bilinear forms (8.51),(8.55) that are of the type

$$\begin{cases} B(\mathbf{D})G \cdot F = 0, \\ A(\mathbf{D})(F \cdot F + \epsilon G \cdot G) = 0, \end{cases} \tag{8.56}$$

where A is even and B either odd (mKdV) or even (sG). If B is odd one can make the rotation $F = f + g, G = \sqrt{-\epsilon}(f - g)$ after which the pair (8.56) becomes

$$\begin{cases} B(\mathbf{D})g \cdot f = 0, \quad (B \text{ odd}) \\ A(\mathbf{D})g \cdot f = 0. \end{cases} \tag{8.57}$$

Let us now construct soliton solutions to the class of Eq. (8.56). For the vacuum we choose $F = 1, G = 0$ and therefore we must have $A(0) = 0$. For the 1SS we may try

$$F = 1 + \varepsilon \alpha e^\eta, G = \varepsilon \beta e^\eta.$$

Direct calculation yields three conditions

$$\beta B(\mathbf{p}) = 0, \quad \alpha \beta B(0) = 0, \quad \alpha A(\mathbf{p}) = 0.$$

Since we cannot impose two dispersion relations at the same time either α or β must vanish. Thus we can in principle have *two different kinds of solitons*

$$\begin{aligned} &\text{type a}: F = 1 + e^{\eta_A}, G = 0, \text{ with dispersion relation } A(\mathbf{p}) = 0, \\ &\text{type b}: F = 1, G = e^{\eta_B}, \quad \text{with dispersion relation } B(\mathbf{p}) = 0. \end{aligned} \tag{8.58}$$

For mKdV and SG the A polynomial is too trivial to make the first kind of soliton interesting; indeed, solutions (8.45),(8.53) are of b-type. However, there are integrable equations for which the dispersion relation $A(\mathbf{p}) = 0$ has nontrivial solutions, so we will now consider the general problem of constructing a 2SS.

a+a: If we want to combine two solitons of type a the starting point must be

$$F = 1 + \varepsilon e^{\eta_1} + \varepsilon e^{\eta_2} + O(\varepsilon^2), \quad G = O(\varepsilon^2), \quad \text{with } A(\mathbf{p}_1) = A(\mathbf{p}_2) = 0.$$

If we take $G = 0$ then the A-equation is KdV type for F and the solution is given by (8.37,8.38) with P replaced by A.

a+b: Now the starting point is

$$F = 1 + \varepsilon e^{\eta_1} + O(\varepsilon^2), \quad G = \varepsilon e^{\eta_2} + O(\varepsilon^2), \quad \text{with } A(\mathbf{p}_1) = B(\mathbf{p}_2) = 0.$$

Studying the equations order by order shows that a 2SS is obtained with

$$F = 1 + e^{\eta_1}, \quad G = e^{\eta_2} + L_{12} e^{\eta_1 + \eta_2}, \quad \text{with } L_{12} = -\frac{B(\mathbf{p}_2 - \mathbf{p}_1)}{B(\mathbf{p}_1 + \mathbf{p}_2)}. \tag{8.59}$$

b+b: In this case we start with

$$F = 1 + O(\varepsilon^2), \quad G = \varepsilon e^{\eta_1} + \varepsilon e^{\eta_2} + O(\varepsilon^2), \quad \text{with } B(\mathbf{p}_1) = B(\mathbf{p}_2) = 0,$$

and the 2SS turns out to have the form

$$F = 1 - K_{12} e^{\eta_1 + \eta_2}, \quad G = e^{\eta_1} + e^{\eta_2}, \quad \text{with } K_{12} = \varepsilon \frac{A(\mathbf{p}_1 - \mathbf{p}_2)}{A(\mathbf{p}_1 + \mathbf{p}_2)}. \tag{8.60}$$

These results indicate that again we can have 2SS for a huge class of equations, since no essential restrictions were given for the two polynomials A and B. And again the situation is quite different with 3SS, which exists only for very specific equations. Here we just point out that by considering the different limits one can see that the only candidate for a combination of three b-type solitons is[4]

$$\begin{cases} G = e^{\eta_1} + e^{\eta_2} + e^{\eta_3} - K_{12} K_{13} K_{23} e^{\eta_1 + \eta_2 + \eta_3}, \\ F = 1 - K_{12} e^{\eta_1 + \eta_2} - K_{13} e^{\eta_1 + \eta_3} - K_{23} e^{\eta_2 + \eta_3}. \end{cases} \tag{8.61}$$

The requirement that (8.61) actually is a solution imposes again severe restrictions on the polynomials A and B.

What if the parameters are complex but so that $\eta_2 = \eta_1^*$? Clearly the 2SS of the b+b case is still real. In more detail, if $\eta_1 = p_r x + \omega_r t + i(p_i x + \omega_i t)$, $\eta_2 = p_r x + \omega_r t - i(p_i x + \omega_i t)$ we get

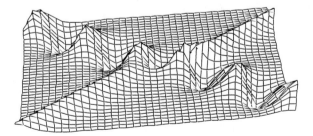

Fig. 8.2 mKdV breather scattering with a soliton. $\left[(8.61) \text{ with } p_1 = 1 + i, p_2 = 1 - i, p_3 = \sqrt{2} \right]$

---

[4] When comparing the $\eta \to -\infty$ and $\eta \to +\infty$ limits, note that (G,F) and $(-F,G)$ correspond to the same soliton when u is defined by (8.49).

$$\frac{G}{F} = \frac{e^{p_r x + \omega_r t} 2\cos(p_i x + \omega_i t)}{1 - K_{12} e^{2(p_r x + \omega_r t)}} = \frac{\cos(p_i x + \omega_i t)}{\sqrt{K_{12}} \sinh(p_r x + \omega_r t + \frac{1}{2}\log(K_{12}))}. \tag{8.62}$$

We see that in addition to moving with constant speed $v = -\omega_r/p_r$ the soliton oscillates, such a solution is called a *breather soliton*. Note that the velocity and the breathing frequency depend on different parameters. In the mKdV case $\omega_r = -p_r(p_r^2 - 3p_i^2)$ so that $v = (p_r^2 - 3p_i^2)$, from which we see that a breather can have a negative velocity. As an illustration we have in Fig. 8.2 breather-soliton scattering constructed using (8.61).

8.5 The Nonlinear Schrödinger Equation

Let us next consider the nonlinear Schrödinger equation (nlS)

$$i\phi_t + \phi_{xx} + 2\epsilon|\phi|^2\phi = 0, \quad \epsilon = \pm 1, \tag{8.63}$$

where the function ϕ is complex.

8.5.1 One-Soliton Solutions

As usual, we start by constructing a traveling wave solution. Now, however, we have to make an ansatz also for the complex phase:

$$\phi(x,t) = e^{i(a(x+bt)+c)} f(x - vt). \tag{8.64}$$

Thus the envelope f travels with speed v, while the phase has its own behavior. From (8.63) one then finds that $a = v/2$ and then the equation reduces to $f'' + 2\epsilon f^3 - (\frac{1}{2}bv + \frac{1}{4}v^2)f = 0$, which can be integrated once to

$$(f')^2 + \epsilon f^4 - (\tfrac{1}{2}bv + \tfrac{1}{4}v^2)f^2 + C = 0. \tag{8.65}$$

If we assume that $f, f' \to 0$ as $|x - vt| \to \infty$, then $C = 0$, and if $\epsilon = +1$ the result is

$$\phi = \frac{\kappa e^{i\left[\frac{v}{2}x + (\kappa^2 - \frac{v^2}{4})t + c\right]}}{\cosh(\kappa(x - vt))}. \tag{8.66}$$

This solution is called a *bright soliton*. Note that we can have this normal soliton solution only if $\epsilon = +1$ in (8.63). There are two free parameters (in addition to the overall position and phase): velocity v and amplitude κ. The fact that amplitude and velocity are both free is in clear contrast with the results for KdV. Indeed, it seems as if there is no dispersion relation but we will later see that there are in fact two complex parameters and the dispersion relation still allows two free real parameters.

The soliton solution (8.66) describes a bright pulse in a dark background, but there is also the possibility of a *dark soliton* in a bright background (envelope-hole soliton), if $\epsilon = -1$. In that case we take $C \neq 0$ in (8.65) and get the solution

$$\phi = \kappa \tanh(\kappa(x - vt))e^{i\left[\frac{v}{2}x - (\frac{v^2}{4} + 2\kappa^2)t\right]}. \tag{8.67}$$

This solution also has two free parameters, but if we want to obtain multisoliton solutions then each soliton must have the same background amplitude and frequency and therefore these parameters are global. Still other possibilities exist ("gray" and "antigray") but for them the ansatz is more general than (8.64).

8.5.2 The Bilinear Approach

The substitution that bilinearizes (8.63) is

$$\phi = g/f, \quad g \text{ complex, } f \text{ real}, \tag{8.68}$$

and yields

$$f\left[(iD_t + D_x^2)g \cdot f\right] - g\left[D_x^2 f \cdot f - \epsilon 2|g|^2\right] = 0. \tag{8.69}$$

For normal solitons we split this into

$$\begin{cases} (iD_t + D_x^2)g \cdot f = 0, \\ \quad D_x^2 f \cdot f = \epsilon 2|g|^2. \end{cases} \tag{8.70}$$

Clearly we can start from the vacuum soliton $f = 1, g = 0$. In the formal expansion for a 1SS the function g will then have a degree of at least 1 and according to the second equation f will go at least to degree 2. Thus if we take $g = e^\eta$ the correction term to f must be proportional to $e^{\eta + \eta^*}$, which leads us to try the ansatz

$$g = e^\eta, \quad f = 1 + ae^{\eta + \eta^*}, \quad \eta = px + \omega t, \quad p \text{ and } \omega \text{ complex}. \tag{8.71}$$

From the first equation of (8.70) we then get the dispersion relation

$$i\omega + p^2 = 0, \tag{8.72}$$

and from the second the phase factor

$$a = \frac{\epsilon}{(p + p^*)^2}. \tag{8.73}$$

If we split η and its parameters into real and imaginary parts:

$$p = p_R + ip_I, \quad \omega = \omega_R + i\omega_I,$$

then the dispersion relation says

$$\omega_R = -2p_R p_I, \quad \omega_I = p_R^2 - p_I^2$$

and we can write

$$\eta = p_R(x - 2p_I t) + i[p_I x + (p_R^2 - p_I^2)t],$$

and finally if $\epsilon = +1$,

$$\phi = \frac{g}{f} = \frac{e^{p_R(x - 2p_I t) + i[p_I x + (p_R^2 - p_I^2)t]}}{1 + \frac{\epsilon}{4p_R^2} e^{2p_R(x - 2p_I t)}} = \frac{p_R e^{i[p_I x + (p_R^2 - p_I^2)t]}}{\cosh(p_R(x - 2p_I t) + d)},$$

which is the result (8.66) obtained before. (If $\epsilon = -1$ we get a singular solution.)
It is clear that the above construction generalizes to the class

$$\begin{cases} B(\mathbf{D}) G \cdot F = 0, \\ A(\mathbf{D}) F \cdot F = \epsilon 2|G|^2, \end{cases} \tag{8.74}$$

so that the 1SS is as before, with

dispersion relation: $B(\mathbf{p}) = B(-\mathbf{p}^*) = 0$, phase factor: $a = \dfrac{\epsilon}{A(\mathbf{p} + \mathbf{p}^*)}$.

The dark soliton mentioned above is related to a different splitting of (8.69):

$$\begin{cases} (iD_t + D_x^2 - 2\rho^2) g \cdot f = 0, \\ (D_x^2 - 2\rho^2) f \cdot f = -2|g|^2. \end{cases} \tag{8.75}$$

Now the 0SS is given by a pure phase

$$g \equiv g_0 = \rho e^\theta, \ f = 1, \ \theta = i(kx - \omega t), \ \omega = k^2 + 2\rho^2, \tag{8.76}$$

and the 1SS by

$$g = g_0(1 + Ze^\eta), \quad f = 1 + e^\eta, \tag{8.77}$$

$$\eta = px - \Omega t, \quad \Omega = p(2k - \sqrt{4\rho^2 - p^2}), \quad Z = \frac{\sqrt{4\rho^2 - p^2} + ip}{\sqrt{4\rho^2 - p^2} - ip}. \tag{8.78}$$

An illustration is given in Fig. 8.3. It is easy to also show that

$$|\phi|^2 = \rho^2 - \frac{p^2/4}{\cosh^2(\frac{1}{2}\eta)},$$

from which we see the "dark" nature of the soliton. If $\rho = p/2$ then $Z = -1$ and we
recover (8.67), if $\rho > p/2$ we have a "gray" soliton.

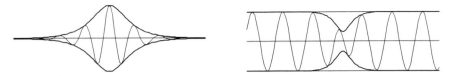

Fig. 8.3 A bright soliton (*left*) and a dark soliton (*right*): *thick line*: $|\phi|$, *thin line*: $\Re\phi$

8.5.3 Two-Soliton Solutions

The bright 2SS is obtainable from the natural extension of (8.71):

$$g = e^{\eta_1} + e^{\eta_2} + a_{12\bar{1}}e^{\eta_1+\eta_2+\eta_1^*} + a_{12\bar{2}}e^{\eta_1+\eta_2+\eta_2^*}, \tag{8.79}$$

$$f = 1 + a_{1\bar{1}}e^{\eta_1+\eta_1^*} + a_{1\bar{2}}e^{\eta_1+\eta_2^*} + a_{\bar{1}2}e^{\eta_1^*+\eta_2} + a_{2\bar{2}}e^{\eta_2+\eta_2^*} + a_{1\bar{1}2\bar{2}}e^{\eta_1+\eta_1^*+\eta_2+\eta_2^*}. \tag{8.80}$$

The principles of forming these expressions were the following: (1) g is odd in e^{η} and in particular there is always one more η than η^* in the exponent; (2) f is even and real; (3) no η twice in the exponent; and (4) treat η and η^* as independent. The various coefficients a can be determined when this ansatz is substituted into the equation. It turns out that the ansatz does not work for arbitrary polynomials A and B, and in that respect the 2SS behaves more like a 4SS (it would contain ε^4 terms of the expansion). For the nlS we get

$$\eta_k = p_k x + i p_k^2 t + \eta_k^0,$$

$$a_{i\bar{j}} = \frac{1}{(p_i + p_j^*)^2}, \quad a_{ij} = (p_i - p_j)^2, \tag{8.81}$$

$$a_{\bar{i}j} = (a_{i\bar{j}})^*, \quad a_{\bar{i}\bar{j}} = (a_{ij})^*, \quad a_{ij\bar{k}} = a_{ij}a_{i\bar{k}}a_{j\bar{k}}, \quad a_{1\bar{1}2\bar{2}} = a_{1\bar{1}}a_{12} \, a_{1\bar{2}}a_{\bar{1}2}a_{\bar{1}\bar{2}}a_{2\bar{2}}.$$

The η-limits of this solution are easily obtained:

$$\Re(\eta_2) \to -\infty : \phi \to \frac{e^{\eta_1}}{1 + a_{1\bar{1}}e^{\eta_1+\eta_1^*}},$$

$$\Re(\eta_1) \to -\infty : \phi \to \frac{e^{\eta_2}}{1 + a_{2\bar{2}}e^{\eta_2+\eta_2^*}},$$

$$\Re(\eta_2) \to +\infty : \phi \to \frac{a_{12\bar{2}}e^{\eta_1}}{a_{2\bar{2}} + a_{1\bar{1}2\bar{2}}e^{\eta_1+\eta_1^*}},$$

$$\Re(\eta_1) \to +\infty : \phi \to \frac{a_{12\bar{1}}e^{\eta_2}}{a_{1\bar{1}} + a_{1\bar{1}2\bar{2}}e^{\eta_2+\eta_2^*}}.$$

The $\Re(\eta) \to -\infty$ limits are of the standard form of (8.66), while for $\Re(\eta) \to +\infty$ there are extra complex factors. Since, e.g., $a_{12\bar{2}}/a_{2\bar{2}} = a_{12}a_{1\bar{2}}$ and $a_{1\bar{1}2\bar{2}}/a_{2\bar{2}} = a_{1\bar{1}}|a_{12}a_{1\bar{2}}|^2$ we obtain the phase shift

$$\eta_i \to \eta_i + \log c_{ij}, \, c_{ij} = a_{ij}a_{ij} = \frac{(p_i - p_j)^2}{(p_i + p_j^*)^2}, \text{ where } ij = 12 \text{ or } 21.$$

The magnitude $|c_{ij}|$ therefore determines the change in the location of soliton i as it scatters with soliton j, in explicit form:

$$|c_{ij}| = \frac{(p_{Ri} - p_{Rj})^2 + (p_{Ii} - p_{Ij})^2}{(p_{Ri} + p_{Rj})^2 + (p_{Ii} - p_{Ij})^2}.$$

Remark 2.

1. For nlS we can easily create breather solutions. As usual it is a bound state of two different solitons traveling with the same speed, therefore we just take $p_{I1} = p_{I2}$ and to avoid singular solutions, $p_{R1}^2 \neq p_{R2}^2$.
2. It is also possible to have multisoliton solutions composed of dark solitons, but the background g_0 must be the same for all of them.
3. The above results generalize easily for the integrable equation [17]

$$i\phi_t + \beta\phi_{xx} + \delta|\phi|^2\phi + i\gamma\phi_{xxx} + i3\alpha|\phi|^2\phi_x = 0, \tag{8.82}$$

if $\alpha\beta = \gamma\delta$.

4. The nlS has also integrable multi-component equations, e.g., Manakov's equation

$$\begin{cases} iq_{1t} + q_{1xx} + (|q_1|^2 + |q_2|^2)q_1 = 0, \\ iq_{2t} + q_{2xx} + (|q_1|^2 + |q_2|^2)q_2 = 0, \end{cases} \tag{8.83}$$

with interesting scattering that looks inelastic [18], see Fig. 8.4.

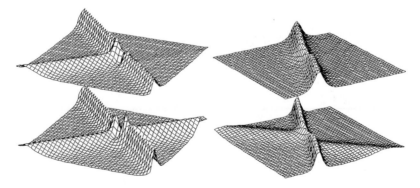

Fig. 8.4 Exotic scattering: On the *left* scattering in the Manakov model (8.83). Note the amplitude shift between components, *top*: $|q_1|$, *bottom*: $|q_2|$. On the *right* scattering of the different solitons in the Hirota–Satsuma model (8.84), in the v field (*above*) we see a sudden phase shift, the u field (*below*) shows the reason for it [(8.88), (8.89), (8.90), (8.91) with $p_1 = p_2 = 1$]

8.5.4 Multicomponent Equations

The nlS-class of bilinear equations (8.74) is also associated with some other equation, including real ones, and they can have several kinds of solitons if both of the polynomials A and B are nonlinear enough.

As an example consider the Hirota–Satsuma equation [19]

$$\begin{cases} u_t + u_{xxx} + 6uu_x - 6vv_x = 0, \\ v_t - 2v_{xxx} - 6uv_x = 0. \end{cases} \tag{8.84}$$

If $v \equiv 0$ the equation reduces to KdV (8.1) and u is then the standard KdV-soliton with dispersion relation $q + p^3 = 0$. However, it is readily verified that (8.84) also has the 1SS

$$u = \frac{2p^2}{\cosh^2(p(x+2p^2t))}, \quad v = \frac{2p^2}{\cosh(p(x+2p^2t))}. \tag{8.85}$$

Note that the KdV-type soliton has velocity p^2, while the soliton (8.85) has velocity $-2p^2$. Thus these different kinds of solitons also travel in different directions.

The system (8.84) is bilinearized with the substitution

$$u = 2\partial_x^2 \log(f), \quad v = g/f, \tag{8.86}$$

which results in

$$\begin{cases} (2D_x^3 - D_t)g \cdot f = 0, \\ (D_x^4 + D_xD_t)f \cdot f = 3g^2. \end{cases} \tag{8.87}$$

There are two kinds of 1SS:

a) $g = 0$, $f = 1 + e^\eta$ (KdV-type, dispersion relation $q = -p^3$),
b) $g = 4p^2e^\eta$, $f = 1 + e^{2\eta}$ (nlS-type but real, dispersion relation $q = 2p^3$).

For the 2SS we get
    a+a: $g = 0$, f as in (8.37,8.35,8.38).
    a+b:

$$g = 4p_2^2 e^{\eta_2}(1 + B_{12}e^{\eta_1}), \tag{8.88}$$

$$f = 1 + e^{\eta_1} + e^{2\eta_2} + B_{12}^2 e^{\eta_1 + 2\eta_2}, \tag{8.89}$$

$$B_{12} = \frac{p_1^2 - 2p_1p_2 + 2p_2^2}{p_1^2 + 2p_1p_2 + 2p_2^2}, \tag{8.90}$$

$$\text{DR}: \quad q_1 = -p_1^3, q_2 = 2p_2^3. \tag{8.91}$$

    b+b:

$$g = 4p_1^2 e^{\eta_1}(1 + A_{12}e^{2\eta_2}) + 4p_2^2 e^{\eta_2}(1 + A_{12}e^{2\eta_1}), \tag{8.92}$$

$$f = 1 + e^{2\eta_1} + C_{12}e^{\eta_1 + \eta_2} + e^{2\eta_2} + A_{12}^2 e^{2\eta_1 + 2\eta_2}, \tag{8.93}$$

$$A_{12} = \frac{(p_1 - p_2)^2}{(p_1 + p_2)^2}, \quad C_{12} = \frac{16 p_1^2 p_2^2}{(p_1 + p_2)^2 (p_1^2 + p_2^2)}, \tag{8.94}$$

$$\text{DR}: \quad q_i = 2 p_i^3. \tag{8.95}$$

The a+a scattering is as for KdV. For the a+b scattering we find

$$\eta_2 \to -\infty \; u \to \frac{p_1^2 / 2}{\cosh^2(\frac{1}{2}\eta_1)} \qquad\qquad v \to 0,$$

$$\eta_2 \to +\infty \; u \to \frac{p_1^2 / 2}{\cosh^2(\frac{1}{2}[\eta_1 + 2\log B_{12}])} \; v \to 0,$$

$$\eta_1 \to -\infty \; u \to \frac{2 p_2^2}{\cosh^2(\eta_2)} \qquad\qquad v \to \frac{2 p_2^2}{\cosh(\eta_2)},$$

$$\eta_1 \to +\infty \; u \to \frac{2 p_2^2}{\cosh^2(\eta_2 + \log(B_{12}))} \quad v \to \frac{2 p_2^2}{\cosh(\eta_2 + \log(B_{12}))}.$$

For b+b scattering the result is

$$\eta_2 \to -\infty \; u \to \frac{2 p_1^2}{\cosh^2(\eta_1)} \qquad\qquad v \to \frac{2 p_1^2}{\cosh(\eta_1)},$$

$$\eta_2 \to +\infty \; u \to \frac{2 p_1^2}{\cosh^2(\eta_1 + \log(A_{12}))} \quad v \to \frac{2 p_1^2}{\cosh(\eta_1 + \log(A_{12}))},$$

$$\eta_1 \to -\infty \; u \to \frac{2 p_2^2}{\cosh^2(\eta_2)} \qquad\qquad v \to \frac{2 p_2^2}{\cosh(\eta_2)},$$

$$\eta_1 \to +\infty \; u \to \frac{2 p_2^2}{\cosh^2(\eta_2 + \log(A_{12}))} \quad v \to \frac{2 p_2^2}{\cosh(\eta_2 + \log(A_{12}))}.$$

The phase shifts obtained for this model follow the usual pattern, in particular the different components u and v get identical phase shifts. If the solitons are of the same type (a+a or b+b) the phase shift is determined by A_{12}; if they are different the phase shift is determined by B_{12}.

The a+b scattering provides us an example of "ghost" solitons: If we look only at the v field we see a standard soliton traveling along and then suddenly experiencing a phase shift as shown in Fig. 8.4 (upper part on the right). If we ignore the u-field we do not see any reason for this behavior, but if u is included (Fig. 8.4, lower part on the right) the behavior can be understood: the a-type soliton associated with η_1 is a ghost soliton. It shows up directly only in the u-field, but its secondary effects can be seen in the v-field.

8.5.5 Dromions

As the final example in the nlS-class let us consider the Davey–Stewartson (DS) equation

$$\begin{cases} i\partial_t \phi + (\partial_x^2 + \partial_y^2)\phi + \phi v = 0, \\ \partial_x \partial_y v = 2(\partial_x^2 + \partial_y^2)|\phi|^2. \end{cases} \tag{8.96}$$

This is a (2+1)-dimensional generalization of the nlS equation (8.63) but the two dimensions get twisted in the second equation.

Normally soliton equations in $(2+1)$ dimensions have just plane wave solutions that are localized only in one direction. However, it was found [20] that DS equation can also have solutions that decay exponentially in *all* spatial directions. These were rather mysterious at first but when the different roles of the ϕ and v fields were recognized [21, 22] it was possible to construct general multi-dromion solutions [23, 24]. The fundamental observation is that there are still underlying plane waves, but they only show up as such in the v-field—in the ϕ-field there is an effect only where two v-plane waves intersect, i.e, where the plane waves suffer a phase shift [25]. Since the plane waves do not show up in the ϕ-field they are called "ghost" solitons.

Equation (8.96) is bilinearized with

$$\phi = G/F, \quad v = 2(\partial_x^2 + \partial_y^2)\log F, \tag{8.97}$$

which yields

$$\begin{cases} (iD_t + D_x^2 + D_y^2)G \cdot F = 0, \\ D_x D_y F \cdot F = 2|G|^2. \end{cases} \tag{8.98}$$

The pair (8.98) has two kinds of 1SSs: standard plane wave solitons of the NLS-type

$$F_{1S} = 1 + K_{11}e^{\eta_1 + \eta_1^*}, \quad G_{1S} = e^{\eta_1}, \tag{8.99}$$

$$\eta_j = p_j x + q_j y + \Omega_j t + \eta_j^0, \quad \Omega_j = i(p_j^2 + q_j^2), \tag{8.100}$$

$$K_{ij} = 1/[(p_i + p_j^*)(q_i + q_j^*)], \tag{8.101}$$

and also "ghost" solitons

$$F_{1D} = 1 + ce^{\eta_1 + \eta_1^*}, G_{1D} = 0, \tag{8.102}$$

with vanishing ϕ. For ghosts, in addition to the dispersion relation (8.100) it is also required that the η_i-plane waves are *parallel to the coordinate axes*, i.e., either $p_i = 0$ or $q_i = 0$.

In order to construct a 1-dromion solution we take two perpendicular ghost solitons, i.e., $\eta = px + ip^2 t$, $\rho = qy + iq^2 t$ and combine them as follows:

$$F = \delta + \alpha e^{\eta + \eta^*} + \beta e^{-\rho - \rho^*} + \gamma e^{\eta + \eta^* - \rho - \rho^*}, \tag{8.103}$$

$$G = \kappa e^{\eta - \rho^*}, \tag{8.104}$$

$$|\kappa|^2 = 4p_R q_R(\alpha\beta - \gamma\delta). \tag{8.105}$$

The constants α, β, γ and δ are real numbers and to avoid singular solutions they should all be nonnegative, furthermore $p_R q_R(\alpha\beta - \gamma\delta) > 0$.

If one compares (8.103),(8.104),(8.105) to the usual 2SS of NLS given in (8.79),(8.80), one notes in particular that now both F and G are even in the

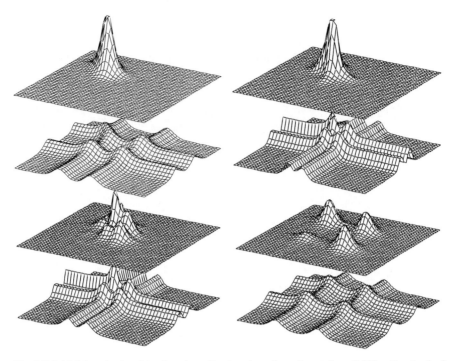

Fig. 8.5 Initial dromion breaking into 4 smaller dromions. In each part the v-field has the standard plane wave structure

exponentials, whereas for normal solitons G is odd. However, if the ghost plane waves resonate so that $\alpha = \beta = 0$ we can express the result in terms of $\tilde{\eta} = \eta - \rho^*$ and obtain a 1SS described in (8.99),(8.100),(8.101).

One can construct multi-dromion solutions with two bilinear approaches, see [23, 24]. Such formulae were used in Fig. 8.5, which shows the splitting of one dromion into four different smaller dromions.

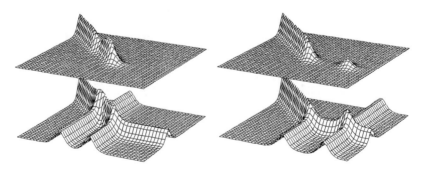

Fig. 8.6 Solitoff-dromion scattering. Note the two resonating plane waves

The dromion concept (underlying ghost plane waves with dromions at the points of their intersection) is a robust concept and other examples are known [26]. Furthermore, the underlying plane waves can be made to resonate and create solitoffs [27], see Fig. 8.6.

8.6 Hierarchies

The above examples illustrate how new dependent variables can be introduced in bilinearization, even more than there were originally. In general the soliton equations are arranged into hierarchies and for higher members in a given hierarchy one also has to introduce new *independent* variables.

8.6.1 A Shallow Water Wave Equation

Consider the following equation:

$$v_{xxxt} + \alpha v_x v_{xt} + \beta v_t v_{xx} - v_{xx} - v_{xt} = 0. \tag{8.106}$$

If $\alpha = \beta = 3$ the equation can be integrated once and the substitution $v = 2\partial_x \log f$ leads to (8.25), but if $\alpha = 4$, $\beta = 2$ this substitution yields first [28]

$$3D_x\left[(D_x^3 D_t - D_x^2 - D_x D_t) f \cdot f\right] \cdot f^2 + \underline{D_t\left(D_x^4 f \cdot f\right) \cdot f^2} - D_x\left[D_x^3 D_t f \cdot f\right] \cdot f^2 = 0. \tag{8.107}$$

We observe that all but the underlined term are of the form $D_x[\dots] \cdot f^2$. To proceed further introduce a new independent variable τ by

$$(D_x^4 + D_x D_\tau) f \cdot f = 0. \tag{8.108}$$

Next we use the identity

$$D_t\left(D_x D_\tau f \cdot f\right) \cdot f^2 = D_x\left[D_t D_\tau f \cdot f\right] \cdot f^2 \tag{8.109}$$

and using (8.109),(8.108) in (8.107) we find that (8.106) is equivalent to

$$\begin{cases} (D_x^4 + D_x D_\tau) f \cdot f = 0, \\ (2D_x^3 D_t - 3D_x^2 - 3D_x D_t - D_t D_\tau) f \cdot f = 0. \end{cases} \tag{8.110}$$

In this case we had originally 1 equation for 1 function of 2 variables, but the bilinear form is given by 2 equations for 1 function of 3 variables. This method of introducing new independent variables is typical for higher members of a hierarchy.

8.6.2 The Kaup/Lax5 Equation

There are two NEEs with dispersion relation $p^5 + \omega = 0$, the Sawada–Kotera equation (8.27) and Kaup's equation (also called Lax5):

$$u_{xxxx} + 20uu_{xxx} + 40u_x u_{xx} + 120u^2 u_x + u_t = 0. \tag{8.111}$$

The substitution $u = \partial_x^2 \log f$ leads to [29]

$$D_x \left[(D_x{}^6 + D_x D_t) f \cdot f \right] \cdot f^2 - 5D_x (D_x{}^4 f \cdot f) \cdot (D_x{}^2 f \cdot f) = 0. \tag{8.112}$$

This can be immediately written in a trilinear form

$$(7T_x^6 + 20T_x^3 T_x^{*3} + 27T_x T_t) F \cdot F \cdot F = 0, \tag{8.113}$$

but does it have a bilinear representation?

As above one introduces a new independent variable τ and the constraint (8.108). Using (8.109) and the further identity

$$3D_x (D_x D_\tau f \cdot f) \cdot (D_x{}^2 f \cdot f) = D_\tau (D_x{}^4 f \cdot f) \cdot f^2 - D_x (D_x{}^3 D_\tau f \cdot f) \cdot f^2 \tag{8.114}$$

one finds that (8.111) is equivalent to

$$\begin{cases} (D_x^4 + D_x D_\tau) f \cdot f = 0, \\ (3D_x{}^6 - 5D_x{}^3 D_\tau + 3D_x D_t - 5D_\tau{}^2) f \cdot f = 0. \end{cases} \tag{8.115}$$

8.6.3 The Jimbo–Miwa Hierarchy

The fundamental classification of integrable equations in bilinear form is based on the Sato theory and was done by Jimbo and Miwa [6]. This classification contains several infinite hierarchies; one of them, the *KP hierarchy*, starts as

$$(D_1^4 - 4D_1 D_3 + 3D_2^2) f \cdot f = 0, \tag{8.116}$$

$$(D_1^3 + 2D_3) D_2 - 3D_1 D_4) f \cdot f = 0, \tag{8.117}$$

$$(D_1^6 - 20D_1^3 D_3 - 80D_3^2 + 144D_1 D_5 - 45D_1^2 D_2^2) f \cdot f = 0, \tag{8.118}$$

$$\vdots$$

Here f is a function of infinite number of variables $x_n, n = 1, 2, 3, \dots$ ($D_k \equiv D_{x_k}$) obeying infinite number of equations. We recognize the KP-equation as the first equation. Note that all equations are weight homogeneous, if D_k is given weight k.

Normal soliton equations are obtained by *reductions* from a hierarchy [30]. There are many ways to reduce the hierarchy to an integrable set of finite number of equations in a finite number of variables. If we just take Eq. (8.116) alone, we get the KP-equation after substituting $f = e^w$. If we take (8.117) under the reduction $D_4 \rightarrow cD_1$ we get the Hirota–Satsuma–Ito equation, and (8.118) under reduction $D_2 \rightarrow 0$ yields the Sawada–Kotera–Ramani equation. Without the above reductions Eqs. (8.117),(8.118) would not be integrable by themselves. In the previous subsections we also showed how the pair (8.116),(8.118), with $D_2 \rightarrow 0$ corresponds to (8.111) and (8.116),(8.117) with $D_2 \rightarrow \epsilon D_1$, $D_4 \rightarrow \epsilon(aD_x + bD_t)$, and $\epsilon \rightarrow 0$ to (8.106); in each case x_3 was the dummy variable.

In general we could have N equations with $N - 1$ extra independent variables. 1SSs are constructed as before, we just have N dispersion relations, but $N - 1$ of them can be used to determine the parameters associated with the extra independent variables leaving one true dispersion relation. 2SSs are still of the form (8.37),(8.38), but now each equation must define the *same* phase factor A_{12} and the phase factor can in fact be used as one of the labels of the hierarchy. The existence of a common 3SS introduces further conditions, which only integrable equations pass.

8.7 Bilinear Bäcklund Transformation

As a further application of the bilinear approach we discuss the Bäcklund transformation (BT), but only for the KdV equation

$$(D_x^4 + D_xD_t) F \cdot F = 0. \tag{8.119}$$

Let us consider the expression

$$F^2[(D_x^4 + D_xD_t) G \cdot G] - G^2[(D_x^4 + D_xD_t) F \cdot F] = 0. \tag{8.120}$$

Clearly, if F satisfies Eq. (8.119) and G satisfies (8.120), then G also satisfies Eq. (8.119).

Let us now rewrite (8.120) in a more useful form. There are lots of identities between bilinear expressions, including the following:

$$f^2(D_xD_t g \cdot g) - g^2(D_xD_t f \cdot f) = 2D_x(D_t g \cdot f) \cdot (fg),$$
$$f^2(D_x^4 g \cdot g) - g^2(D_x^4 f \cdot f) = 2D_x(D_x^3 g \cdot f) \cdot (fg) + 6D_x(D_x^2 g \cdot f) \cdot (D_x f \cdot g).$$

When these are applied to Eq. (8.120) it can be written as

$$2D_x[(D_x^3 + D_t) G \cdot F] \cdot [FG] + 6D_x[D_x^2 G \cdot F] \cdot [D_x F \cdot G] = 0. \tag{8.121}$$

This does not yet produce a BT because the free parameter is still missing. However, when the following obvious identity

$$-6\lambda D_x[D_x G \cdot F] \cdot [FG] + 6\lambda D_x[GF] \cdot [D_x F \cdot G] = 0$$

is added to (8.121) we get

$$2D_x[(D_x^3 + D_t - 3\lambda D_x)G \cdot F] \cdot [FG] + 6D_x[(D_x^2 + \lambda)G \cdot F] \cdot [D_x F \cdot G] = 0,$$

on the basis of which we get the BT

$$\begin{cases} (D_x^3 + D_t - 3\lambda D_x)G \cdot F = 0, \\ (D_x^2 + \lambda)G \cdot F = 0. \end{cases} \tag{8.122}$$

It is still necessary to prove that if F satisfies (8.119) then G can in fact be solved from (8.122), i.e., that the equations are consistent. Dividing both Eqs. (8.122) with F they can be written as

$$L_1(G) := \left[\partial_x^3 - 3A\partial_x^2 + 3B\partial_x - C - 3\lambda(\partial_x - A) - F^{-1}F_t + \partial_t\right]G = 0,$$
$$L_2(G) := [\partial_x^2 - 2A\partial_x + B + \lambda]G = 0,$$

where $A = F^{-1}F_x$, $B = F^{-1}F_{xx}$, $C = F^{-1}F_{xxx}$. Then we find

$$(L_1 L_2 - L_2 L_1)(G) = -3\partial_x[F^{-2}(D_x^2 F \cdot F)]L_2(G) + \partial_x[F^{-2}(D_x^4 + D_x D_t)F \cdot F]G,$$

which means that Eq. (8.122) are consistent if F solves (8.119). (It should be noted that not all pairs obtained by rearranging bilinear expressions are consistent [31].)

From the result (8.122) we note that the unknown function G appears *linearly*. This is a general property: The original bilinear equations are replaced by twice as many equations, in each of which the old and new solutions both appear linearly and provide us a Lax pair.

8.8 The Three-Soliton Condition as an Integrability Test

8.8.1 Defining the Class of Equations and the Integrability Test

For all searches of integrable systems one has to define the *method of testing* integrability and the *class of equations* which are tested. Naturally the testing method influences the choice of the class of equations.

For example, in the searches based on symmetry methods (see the lectures of Mikhailov, Sanders, Shabat, Sokolov and Wang in this book) one class of equations is

$$u_t = f(u, u_x, u_{xx}, \ldots).$$

Thus the nonlinearity of f is not restricted in any way, but it is explicitly assumed that time derivatives appear only as u_t which can be separated. This assumption is natural for the chosen method. (Of course this assumption can be and has been extended, e.g., to include multi-component models.)

Here we will discuss searches based on Hirota's bilinear method. The class of equations is then defined, e.g., as

$$P(D_x, D_t, D_y, \dots) F \cdot F = 0,$$

and the integrability test is *the existence of multisoliton solutions*, see Sect. 8.3.5. The dispersion relation satisfied by the solitons is given directly by the chosen polynomial: $P(p, q, \dots) = 0$. Note that now the number of independent variables is not restricted in any way, but the nonlinearity is: it is whatever comes after substituting $F = e^w$ in the equation.[5] After this substitution the linear part of the equation will be $P(\partial_x, \partial_y, \dots)w$ and therefore for a given linear part corresponds to a particular nonlinear part (within a given class of equations). To repeat: the dispersion relation is truly arbitrary, but once it is chosen it fixes the nonlinearity.

It is important to note that we let the *equation itself* determine the dispersion relation and *any* solution of the dispersion relation can be used to define a soliton. Another approach [32] is to fix the dispersion relation beforehand, in which case it is only necessary that the dispersion relation is a factor of $P(p, q, \dots)$. One might then impose the three-soliton condition only on solitons having this predetermined dispersion relation (weak condition), as opposed to all possible solitons (strong condition).

8.8.2 About the Search Process

In order to find all KdV-type solitons equations we have to find all polynomials that solve (8.43). Such a search was done in [33] using symbolic algebra (for an earlier numerical work see [34]).

The problem is best studied in the framework of commutative algebra. Equation (8.43) needs to be true only on the affine manifold $\mathcal{M}(P) = \{(\mathbf{p}_1, \mathbf{p}_2, \mathbf{p}_3) | P(\mathbf{p}_1) = P(\mathbf{p}_2) = P(\mathbf{p}_3) = 0\}$. If P is reducible we can write $P = \prod Q_i^{n_i}$, where the Q_i are irreducible, and then the affine manifold is actually defined by the radical $\sqrt{P} = \prod Q_i$. The 3SC (8.43) then means that we must have $S_3(P) = \sum_{i=1}^{3} R_i(\mathbf{p}_1, \mathbf{p}_2, \mathbf{p}_3) \sqrt{P}(\mathbf{p}_i)$, $\forall \mathbf{p}_i$, for some polynomials R_i. One way to solve the problem is then to try to find the R_i but that is very cumbersome.

In practice the best way to implement the dispersion relation (or its radical) is to use a substitution rule. After fixing an ordering for monomials (e.g., first by total degree and then lexicographically) we take the leading monomial $L(P)$ of $\sqrt{P}$ and substitute for it the negative of the rest of $\sqrt{P}$. This will be done repeatedly until no term has a factor $L(P)$, and the process will finish in a finite number of steps. After this the resulting polynomial should vanish identically. In REDUCE [35] the substitution can be conveniently implemented by a LET-rule, with the built-in property that all substitutions are repeated as often as possible. We used the following code for this purpose ($R = \sqrt{P}$):

---

[5] Note, however, the possibility of introducing new variables as in (8.2) and integrating the equation.

```
PROCEDURE TEST3SC(P,R);BEGIN
S:=PART(R,1);
K:=-R+S;
S1:=SUB(X=X1,Y=Y1,T=T1,S);
S2:=SUB(X=X2,Y=Y2,T=T2,S);
S3:=SUB(X=X3,Y=Y3,T=T3,S);
K1:=SUB(X=X1,Y=Y1,T=T1,K);
K2:=SUB(X=X2,Y=Y2,T=T2,K);
K3:=SUB(X=X3,Y=Y3,T=T3,K);
LETS(S1,K1);
LETS(S2,K2);
LETS(S3,K3);
...computation of $S_3(P)$...
END;
```

```
PROCEDURE LETS(S,K);LET S=K;
```

The search process was organized in increasing maximum degree of P and in increasing the number of dimensions. The form of the 3SC is such that if it holds for some P it must also hold for any projection to lower dimensions. For example, if $P(x,y,z)$ passes the test then so will any $P(x,y,\alpha x+\beta y)$. Thus at a fixed degree the necessary condition for increasing the dimension, for example from 2 to 3, is that the starting P has free constants in such a form that they could be combined as $\alpha x+\beta y$. In that case we need to test only those polynomials where the new variable z appears in place of (some of) these combinations.

8.8.3 Results

8.8.3.1 KdV Class

In addition to the equations known before, (8.22),(8.26),(8.29), and their reductions, one new genuinely nonlinear equation was found [33] in the KdV class

$$(D_x^4 - D_x D_t^3 + aD_x^2 + bD_x D_t + cD_t^2)F \cdot F = 0, \tag{8.123}$$

which seems to be a linear combination of KdV and HSI equations. This equation also has 4SS and passes the Painlevé test [36], but nothing else is known about it. Another intriguing result is

$$D_x D_y D_z D_t F \cdot F = 0, \tag{8.124}$$

which satisfies the 3SC but *not* the 4SC.

We have also made a search in the trilinear class $P(T,T^*)F \cdot F \cdot F = 0$ (using the Painlevé test [11, 12]), here we just mention one result for which we have also verified the existence of 3SS:

$$\left(T_y(T_x^{*4} + 8T_x^3 T_x^* + 9T_y^2) + 9T_x^2 T_t\right) F \cdot F \cdot F = 0. \tag{8.125}$$

8.8.3.2 mKdV/sG Class

This class was studied in [37–39]. The condition was again that any set of three solitons could be combined into a 3SS. The final result contained seven equations and five were of mKdV type. Three of them have a nonlinear B polynomial but a factorizable A part (and hence only one kind of soliton with B providing the dispersion relation):

$$\begin{cases} (aD_x^7 + bD_x^5 + D_x^2 D_y + D_t) g \cdot f = 0, \\ \qquad\qquad\qquad D_x^2 g \cdot f = 0, \end{cases} \tag{8.126}$$

$$\begin{cases} (aD_x^3 + bD_y^3 + D_t) g \cdot f = 0, \\ \qquad\qquad D_x D_y g \cdot f = 0, \end{cases} \tag{8.127}$$

$$\begin{cases} (D_x D_y D_t + aD_x + bD_y) g \cdot f = 0, \\ \qquad\qquad\qquad D_x D_y g \cdot f = 0. \end{cases} \tag{8.128}$$

In two cases both A and B are nonlinear enough to support solitons; note that the B polynomials are the same and that the A parts have already appeared in the KdV list:

$$\begin{cases} (D_x^3 + D_y) g \cdot f = 0, \\ (D_x^3 D_t + aD_x^2 + D_t D_y) g \cdot f = 0, \end{cases} \tag{8.129}$$

$$\begin{cases} (D_x^3 + D_y) g \cdot f = 0, \\ (D_x^6 + 5D_x^3 D_y - 5D_y^2 + D_t D_x) g \cdot f = 0. \end{cases} \tag{8.130}$$

Two equations of sine-Gordon type were also found:

$$\begin{cases} (D_x D_t + b) G \cdot F = 0, \\ (D_x^3 D_t + 3bD_x^2 + D_t D_y)(F \cdot F + G \cdot G) = 0, \end{cases} \tag{8.131}$$

$$\begin{cases} (aD_x^3 D_t + D_t D_y + b) G \cdot F = 0, \\ \qquad\qquad D_x D_t (F \cdot F + G \cdot G) = 0. \end{cases} \tag{8.132}$$

Of course the various reductions of these equations are also integrable.

8.8.3.3 nlS Class

The class is defined by (8.74) and the condition was the existence of a 2SS, as discussed in Sect. 8.5.3. Three equations were found in the search of Refs. [39, 40]:

$$\begin{cases} (D_x^2 + iD_y + c) G \cdot F = 0, \\ (a(D_x^4 - 3D_y^2) + D_x D_t) F \cdot F = |G|^2, \end{cases} \tag{8.133}$$

$$\begin{cases} (i\alpha D_x^3 + 3cD_x^2 + i(bD_x - 2dD_t) + g) G \cdot F = 0, \\ (\alpha D_x^3 D_t + aD_x^2 + (b + 3c^2)D_x D_t + dD_t^2) F \cdot F = |G|^2, \end{cases} \tag{8.134}$$

$$\begin{cases} (i\alpha D_x^3 + 3D_x D_y - 2iD_t + c) G \cdot F = 0, \\ (a(\alpha^2 D_x^4 - 3D_y^2 + 4\alpha D_x D_t) + bD_x^2) F \cdot F = |G|^2. \end{cases} \tag{8.135}$$

Perhaps the most interesting new equation above is the combination in (8.135) of the two most important $(2+1)$-dimensional equations: Davey–Stewartson and Kadomtsev–Petviashvili equations [39]. In the special case $b = c = 0$ its N-soliton solutions have been discussed in [41].

8.9 From Bilinear to Nonlinear

In many applications it is useful if the leading derivatives in the evolution equation appear in terms linear in the dependent variables, while the nonlinear terms contain lower order derivatives. Starting from bilinear equations one can obtain such a form by a substitution like $F = e^w$. We will now discuss this "nonlinearization" of bilinear equations.

8.9.1 KdV-Type Equations

For KdV-type equations (involving only one dependent variable F) the standard substitution is $F = e^w$. Now since for soliton solutions F is a sum of exponentials, w grows linearly in some directions. In order to obtain a dependent variable that looks like a soliton it is necessary to make a further change of variables, e.g., $u = w_{xx}$. Then for soliton solutions u looks like a plane wave and in terms of u we have a bona-fide nonlinear evolution equation.

For example in the case of the $2+1$-dimensional HSI equation (8.26) we first get

$$w_{txxx} + 6w_{tx}w_{xx} + aw_{xx} + w_{ty} = 0. \tag{8.136}$$

The x-derivative of this equation can we written as

$$v_{txxx} + 3v_{tx}v_x + 3v_t v_{xx} + av_{xx} + v_{ty} = 0 \tag{8.137}$$

using $v = 2w_x$. One more derivative is necessary and thus we introduce $u = v_x = 2w_{xx}$. But in order to write the x-derivative of (8.137) without integrals we have to write it as a coupled system

$$u_{txxx} + 3u_{tx}u + 6u_t u_x + 3v_t u_{xx} + au_{xx} + u_{ty} = 0, \; v_x = u. \tag{8.138}$$

(In the $1+1$-dimensional reduction $v_{ty} \to v_{tx}$ we could have made the substitution $v_x = u$ directly on (8.137) with the result (8.24) (up to scalings).)

Putting $F = e^w$ in the trilinear equation (8.125) yields

$$w_{xxxy} + 8w_{xxy}w_{xx} + 4w_{xy}w_{xxx} + w_{yyy} + w_{xxt} = 0. \tag{8.139}$$

8.9.2 mKdV-Type Equations

For the mKdV class (8.57) the generic nonlinearizing substitution is

$$f = e^{r+w}, g = e^{r-w}.$$

As an example let us consider (8.126) in the special case $a = b = 0$; the equations then read

$$w_{xxy} + 4w_x^2 w_y + 4r_{xy} w_x + 2r_{xx} w_y + w_t = 0,$$
$$r_{xx} + 2w_x^2 = 0.$$

In the limit $y \to x$ we could eliminate r_{xx} from the first equation and get the potential mKdV equation (8.48). In the general case we introduce new variables $v = 2w_x$, $q = 2r_x$, take an x-derivative of the first equation, eliminate q_x and get

$$v_{xxy} - 4v^2 v_y + 2q_y v_x + v_t = 0, \quad q_x + v^2 = 0. \tag{8.140}$$

For a discussion of the nonlinear versions of (8.127), (8.128), see [42].

8.9.3 nlS-Type Equations

For the nonlinear Schrödinger equation the canonical first step in the transformation from the bilinear form to the nonlinear form is

$$F = e^w, G = \phi e^w, \ w \ \text{real}, \ \phi \ \text{complex}, \tag{8.141}$$

cf. (8.68). Again w will grow linearly in some directions and therefore further changes of variables are necessary. We will now derive the nonlinear forms for the search results (8.133),(8.134),(8.135).

Let us first consider (8.133). The transformation (8.141) yields

$$\begin{cases} \phi_{xx} + i\phi_y + c\phi + 2w_{xx}\phi = 0, \\ a(2w_{xxxx} + 12w_{xx}^2 - 6w_{yy}) + 2w_{xt} = |\phi|^2. \end{cases} \tag{8.142}$$

but by operating ∂_x^2 on the second equation allows us to introduce $u = 2w_{xx}$ and obtain Melnikov's equation [43]

$$\begin{cases} \phi_{xx} + i\phi_y + c\phi + u\phi = 0, \\ a[u_{xxxx} + 3(u^2)_{xx} - 3u_{yy}] + u_{xt} = (|\phi|^2)_{xx}. \end{cases} \tag{8.143}$$

This equation may be interpreted as a combination of KP and nlS: (1) If $\phi = 0$ the second equation yields KPII, while (2) if $a = 0$ we get a kind of (2+1)-dimensional nlS, which reduces to the usual nlS if $t \to x$. Note that the first equation has only two independent variables, x and y, while the second has x, t and y.

Equation (8.134) is a combination of the HSI equation (8.24) and Hirota's generalization of nlS [4]. Transformation (8.141) yields now

$$\begin{cases} i\alpha(\phi_{xxx} + 6w_{xx}\phi_x) + 3c(\phi_{xx} + 2w_{xx}\phi) + i(b\phi_x - 2d\phi_t) + \gamma\phi = 0, \\ 2\alpha(w_{xxxt} + 6w_{xx}w_{xt}) + 2aw_{xx} + 2(b + 3c^2)w_{xt} + dw_{tt} = |\phi|^2. \end{cases} \tag{8.144}$$

(1) In the limit $\phi = 0, d = 0$ we obtain (8.24), while (2) HnlS is obtained in the singular limit: first put $c = b = \gamma = 0, d = -\frac{1}{2}\alpha$, then divide the first equation by α

and put $\alpha = 0$ in the second. The result is $q_t + q_{xxx} + 3|q|^2 q_x = 0$, which is a special case of (8.82).

Equation (8.135) is a combination of KP and DS. The substitution (8.141) first yields

$$\begin{cases} i\alpha(\phi_{xxx} + 6\phi_x w_{xx}) + 3(\phi_{xy} + 2w_{xy}\phi) - 2i\phi_t + c\phi = 0, \\ a[2\alpha^2(w_{xxxx} + 6w_{xx}^2) - 6w_{yy} + 8\alpha w_{xt}] + 2bw_{xx} = |\phi|^2. \end{cases} \tag{8.145}$$

There are again two different limits to consider:

1. If we put $\phi = 0$, $\alpha = 1$, $a = 1$, $b = 0$, operate on the second equation by ∂_x^2 and define $u = 2w_{xx}$ then we get KPII.
2. If we take $\alpha = 0$, $c = 0$, operate on the second equation by $\partial_x \partial_y$ and define $u = w_{xy}$ then we obtain

$$\begin{cases} 3\phi_{xy} - 2i\phi_t + 6u\phi = 0, \\ -6au_{yy} + bu_{xx} = (|\phi|^2)_{xy}. \end{cases} \tag{8.146}$$

For $b = -6a = 1/\delta$, $u = -q + \frac{1}{2}\delta|\phi|^2$ and after a 45° rotation in the (x,y)-plane this equation gets the hyperbolic–elliptic form of the DS equation for ϕ and q.

In contrast to (8.143) both equations in (8.145) have three variables.

8.10 Conclusions

In these lectures we have discussed Hirota's method from the ground up, i.e., starting with nonlinear evolution equations, finding their bilinear forms and then their multisoliton solutions. We have also described how for a large class of equations one can construct one- and two-soliton solutions, while the existence of three-soliton solutions imposes severe restrictions which can be used to search for integrable equations.

Hirota's direct method is one of the important tools in the study of nonlinear evolution equations, it is particularly effective for constructing multisoliton solutions and should therefore be in the toolbox of anyone who works with solitons.

Acknowledgments The author would like to thank J. Satsuma for hospitality at the University of Tokyo, where this lecture course was finished.

References

1. R. Hirota, Exact Solution of the Korteweg-de Vries Equation for Multiple Collisions of Solitons, Phys. Rev. Lett. 27, 1192–1194, 1971.

2. R. Hirota, Exact Solution of the modified Korteweg-de Vries Equation for Multiple Collisions of Solitons, J. Phys. Soc. Japan 33, 1456–1459, 1972.
3. R. Hirota, Exact Solution of the Sine-Gordon Equation for Multiple Collisions of Solitons, J.Phys. Soc. Japan 33, 1459–1463, 1972.
4. R. Hirota, Exact envelope-soliton solutions of a nonlinear wave equation, J. Math. Phys. 14, 805–809, 1973.
5. R. Hirota, A New Form of Bäcklund Transformations and Its Relation to the Inverse Scattering Problem Progr, Theor. Phys. 52, 1498–1512, 1974.
6. M. Jimbo and T. Miwa, Solitons and Infinite Dimensional Lie Algebras Publ. RIMS, Kyoto Univ. 19, 943–1001, 1983.
7. R. Hirota, Direct Method of Finding Exact Solutions of Nonlinear Evolution Equations, in Bäcklund Transformations, the Inverse Scattering Method, Solitons, and Their Applications, M. Miura (ed.), Lecture Notes in Mathematics, vol 515, Springer, 40–68, 1976.
8. R. Hirota and J. Satsuma, A Variety of Nonlinear Network Equations Generated from the Bäcklund Transformation for the Toda Lattice. Progress Theoretical Phys., Suppl. 59 64–100, 1976.
9. R. Hirota, Direct methods in soliton theory. in Solitons, R.K. Bullough and P.J. Caudrey (eds.), Springer, 157–176, 1980.
10. J. Hietarinta, Hirota's bilinear method and partial integrability, in Partially Integrable Equations in Physics, R. Conte and N. Boccara (eds.), Kluwer, 459–478, 1990.
11. B. Grammaticos, A. Ramani, and J. Hietarinta, Multilinear operators: the natural extension of Hirota's bilinear formalism, Phys. Lett. A 190, 65–70, 1994.
12. J. Hietarinta, B. Grammaticos, and A. Ramani, Integrable Trilinear PDE's, in Nonlinear Evolution Equations & Dynamical Systems, NEEDS '94, V.G. Makhankov, A.R. Bishop and D.D. Holm (eds.), World Scientific, 54–63, 1995.
13. F. Kako and N. Yajima, Interaction of Ion-Acoustic Solitons in Two-Dimensional Space, J. Phys. Soc. Japan 49, 2063–2071, 1980
14. R. Hirota and M. Ito, Resonance of Solitons in One Dimension, J. Phys. Soc. Japan 52, 744–748, 1983
15. K. Ohkuma and M. Wadati, The Kadomtsev-Petviashvili Equation: the Trace Method and the Soliton Resonances, J. Phys. Soc. Japan 52, 749–760, 1983.
16. R. Hirota, Exact N-soliton solutions of wave equation of long waves in shallowwater in non-linear lattices, J. Math. Phys. 14, 810–814, 1973.
17. R. Hirota, Exact envelope-soliton solutions of a nonlinear wave equation, J. Math. Phys. 14, 805–809, 1973.
18. R. Radhakrishnan, M. Lakshmanan and J. Hietarinta, Inelastic collision and switching of coupled bright solitons in optical fibers, Phys. Rev. E 56, 2213–2216, 1997.
19. R. Hirota and J. Satsuma, Soliton solutions of a coupled Korteweg-de Vries equation, Phys. Lett. A 85, 407–408, 1981
20. M. Boiti, J.J-P. Leon, L. Martina, and F. Pempinelli, Scattering of localized solitons on the plane, Phys. Lett. A 132, 432–439, 1988.
21. A.S. Fokas and P.M. Santini, Coherent Structures in Multidimensions, Phys. Rev. Lett. 63, 1329–1333, 1989
22. A.S. Fokas and P.M. Santini, Dromions and a boundary value problem for the Davey-Stewartson 1 equation, Physica D 44, 99–130, 1990.
23. J. Hietarinta and R. Hirota, Multidromion solutions to the Davey-Stewartson equation, Phys. Lett. A 145, 237–244, 1990.
24. C. Gilson and J.J.C. Nimmo, A direct method for dromion solution of the Davey- Stewartson equations and their asymptotic properties, Proc. R. Soc. Lond. A 435, 339–357, 1991.
25. J. Hietarinta and J. Ruokolainen, Dromions – The movie. (video animation, see http:// users.utu.fi/hietarin/dromions) 1990.
26. J. Hietarinta, One-dromion solutions for generic classes of equations, J, Phys. Lett. A 149, 113–118, 1990.

27. C. Gilson, Resonant behaviour in the Davey–Stewartson equation, Phys. Lett. A 161, 423–428, 1992.
28. R. Hirota and J. Satsuma, N-Soliton Solutions of Model Equations for Shallow Water Waves, J. Phys. Soc. Japan 40, 611–612, 1976
29. J. Satsuma and D.J. Kaup, A Bäcklund transformation for a Higher Order Korteweg-de Vries Equation, J. Phys. Soc. Japan 43, 692–697, 1977
30. R. Hirota, Reduction of soliton equations in bilinear form, Physica D 18, 161–170, 1986.
31. J. Hietarinta, Bäcklund transformations from the bilinear viewpoint, CRM Proceedings and Lecture Notes 29 245–251, 2001.
32. A. Newell and Z. Yunbo, The Hirota conditions, J. Math. Phys. 27, 2016–20121, 1986.
33. J. Hietarinta, A search for bilinear equations passing Hirota's three-soliton condition. I. KdV-type bilinear equations, J. Math. Phys. 28, 1732–1742, 1987.
34. M. Ito, An Extension of Nonlinear Evolution Equations of the K-dV (mK-dV) Type to Higher Orders, J. Phys. Soc. Jpn. 49, 771–778, 1980.
35. A.C. Hearn, REDUCE User's Manual Version 3.2, 1985.
36. B. Grammaticos, A. Ramani, and J. Hietarinta, A search for integrable bilinear equations: the Painlevé approach, J. Math. Phys. 31, 2572–2578, 1990.
37. J. Hietarinta, A search of bilinear equations passing Hirota's three-soliton condition: II. mKdV-type bilinear equations, J. Math. Phys. 28, 2094–2101, 1987
38. J. Hietarinta, A search of bilinear equations passing Hirota's three-soliton condition: III. sine-Gordon-type bilinear equations, J. Math. Phys. 28, 2586–2592, 1987.
39. J. Hietarinta, Recent results from the search for bilinear equations having threesoliton solutions, in Nonlinear evolution equations: integrability and spectral methods, A. Degasperis, A.P. Fordy and M. Lakshmanan (eds.), Manchester U.P., 307–317, 1990.
40. J. Hietarinta, A search of bilinear equations passing Hirota's three-soliton condition: IV. Complex bilinear equations, J. Math. Phys. 29, 628–635, 1988.
41. R. Hirota and Y. Ohta, Hierarchies of Coupled Soliton Equations. I, J. Phys. Soc. Jpn. 60, 798–809, 1991.
42. J.J.C. Nimmo, Darboux transformations in two dimensions. in Applications of Analytic and Geometric Methods to Nonlinear Differential Equations, P. Clarkson (ed.), Kluwer Academic, 183–192, 1992.
43. V.K. Melnikov, A direct method for deriving a multisoliton solution for the problem of interaction of waves on the x, y plane, Commun. Math. Phys. 112, 639–652, 1987.

Chapter 9
Integrability of the Quantum XXZ Hamiltonian

T. Miwa

9.1 Integrability

In classical mechanics, the integrability of a Hamiltonian H in the Liouville sense is the existence of enough many conserved (i.e., Poisson commuting with the Hamiltonian) quantities H_n. For consistency reason they are mutually commuting:

$$\{H_m, H_n\} = 0, \quad H = H_0. \tag{9.1}$$

The orbits are determined by the algebraic equations $H_n = \lambda_n$.

If the system has an infinite degrees of freedom (e.g., the KdV equation), it is not at all obvious how to find explicit solutions for integrable Hamiltonians. Still the existence of infinitely many conserved quantities is a key to finding them.

In quantum mechanics, the Hamiltonian is an operator acting on some physical space. Equation (9.1) is replaced by Lie bracket relations in the operator algebra:

$$[H_m, H_n] = 0. \tag{9.2}$$

The physical states are by definition simultaneous eigenvectors of these commuting operators:

$$H_n|v\rangle = \lambda_n|v\rangle.$$

Again, it is not obvious how to find all the simultaneous eigenvectors of these commuting operators.

Let us consider the simplest example, i.e., the harmonic oscillator:

$$H = -\frac{1}{2}\left(\frac{d^2}{dx^2} - x^2\right). \tag{9.3}$$

The degree of freedom is 1 because the space on which H acts is the space of functions in one variable. We do not need any further commuting operators in this case. Integrability of this H lies in the algebraic structure which makes an explicit diago-

T. Miwa (✉)
Department of Mathematics, Kyoto University, Kyoto, Japan,
tetsuji@kusm.kyoto-u.ac.jp

Miwa, T.: *Integrability of the Quantum XXZ Hamiltonian.* Lect. Notes Phys. **767**, 315–323 (2009)
DOI 10.1007/978-3-540-88111-7_9

nalization of H possible. Let us recall this construction. We introduce operators

$$P = -\frac{1}{2}\left(\frac{d}{dx} + x\right), \quad Q = \frac{d}{dx} - x$$

so that

$$H = QP + \frac{1}{2},$$

and P and Q generate the three-dimensional Heisenberg algebra with the relation

$$[P,Q] = 1.$$

The structure of the physical space is clearly understood by means of this algebra. One can find the eigenvector corresponding to the lowest eigenvalue by solving the equation

$$Pf(x) = 0, \quad f(x) = e^{-\frac{1}{2}x^2}.$$

Applying the operator Q to $f(x)$, one can create other eigenvectors. This is because H and P,Q satisfy the commutation relations

$$[H,P] = -P, \quad [H,Q] = Q.$$

Namely, P changes the eigenvalue of H by -1 and Q by 1. We can think of this fact as follows: the operator P annihilates a particle, and Q creates it.

Now we introduce the quantum Hamiltonian acting on the tensor product of $\mathbb{C}^2$, which is our main subject:

$$H = -\frac{1}{2}\sum_n \left(\sigma_n^x \sigma_{n+1}^x + \sigma_n^y \sigma_{n+1}^y + \Delta \sigma_n^z \sigma_{n+1}^z\right). \tag{9.4}$$

Here $\sigma_n^x, \sigma_n^y, \sigma_n^z$ are the Pauli matrices acting on the nth tensor component. If we consider finite, say N-fold, tensor product with the periodic condition, one can find one parameter family of operators $T(\zeta)$ satisfying

$$[T(\zeta_1), T(\zeta_2)] = 0, \quad T(\zeta) = T(1)(1 + H(\zeta - 1) + \cdots). \tag{9.5}$$

In particular, we have $[H, T(\zeta)] = 0$. Suppose that a matrix

$$R = \left(R_{\varepsilon\tau}^{\varepsilon'\tau'}(\zeta)\right)_{\varepsilon,\tau,\varepsilon',\tau'=0,1}$$

acting on $\mathbb{C}^2 \otimes \mathbb{C}^2$ satisfies the cubic relations

$$\sum_{\varepsilon,\tau,\eta} R_{\varepsilon''\tau''}^{\varepsilon\tau}(\zeta_1/\zeta_2) R_{\varepsilon\eta''}^{\varepsilon'\eta}(\zeta_1/\zeta_3) R_{\tau\eta}^{\tau'\eta'}(\zeta_2/\zeta_3) \tag{9.6}$$

$$= \sum_{\varepsilon,\tau,\eta} R_{\tau''\eta''}^{\tau\eta}(\zeta_1/\zeta_2) R_{\varepsilon''\eta}^{\varepsilon\eta'}(\zeta_1/\zeta_3) R_{\varepsilon,\tau}^{\varepsilon'\tau'}(\zeta_2/\zeta_3).$$

This is called the Yang–Baxter equation.

Define $T = \left(T^{\varepsilon'_1 \cdots \varepsilon'_N}_{\varepsilon''_1 \cdots \varepsilon''_N}(\zeta)\right)_{\varepsilon'_i, \varepsilon''_i = 0,1}$ acting on $\otimes^N \mathbb{C}^2$ by

$$T^{\varepsilon'_1 \cdots \varepsilon'_N}_{\varepsilon''_1 \cdots \varepsilon''_N}(\zeta) = \sum_{\tau_1, \ldots, \tau_N = 0,1} R^{\tau_2 \varepsilon'_1}_{\tau_1 \varepsilon''_1}(\zeta) R^{\tau_3 \varepsilon'_2}_{\tau_2 \varepsilon''_2}(\zeta) \cdots R^{\tau_1 \varepsilon'_N}_{\tau_N \varepsilon''_N}(\zeta). \tag{9.7}$$

Then, one can show the commutativity (9.5). We also remark that $T(1)$ is the shift operator:

$$T(1)\left(v_{\varepsilon_1} \otimes v_{\varepsilon_2} \otimes \cdots \otimes v_{\varepsilon_N}\right) = v_{\varepsilon_2} \otimes v_{\varepsilon_3} \otimes \cdots \otimes v_{\varepsilon_1}.$$

Drinfeld and Jimbo clarified the representation theoretical meaning of the Yang–Baxter equation (9.6) by introducing $U_q(\widehat{g})$, which is a q-deformation of the universal enveloping algebra of the affine Lie algebra

$$\widehat{g} = g \otimes \mathbb{C}[t, t^{-1}] \otimes \mathbb{C}K.$$

Here we explain this in the simplest setting of the two-dimensional representation of $U_q(\widehat{sl}_2)$. The algebra is generated by six generators $e_i, f_i, t_i = q^{h_i}$ $(i = 0, 1)$. They satisfy the commutation relations (in fact, a little bit more, but we omit them):

$$[h_i, h_j] = 0, \quad [e_i, f_j] = \delta_{ij} \frac{t_i - t_i^{-1}}{q - q^{-1}},$$

$$[h_i, e_j] = \begin{cases} 2e_j \text{ if } i = j \\ -2e_j \text{ if } i \neq j, \end{cases} \qquad [h_i, f_j] = \begin{cases} -2f_j \text{ if } i = j \\ 2f_j \text{ if } i \neq j. \end{cases}$$

The algebra has a two-dimensional representation

$$e_0 = \begin{pmatrix} 0 & 0 \\ 1 & 0 \end{pmatrix}, \quad f_0 = \begin{pmatrix} 0 & 1 \\ 0 & 0 \end{pmatrix}, \quad h_0 = \begin{pmatrix} 1 & 0 \\ 0 & -1 \end{pmatrix},$$

$$e_1 = \begin{pmatrix} 0 & 0 \\ 1 & 0 \end{pmatrix}, \quad f_1 = \begin{pmatrix} 0 & 1 \\ 0 & 0 \end{pmatrix}, \quad h_1 = \begin{pmatrix} -1 & 0 \\ 0 & 1 \end{pmatrix}.$$

In this representation, the central element $K = h_0 + h_1$ is zero.

One can define the tensor product of two representations by using the algebra map

$$\Delta : U_q(\widehat{sl}_2) \rightarrow U_q(\widehat{sl}_2) \otimes U_q(\widehat{sl}_2)$$

given by

$$\Delta(e_i) = e_i \otimes 1 + t_i \otimes e_i,$$
$$\Delta(f_i) = f_i \otimes t_i^{-1} + 1 \otimes f_i,$$
$$\Delta(h_i) = h_i \otimes 1 + 1 \otimes h_i.$$

The algebra admits one-parameter family of automorphisms:

$$\rho_\zeta : U_q(\widehat{sl}_2) \to U_q(\widehat{sl}_2)$$

given by

$$\rho_\zeta(e_i) = \zeta e_i, \quad \rho_\zeta(f_i) = \zeta^{-1} f_i, \quad \rho_\zeta(h_i) = h_i.$$

Therefore, one can define one-parameter family of two-dimensional representations, which we denote by $(\mathbb{C}^2)_\zeta = \mathbb{C}v_0 \oplus \mathbb{C}v_1$. When $\zeta = 1$, we denote $(\mathbb{C}^2)_1 = \mathbb{C}^2$.

For generic values of ζ_1, ζ_2 the tensor product $(\mathbb{C}^2)_{\zeta_1} \otimes (\mathbb{C}^2)_{\zeta_2}$ is irreducible, and it is isomorphic to $(\mathbb{C}^2)_{\zeta_2} \otimes (\mathbb{C}^2)_{\zeta_1}$. The unique isomorphism (called the intertwiner)

$$R : (\mathbb{C}^2)_{\zeta_1} \otimes (\mathbb{C}^2)_{\zeta_2} \to (\mathbb{C}^2)_{\zeta_2} \otimes (\mathbb{C}^2)_{\zeta_1}$$

is written as

$$R(v_{\varepsilon'} \otimes v_{\tau'}) = \sum_{\varepsilon'', \tau''=0,1} (v_{\tau''} \otimes v_{\varepsilon''}) R^{\varepsilon' \tau'}_{\varepsilon'' \tau''}.$$

The R-matrix has the dependence on ζ_1, ζ_2 through $\zeta_1/\zeta_2 : R = R(\zeta_1/\zeta_2)$.

The Yang–Baxter equation is a consequence of the uniqueness of the intertwiner

$$(\mathbb{C}^2)_{\zeta_1} \otimes (\mathbb{C}^2)_{\zeta_2} \otimes (\mathbb{C}^2)_{\zeta_3} \to (\mathbb{C}^2)_{\zeta_3} \otimes (\mathbb{C}^2)_{\zeta_2} \otimes (\mathbb{C}^2)_{\zeta_1}.$$

The main problem is the diagonalization of the family of transfer matrices. The quantum inverse scattering method was designed for this problem. The idea is similar to the algebraic method for the harmonic oscillator. Namely, we proceed as follows:

(i) find an eigenvector (reference state) Ω_0;
(ii) construct an operator $B(\xi)$, which creates eigenvectors in the form

$$B(\xi_1) \cdots B(\xi_n)\Omega_0 \tag{9.8}$$

for an appropriate choice of $\xi_1, \ldots, \xi_n$;
(iii) obtain the eigenvalue of $T(\zeta)$ in the form $\prod_{i=1}^n \tau(\zeta/\xi_i)$.

For the XXZ Hamiltonian, we can take $\Omega_0 = \underbrace{v_0 \otimes \cdots \otimes v_0}_{N}$ for which $T(\zeta)\Omega_0 = \Omega_0$. This is because the total spin operator $S^z = \sum_n \sigma_n^z$ commutes with H and the subspace where $S_z = N$ is one-dimensional.

The operator $B(\zeta) = T_{10}(\zeta)$ is given by

$$(T_{ab})^{\varepsilon_1' \cdots \varepsilon_N'}_{\varepsilon_1'' \cdots \varepsilon_N''}(\zeta) = \sum_{\tau_1, \ldots, \tau_N=0,1} R^{\tau_2 \varepsilon_1'}_{a \varepsilon_1''}(\zeta) R^{\tau_3 \varepsilon_2'}_{\tau_2 \varepsilon_2''}(\zeta) \cdots R^{b \varepsilon_N'}_{\tau_N \varepsilon_N''}(\zeta). \tag{9.9}$$

Note that the transfer matrix (9.7) is given by $T = T_{00} + T_{11}$.

If $B(\zeta)$ satisfies the relation

$$T(\zeta)B(\xi) = \tau(\zeta/\xi)B(\xi)T(\zeta),$$

then (ii) and (iii) are true for arbitrary choice of $\xi_1, \ldots, \xi_n$. This is not possible because we are working on the finite-dimensional vector space $\mathbb{C}^2 \otimes \cdots \otimes \mathbb{C}^2$.

The set of parameters $\xi_1, \ldots, \xi_n$ must satisfy a set of algebraic equations so that the vector (9.8) really gives an eigenvector. This is called the Bethe Ansatz equations. It is difficult to solve the equations for a finite value of N.

Another difficulty in comparison to the harmonic oscillator case is that the eigenvector Ω_0 does not necessarily give the lowest eigenvalue of H. In the region $\Delta = (q + q^{-1})/2 < -1$, the lowest eigenvalue belongs to the subspace where $S_Z = 0$. Therefore, we must take n in (9.8) to be large ($\sim N/2$) in order to obtain eigenvectors whose eigenvalues are close to the lowest.

For finite N it is not possible to overcome these difficulties completely. However, in the infinite limit $N \to \infty$ and in the sectors where S_Z is finite, the situation simplifies and (iii) is valid without limitation for $\xi_1, \ldots, \xi_n$ (except $|\xi_i| = 1$).

We will explain a method for obtaining these eigenvectors, which is based on the representation theory of $U_q(\widehat{sl}_2)$. This method is very different from the quantum inverse scattering method. In fact, we find a "simple" lowest eigenvector $|vac\rangle$, called the vacuum vector, of the renormalized transfer matrix $T^{(\infty)}(\zeta)$ in the limit $N = \infty$ such that $T^{(\infty)}(\zeta)|vac\rangle = |vac\rangle$. Then, we create other eigenvectors by using operators $\psi_\varepsilon(\xi)$ ($\varepsilon = 0, 1$) which satisfy the commutation relations

$$T^{(\infty)}(\zeta)\psi_\varepsilon(\xi) = \tau(\zeta/\xi)\psi_\varepsilon(\xi)T^{(\infty)}(\zeta). \tag{9.10}$$

Namely, our method is a generalization of the algebraic method for the harmonic oscillator.

Our approach is, however, a so-called bootstrap approach. It means we raise ourselves in the air by pulling up the bootstraps of our shoes by our hands. By heuristic argument we derive algebraic relations for renormalized operators that are valid only in the limit $N = \infty$. We will find the operators satisfying these relations by using representation theory. The second step is rigorous, but the first step is like pulling ourselves by bootstraps.

9.2 Symmetry

We prepare more facts from the representation theory of the algebra $U_q(\widehat{sl}_2)$. One can add an operator D, to the algebra, that satisfies

$$[D, e_i] = -e_i, [D, f_i] = f_i, [D, h_i] = 0.$$

There exists representations of $U_q(\widehat{sl}_2)$, which we denote by $\mathscr{H}_i$ ($i = 0, 1$) with the following characterization (i–iv):

(i) the vector space $\mathscr{H}_i$ is graded by the eigenspaces of D

$$\mathscr{H}_i = \oplus_l (\mathscr{H}_i)_{l=0}^\infty \text{ where } D = l \text{ on } (\mathscr{H}_i)_l;$$

(ii) the space $(\mathcal{H}_i)_0$ is one-dimensional and it is spanned by a vector v_i called the highest weight vector;

(iii) the highest weight vector satisfies

$$e_j v_i = 0, \quad h_j v_i = \delta_{ji} v_i;$$

in particular, $K = 1$ on $\mathcal{H}_i$;

(iv) the vector v_i is cyclic, i.e., the space $\mathcal{H}_i$ is generated by v_i by the actions of f_0, f_1;

(v) the action of e_0, e_1, f_0, f_1 are locally nilpotent, i.e., for any vector $v \in \mathcal{H}_i, x^N v = 0$ for $x = e_0, e_1, f_0, f_1$ and sufficiently large N.

In application to the XXZ Hamiltonian, the following property is most important:

(vi) there exists an intertwiner of the representations:

$$\Phi(\zeta) : \mathcal{H}_i \to \mathcal{H}_{1-i} \otimes (\mathbb{C}^2)_\zeta. \tag{9.11}$$

Namely, the following commutativity of the mappings holds:

$$\Delta(x) \circ \Phi(\zeta) = \Phi(\zeta) \circ x \text{ for all } x \in U_q(\widehat{\mathrm{sl}}_2).$$

The intertwiner is unique up to normalization. For simplicity, we do not bother this normalization in this chapter. Therefore, formulas are written up to finite scalar.

The key idea is the following hypothesis. We identify the semi-infinite tensor product with $\mathcal{H} = \mathcal{H}_0 \oplus \mathcal{H}_1$:

$$\cdots \otimes \mathbb{C}^2 \otimes \mathbb{C}^2 \simeq \mathcal{H}, \tag{9.12}$$

and the action of the semi-infinite transfer matrix with the intertwiner $\Phi(\zeta)$:

$$T^{(\infty/2)}(\zeta)\left(\cdots \otimes v_{\varepsilon_2'} \otimes v_{\varepsilon_1'}\right)$$
$$= \sum_{\tau_1, \tau_2, \dots} \left(\cdots \otimes v_{\varepsilon_2''} \otimes v_{\varepsilon_1''} \otimes v_\varepsilon\right) R_{\varepsilon \varepsilon_1''}^{\tau_1 \varepsilon_1'}(\zeta) R_{\tau_1 \varepsilon_2''}^{\tau_2 \varepsilon_2'}(\zeta) \cdots.$$

The reason of the hypothesis (9.12) is as follows. Repeated application of the intertwiners $\Phi(1)$ gives the isomorphisms of representations.

$$\mathcal{H} \overset{\sim}{\to} \mathcal{H} \otimes \mathbb{C}^2 \overset{\sim}{\to} \mathcal{H} \otimes \mathbb{C}^2 \otimes \mathbb{C}^2 \overset{\sim}{\to} \mathcal{H} \otimes \mathbb{C}^2 \otimes \mathbb{C}^2 \otimes \mathbb{C}^2 \overset{\sim}{\to} \cdots.$$

In the infinite limit, we have (9.12). Of course this is not a proof of any mathematical statement. However, it gives us a possibility of constructing the renormalized Hamiltonian in the language of $U_q(\widehat{\mathrm{sl}}_2)$.

Now, the question is how to understand the whole infinite tensor product

$$\cdots \otimes \mathbb{C}^2 \otimes \mathbb{C}^2 \otimes \mathbb{C}^2 \otimes \mathbb{C}^2 \cdots.$$

Define an anti-automorphism of $U_q(\widehat{\mathrm{sl}}_2)$:

$$b : U_q(\widehat{\mathfrak{sl}}_2) \to U_q(\widehat{\mathfrak{sl}}_2),$$
$$b(e_i) = q t_i^{-1} e_i, \quad b(f_i) = q^{-1} f_i t_i, \quad b(t_i) = t_i^{-1}.$$

Let M be a $U_q(\widehat{\mathfrak{sl}}_2)$-module. The anti-automorphism b defines an action of $U_q(\widehat{\mathfrak{sl}}_2)$ on the dual space. Let us denote the dual space with this action by M^{*b}. The dual space of tensor product is given by the opposite tensor product of dual spaces:

$$(M_1 \otimes M_2)^{*b} \simeq M_2^{*b} \otimes M_1^{*b}.$$

In particular, we have an isomorphism

$$(\mathbb{C}^2)_\zeta \simeq (\mathbb{C}^2)_\zeta^{*b}, \quad v_\varepsilon \mapsto v_{-\varepsilon}^*,$$

where v_ε^* is the dual basis to v_ε. We identified one half of the infinite tensor product with $\mathscr{H}$ (9.12). The other half of infinite tensor product is identified with $\mathscr{H}^{*b}$:

$$\mathbb{C}^2 \otimes \mathbb{C}^2 \otimes \cdots \simeq (\mathbb{C}^2)^{*b} \otimes (\mathbb{C}^2)^{*b} \otimes \cdots \simeq (\cdots \otimes \mathbb{C}^2 \otimes \mathbb{C}^2)^{*b} \simeq \mathscr{H}^{*b}.$$

Thus we have a representation theoretical picture

$$\cdots \otimes \mathbb{C}^2 \otimes \mathbb{C}^2 \otimes \mathbb{C}^2 \otimes \mathbb{C}^2 \cdots \simeq \mathscr{H} \otimes \mathscr{H}^{*b}.$$

All these are heuristic arguments, and we need some justification (if not a mathematical proof). Especially, we should answer why the semi-infinite space is a direct sum of two irreducible representations.

We restrict our consideration to the region of the parameter q with $-1 < q < 0$, or equivalently $\Delta < -1$. In the region $|\Delta| < 1$, i.e., $|q| = 1$, the representation theoretical interpretation is not valid. In the extreme limit $q = 0_-$, i.e., $\Delta = -\infty$, the Hamiltonian simplifies. Namely, only the σ^z terms survive:

$$H(-\infty) \sim \sum_n \sigma_n^z \sigma_{n+1}^z.$$

This is already diagonal. The lowest eigenvalue of this diagonal Hamiltonian is given by one of the following two vectors:

$$\cdots \otimes v_\mp \otimes v_\pm \otimes v_\mp \otimes v_\pm \cdots.$$

The eigenvalue is $-\infty$. We renormalize this value to zero. Then, all the other finite eigenvalues are obtained from the simple tensor product

$$\cdots \otimes v_{\varepsilon_2} \otimes v_{\varepsilon_1} \otimes v_{\varepsilon_0} \otimes v_{\varepsilon_{-1}} \cdots,$$

where $\varepsilon_n = (-1)^{i+n}$ $(i = 0, 1)$ for $n \to \infty$, and $\varepsilon_n = (-1)^{j+n}$ $(j = 0, 1)$ for $n \to -\infty$, separately. The choice of $i = 0, 1$ corresponds to $\mathscr{H}_i$, and $j = 0, 1$ to $\mathscr{H}_j^{*b}$. We denote the boundary condition in this sense by (i, j).

Let us interpret the transfer matrix $T(\zeta)$ in the limit $N = \infty$. Recall that $T(1)$ is the shift operator. Therefore, $T(\zeta)$, in general, changes the boundary condition (i, j) to $(1 - i, 1 - j)$. Therefore, we want to find the vacuum vectors $|vac\rangle_i$ satisfying the boundary condition (i, i) such that $T(\zeta)|vac\rangle_i = t_0(\zeta)|vac\rangle_{1-i}$. In the limit $N = \infty$, the eigenvalue $t_0(\zeta)$ is actually divergent. We need renormalization. Let us define the renormalized transfer matrix $T^{(\infty)}(\zeta)$ starting from the identification (9.12). As an operator in $\mathrm{End}(\mathscr{H} \otimes \mathscr{H}^*)$ we have

$$T^{(\infty)}(\zeta) = \sum_\varepsilon \Phi_\varepsilon(\zeta) \otimes \Phi_{-\varepsilon}(\zeta)^t.$$

Here

$$\Phi(\zeta)^t : (\mathbb{C}^2)_\zeta \otimes \mathscr{H}^* \simeq (\mathscr{H} \otimes (\mathbb{C}^2)_\zeta)^* \to \mathscr{H}^*$$

is the transpose of $\Phi(\zeta)$. We think of $T^{(\infty)}(\zeta)$ as

$$T^{(\infty)}(\zeta) : \mathscr{H} \otimes \mathscr{H}^* \overset{\Phi(\zeta) \otimes \mathrm{id}}{\to} \mathscr{H} \otimes (\mathbb{C}^2)_\zeta \otimes \mathscr{H}^* \overset{\mathrm{id} \otimes \Phi(\zeta)^t}{\to} \mathscr{H} \otimes \mathscr{H}^*.$$

It is convenient to use the canonical isomorphism

$$\mathscr{H} \otimes \mathscr{H}^* \simeq \mathrm{End}(\mathscr{H}).$$

An operator $A \otimes B \in \mathrm{End}(\mathscr{H}) \otimes \mathrm{End}(\mathscr{H}^*) \simeq \mathrm{End}(\mathscr{H} \otimes \mathscr{H}^*)$ is translated to an operator in $\mathrm{End}(\mathrm{End}(\mathscr{H}))$ mapping $f \in \mathrm{End}(\mathscr{H})$ to $A \circ f \circ B \in \mathrm{End}(\mathscr{H})$. Thus, the renormalized transfer matrix is defined as

$$T^{(\infty)}(\zeta)f = \sum_\varepsilon \Phi_\varepsilon(\zeta) \circ f \circ \Phi_{-\varepsilon}(\zeta).$$

Now, we proceed to the diagonalization of $T^{(\infty)}(\zeta)$. We need a few more facts from representation theory:

(vii) the intertwiners obey the homogeneity with respect to the grading by D:

$$\xi^{-D} \Phi_\varepsilon(\zeta) \xi^D = \Phi_\varepsilon(\zeta/\xi);$$

(viii) and also the "unitarity" relation

$$\sum_\varepsilon \Phi_\varepsilon(-q^{-1}\zeta) \Phi_{-\varepsilon}(\zeta) = \mathrm{id};$$

(ix) another kind of intertwiner exists:

$$\Psi^*(\xi) : (\mathbb{C}^2)_\xi \otimes \mathscr{H} \to \mathscr{H}$$

(x) and satisfies the commutation relation

$$\Phi_\varepsilon(\zeta) \Psi^*_{\varepsilon'}(\xi) = \tau(\zeta/\xi) \Psi^*_{\varepsilon'}(\xi) \Phi_\varepsilon(\zeta).$$

Now we define $|vac\rangle_i \in \mathrm{End}(\mathscr{H}_i)$ by

$$|vac\rangle_i = (-q)^D.$$

Then we have

$$
\begin{aligned}
T^{(\infty)}(\zeta)|vac\rangle_i &= \sum_\varepsilon \Phi_\varepsilon(\zeta) \circ (-q)^D \circ \Phi_{-\varepsilon}(\zeta) \\
&= (-q)^D \circ \sum_\varepsilon \Phi_\varepsilon(-q^{-1}\zeta) \circ \Phi_{-\varepsilon}(\zeta) \\
&= (-q)^D \\
&= |vac\rangle_{1-i}.
\end{aligned}
$$

We have obtained the vacuum vectors.

The particles are created by the operators $\psi_\varepsilon(\xi) \in \mathrm{End}(\mathrm{End}(\mathscr{H}))$ defined by

$$\psi_\varepsilon(\xi)f = \Psi_\varepsilon(\xi) \circ f.$$

We only check (9.10), and it is straightforward from the commutation relation (x).

The method we have explained in this chapter is originally obtained in [1]. A full and expository account is available in [2].

References

1. B. davies, O. Foda, M. Jimbo, T. Miwa, and A. Nakayashiki, Diagonalization of the XXZ Hamiltonian by vertex operators, Commun. Math. Phys. 151, 89–153, 1993.
2. M. Jimbo and T. Miwa, Algebraic Analysis of Solvable Lattice Models CBMS Regional Conference Series in Mathematics No. 86, AMS, 1995.

Index